Lecture Notes in Mathematics

Edited by A. Dold, B. Eckmann and F. Takens

1425

R.A. Piccinini (Ed.)

Groups of Self-Equivalences and Related Topics

Proceedings of a Conference
held in Montreal, Canada, Aug. 8–12, 1988

Springer-Verlag
Berlin Heidelberg New York London Paris Tokyo Hong Kong

Editor

Renzo A. Piccinini
Department of Mathematics and Statistics
Memorial University of Newfoundland
St. John's, Newfoundland, Canada A1C 5S7

Mathematics Subject Classification (1980): 55P10, 55P50, 55R05, 55R20,
57M10, 57N10

ISBN 3-540-52658-7 Springer-Verlag Berlin Heidelberg New York
ISBN 0-387-52658-7 Springer-Verlag New York Berlin Heidelberg

Printing and binding: Druckhaus Beltz, Hemsbach/Bergstr.
2146/3140-543210 – Printed on acid-free paper

FOREWORD

During the week of August 8, 1988, a conference on spaces of self-homotopy equivalences and related topics was held at the Centre de Recherches Mathématiques, Université de Montréal, in Montréal, Canada. The conference was attended by some 30 mathematicians from Canada, Denmark, England, France, Germany, Japan and the United States. This volume contains the Proceedings thereof, covering most of the talks presented at the Centre. It also contains two specially written articles (a survey on groups of self-equivalences and a paper presenting seventeen open problems). The volume is completed by a list of papers on groups of self-equivalences and related areas.

The conference was made possible thanks to a most generous contribution from the Centre des Recherches Mathématiques; their superb organization and most helpful staff were also a key point in the success of the conference. The organizers wish to take this opportunity to thank, both for themselves and also in the name of the participants, the Director of the Centre, Prof. Frank Clarke, and his staff, for their most gracious hospitality.

Each paper in this volume was carefully refereed and sincere thanks are due to all those who helped in this. While the referees must, of course, remain unknown to the public, those who helped me on the technical side can be thanked and named; they are: Edgar Goodaire, Bruce Shawyer and Kathleen Stewart.

Renzo A.Piccinini

Contents

Equivalent homotopy theories and groups of self-equivalences *

Peter Booth
Memorial University of Newfoundland
St. John's, Canada A1C 5S7

Let G be a topological group with classifying space BG, $r: Z \to C$ be a principal G-bundle, $k: C \to BG$ a classifying map for r and $q(k): \bar{C} \to BG$ "k made into a fibration" via the standard factorization of an arbitrary map into the composite of a homotopy equivalence and a Hurewicz fibration [Sp, 2.8,9]. If $p: X \to B$ is a Dold fibration, i.e. if it satifies the WCHP of [Do], then $\mathcal{E}_B(X)$ will denote the group, under composition, of over B homotopy classes of fibre homotopy equivalences ($FHEs$) of X into itself, all understood in the free sense.

Abbreviating G-equivariant to G-, it is well known that G-homotopy theory is equivalent to the homotopy theory of fibrations over the classifying space BG [DDK, DK, Fu1, Fu2, SV]. Taking $\mathcal{W}$ to denote the class of all spaces having the homotopy types of CW-complexes, an application of this result asserts that the group $\mathcal{E}^G(Z)$ of G-homotopy classes of G-homotopy equivalences of Z into itself, with the operation of composition, is isomorphic to $\mathcal{E}_{BG}(\bar{C})$.

We generalize the above, showing that principal bundles can be replaced by other types of fibrations such as principal fibrations, Dold fibrations and ex-fibrations (= sectioned fibrations). If $q: Y \to B$ and $r: Z \to C$ are maps then a *pairwise map* (g, h) from q to r consists of a map $g: Y \to Z$ and a map $h: B \to C$ such that $rg = hq$.

Our basic result on categories (theorem 2.1) asserts that the homotopy theories of:

(i) categories of suitable types of fibrations, together with structure preserving pairwise maps, and

(ii) categories of Dold fibrations over appropriate classifying spaces, together with maps over these spaces,

are equivalent. Our basic result on groups of self-equivalences, corollary 2.2, then follows: certain groups $\mathcal{E}^{\mathcal{F}}(Z)$ of structure preserving pairwise self-homotopy equivalences of fibrations are isomorphic to groups of self-$FHEs$ of corresponding Dold fibrations over appropriate classifying spaces.

There are several ways in which this corollary can yield information about more familiar (not "visibly" pairwise) groups of self-equivalences: sometimes $\mathcal{E}^{\mathcal{F}}(Z)$ is computable so information about groups of self-$FHEs$ is obtained (e.g. in 4.1.2), we have already implied that $\mathcal{E}^{\mathcal{F}}(Z)$ can simplify to $\mathcal{E}^G(Z)$ (this occurs in 4.1.3), and when r is just assumed to be a Hurewicz fibration then $\mathcal{E}^{\mathcal{F}}(Z)$ can sometimes coincide with $\mathcal{E}(Z)$ (see e.g. 4.3.2). We now summarize these three applications.

(4.1.2) If $G \in \mathcal{W}$ is a *grouplike topological monoid* (a topological monoid for which $\pi_0(G)$ with the obvious induced binary operation is a group) and $p_G: EG \to BG$ an associated universal principal G-fibration then $\mathcal{E}_{BG}(EG) \cong \pi_0(G)$. This extends a similar known result for universal covering spaces [Sp, 2.6.4].

*This paper is in final form and no version of it will be submitted for publication elsewhere

(4.1.3) If $r: Z \to S^n$ is a principal G-fibration with $n > 2$ this result characterizes up to isomorphism, or in cases where the order of the characteristic element of r is 2 up to extension, the G-analogue of $\mathcal{E}_{S^n}(Z)$, thereby expanding on [Ts, th. 2.1].

(4.3.2) Using the standard $\mathcal{E}(Z)$ notation for groups of based homotopy classes of based self homotopy equivalences of a based space Z, we show that if Z is a suitable space with two non-zero homotopy groups then $\mathcal{E}(Z)$ is isomorphic to the fundamental group of a certain space of maps between classifying spaces. The group $\mathcal{E}(Z)$ for such spaces has previously been analyzed up to extension, a basic reference is [Sh].

In order to avoid the repetition inherent in developing parallel theories for different types of fibration we phrase our main results (section 2), techniques for analyzing $\mathcal{E}^{\mathcal{F}}(Z)$ especially when C is a sphere (section 3), and the proof of theorem 2.1 (sections 6 and 7) in the unifying language of $\mathcal{F}$-fibrations. Versions that refer to particular types of fibrations are given in section 4.

We now mention a difficulty that has to be overcome before we produce a proof of theorem 2.1. Returning to the case where G is a topological group, a right G-space will be said to be a free G -space over $\mathcal{W}$ if it determines a principal G-bundle $r: Z \to Z/G = C$, where $C \in \mathcal{W}$; $\boldsymbol{FrGSp}$ will denote the category of all free G-spaces over $\mathcal{W}$ and G-maps between them. If B is a space then $\boldsymbol{Dfib}_B$ will denote the category whose objects are Dold fibrations over B (i.e. with range B) and with fibres in $\mathcal{W}$, and morphisms are over B maps between the total spaces of such fibrations.

Given a principal G-bundle $q: Y \to B$ and a right G-space Z there is an associated fibre bundle $q_Z: Y \times_G Z \to B$ [Bor, p. 36] and a functor $\phi(q): \boldsymbol{FrGSp} \to \boldsymbol{Dfib}_B$, which takes Z to q_Z and carries morphisms to morphisms in the obvious fashion. Taking q to be $p_G: EG \to BG$, a universal principal G-bundle [Mil, section 3], the assertion of the equivalence of G-homotopy theory and "over BG" homotopy theory will be made precise in section 4.2 by stating that $\phi = \phi(p_G)$, the functer constructing the associated bundles $(p_G)_Z: EG \times_G Z \to BG$, is in a certain sense a categorical homotopy equivalence from $\boldsymbol{FrGSp}$ to $\boldsymbol{Dfib}_{BG}$ (for a simplicial version of this see [DDK, cor. 2.5]).

A key difficulty in our project of generalizing this argument is that the above definition of ϕ uses the $Y \times_G Z$ construction, and a method of carrying this out when G is not a topological group needs to be found. Our solution, explained in section 5, is based on the observation that $Y \times_G Z$ is a function space!

We work in the context of the category $\boldsymbol{CgTop}$ of compactly generated spaces [Vo, th. 5.1], i.e. spaces having the final (= weak) topology relative to all incoming maps from compact Hausdorff spaces. In the case of Hausdorff spaces this agrees with Steenrod's concept of compactly generated spaces [St]. Any space can be *cg-ified*, i.e. made into a compactly generated space, by retopologizing it with this final topology. Mapping spaces will be assumed to be compactly generated, that is they will carry the *cg*-ification of the compact-open topology. These include $Map(Y, Z)$, $Map_0(Y, Z)$, $\mathcal{H}(Y)$ and $\mathcal{H}_0(Y)$, the spaces of free and based maps of Y to Z and the spaces, and topological monoids under composition, of free and based self homotopy equivalences of Y, respectively. If $p: X \to B$ and $q: Y \to B$ are maps then $Map_B(X, Y)$ will denote the space of over B maps from X to Y, i.e. those (free) maps $f: X \to Y$ for which $qf = p$, and $\mathcal{H}_B(X)$ the space, and topolgical monoid under composition, of (free) self-FHE's of X. We recall the exponential law for mapping spaces: if X, Y and Z are spaces then there is a bijective correspondence between the set of maps $f: X \to Map(Y, Z)$ and the set of maps $g: X \times Y \to Z$ defined by $g(x, y) = f(x)(y)$, for all $x \in X, y \in Y$ [Vo, theorem 3.6].

1 Enriched fibrations

In what follows $\mathcal{F}$ will denote a *category of enriched spaces*, i.e. with objects that are spaces which may carry some extra structure and morphisms maps that preserve any such structure;

it will be assumed to possess a distinguished object F. Then $\mathcal{F}(F)$ will denote the topological monoid, under composition, of morphisms of F into itself. *We assume that both F and $\mathcal{F}(F)$ are in $\mathcal{W}$.* Our discussion will focus on the following examples, others are given in [May] and [Bol].

(1.1) Let G be a grouplike topological monoid. We define $\mathcal{G}$ to be the category whose objects are those right G-spaces that are G-homotopy equivalent to G, and morphisms are G-homotopy equivalences between such spaces. Recalling that the map $\mathcal{G}(G) \to G$ that evaluates at the identity is a multiplicative homeomorphism, its inverse being the "right adjoint" to the operation $G \times G \to G$, we see that the required assumption that G and $\mathcal{G}(G)$ are in $\mathcal{W}$ reduces to $G \in \mathcal{W}$.

(1.2) If F is a given space then we define $\mathcal{H}$ to be the category of all spaces homotopy equivalent to F, and all homotopy equivalences between such spaces. The condition $\mathcal{H}(F) \in \mathcal{W}$ is satisfied if F has the homotopy type of a CW-complex [Mi2, th. 3] or F is a $K(\pi, n)$-space in $\mathcal{W}$ where π is finitely generated and $n \geq 0$ [Ka(P), cor 1.4 (2)].

(1.3) Let $(F, *)$ be a pointed space, i.e. a space with a base point. We will use $\mathcal{H}_0$ to denote the category of all pointed spaces that are pointed homotopy equivalent to $(F, *)$ and all pointed homotopy equivalences between these spaces. The conditions for $\mathcal{H}_0(F, *) \in \mathcal{W}$ are similar to those for $\mathcal{H}(F) \in \mathcal{W}$.

An *$\mathcal{F}$-space over B* is a map $q: Y \to B$ such that for each $b \in B$ the fibre $q^{-1}(b)$ carries the structure of an object of $\mathcal{F}$.

If $p: X \to B$ is map then $X \sqcap Y$ will denote the *fibred product* (or pullback) *space* with underlying set $\{(x, y) \in X \times Y | p(x) = q(y)\}$; the projection $p^*q: X \sqcap Y \to X$ is an *induced $\mathcal{F}$-space* over X. The projection $X \sqcap Y \to Y$ will be denoted by q^*p. These notations are useful because they specify both the spaces and both the maps, respectively, that are used in the constructions.

If $q: Y \to B$ and $r: Z \to C$ are $\mathcal{F}$-spaces then a pairwise map $\langle g, h \rangle$ from q to r will be said to be an *$\mathcal{F}$-pairwise map*, if for each $b \in B$, the map $g | q^{-1}(b): q^{-1}(b) \to r^{-1}(h(b))$ is a morphism of $\mathcal{F}$. Now q is surjective so g determines h; hence we can refer to $\mathcal{F}$-pairwise maps $g: Y \to Z$, and view the space $\mathcal{F}Pws(Y, Z)$ of such maps as being topologized as a mapping space of maps from Y to Z. The above leads naturally to the concepts of *$\mathcal{F}$-pairwise homotopy* (= $\mathcal{F}$-pairwise map from $q \times 1_I$ to r) and *$\mathcal{F}$-pairwise homotopy equivalence* (= $\mathcal{F}PHE$); we will use $\mathcal{F}PHE(Y)$ to denote the space and topological monoid of $\mathcal{F}$-pairwise self-homotopy equivalences of Y. Taking $B = C$ and fixing h to be 1_B the above defines the concepts of *$\mathcal{F}$-map over B*, *$\mathcal{F}$-homotopy over B* and $\mathcal{F}FHE$. The group $\pi_0(\mathcal{F}PHE(Y))$ of $\mathcal{F}$-pairwise homotopy classes of self $\mathcal{F}PHE$s of Y, under composition, will be denoted by $\mathcal{E}^{\mathcal{F}}(Y)$.

We define $Prin_F Y$ to be the space of all $\mathcal{F}$-morphisms from F to individual fibres of q, i.e. $Prin_F Y = \cup_{b \in B} \mathcal{F}(F, \{q^{-1}(b)\})$ regarded as a subspace of $Map(F, Y)$, and $prin_F q: Prin_F Y \to B$ to be the obvious projection map.

Lemma 1.4. Given a map $p: X \to B$ and an $\mathcal{F}$-space $q: Y \to B$ there is a homeomorphism $\xi: X \sqcap Prin_F Y \to Prin_F(X \sqcap Y)$ over X taking (x, g) to $i_x \circ g$, where g is a map $F \to q^{-1}(p(x))$ and $i_x: q^{-1}(p(x)) \to \{x\} \times q^{-1}(p(x))$ denotes the canonical homeomorphism.

Proof. This follows, using the exponential law for mapping spaces, by noticing that functions into $X \sqcap Prin_F Y$ are continuous if and only if the corresponding functions into $Prin_F(X \sqcap Y)$ are continuous.

Suppose the $\mathcal{F}$-space $r: Z \to C$ has the property that for all $\mathcal{F}$-spaces $q: Y \to B$, all homotopies $H: B \times I \to C$ and all maps $g: Y \times \{0\} \to Z$ such that $\langle g, H | B \times \{0\} \rangle$ is an $\mathcal{F}$-pairwise map from q to r under the identifications $Y = Y \times \{0\}$ and $B = B \times \{0\}$, there is a homotopy $G: Y \times I \to Z$ extending g such that $\langle G, H \rangle$ is an $\mathcal{F}$-pairwise map from $q \times 1_I$ to r. Then r will be said to satisfy the *$\mathcal{F}$-covering homotopy property* ($\mathcal{F}CHP$). If r satisfies this condition for those homotopies H that are stationary on $B \times \left[0, \frac{1}{2}\right]$, i.e. with $H(b, t) = H(b, 0)$ whenever $0 \leq t \leq \frac{1}{2}$, then r will be said to have the *$\mathcal{F}$-weak covering homotopy property*

($\mathcal{F}WCHP$).

Let $\mathcal{U}$ be a cover of C. An $\mathcal{F}$-space $r: Z \to C$ will be said to be *locally $\mathcal{F}$-fibre homotopy trivial* (= $L\mathcal{F}FHT$) relative to $\mathcal{U}$ if for each $U \in \mathcal{U}$ the $\mathcal{F}$-space $r|r^{-1}(U): r^{-1}(U) \to U$ is $\mathcal{F}FHE$ to the trivial $\mathcal{F}$-space and projection $F \times U \to U$. It is shown in [Bol] that:

(1.5) in cases where $C \in \mathcal{W}$ the $\mathcal{F}$-space $r: Z \to C$ satisfies the $\mathcal{F}WCHP$ if and only if it is numerably $L\mathcal{F}FHT$, i.e. if it is $L\mathcal{F}FHT$ relative to a numerable cover $\mathcal{U}$ of C.

Our main argument will apply in the context of either of the following *admissible* theories of $\mathcal{F}$-fibrations; the former is developed in [May], the latter in [Bol].

(1.6) $\mathcal{F}$ is either $\mathcal{H}$ or $\mathcal{H}_0$, and $\mathcal{F}$-fibrations are defined to be $\mathcal{F}$-spaces satisfying the $\mathcal{F}CHP$. We do not take $\mathcal{F}$ to be $\mathcal{G}$ in this case for reasons given in 4.1 below.

(1.7) $\mathcal{F}$ is either $\mathcal{G}$, $\mathcal{H}$ or $\mathcal{H}_0$, and $\mathcal{F}$-fibrations are defined to be $\mathcal{F}$-spaces satisfying the $\mathcal{F}WCHP$.

For each of these admissible theories we have:

(1.8) the class of $\mathcal{F}$-fibrations is closed under the formation of induced $\mathcal{F}$-spaces [May, prop. 2.5],

(1.9) if $q: Y \to B$ is an $\mathcal{F}$-fibration and $p: X \to B$ is a homotopy equivalence then the $\mathcal{F}$-pairwise map (q^*p, p) from the induced $\mathcal{F}$-fibration $p^*q: X \sqcap Y \to X$ to q is an $\mathcal{F}PHE$ [Bo2],

(1.10) if $q: Y \to B$ is an $\mathcal{F}$-fibration with $B \in \mathcal{W}$ then $prin_F q$ is a Dold fibration (for it is $L\mathcal{F}FHT$ by 1.5 and [Do, th. 5:12] applies).

(1.11) There is an $\mathcal{F}$-fibration $p_{\mathcal{F}}: E\mathcal{F} \to B\mathcal{F}$, with $B\mathcal{F}$ a CW-complex, that is universal in the following senses:

(1.11a) $Prin_F(E\mathcal{F})$ is weakly contractible, i.e. $\pi_j(Prin_F(E\mathcal{F})) = 0$ for all non-negative integers j ([May] p.48 bottom line and p.50 lines 2 and 3 from the bottom) and

(1.11b) the rule that takes f to $f^*(p_{\mathcal{F}})$, where $f: B \to B\mathcal{F}$ is any map and $B \in \mathcal{W}$, defines a natural bijection from $[B, B\mathcal{F}]$, the set of free homotopy classes of maps from B to $B\mathcal{F}$, to the set of $\mathcal{F}FHE$ classes of $\mathcal{F}$-fibrations over B [May, p.49].

Also we have

(1.11c) $B\mathcal{F} = B\mathcal{F}(F)$, the classifying space for the topological monoid $\mathcal{F}(F)$ ([May, th.9.2] as $G = \mathcal{F}(F)$).

(1.12) If $\mathcal{F}(F) \in \mathcal{W}$ then $Prin_F(E\mathcal{F})$ is contractible. The proof of this involves noticing that [Sc, th. 2] and 1.10 ensure that $Prin_F(E\mathcal{F}) \in \mathcal{W}$ and 1.11a implies that this space has the weak homotopy type of a point; it follows by [Sp, 7.6.24] that it has the homotopy type of a point.

We assume, from this point on, that $(\mathcal{F}, F)$ is a category of enriched spaces with an associated admissible theory of $\mathcal{F}$-fibrations.

2 Main results

Let $\mathcal{F}\boldsymbol{fib}$ denote the category whose objects are $\mathcal{F}$-fibrations with base spaces in $\mathcal{W}$ and morphisms are $\mathcal{F}$-pairwise maps between them. We define $\theta: \boldsymbol{Dfib}_{B\mathcal{F}} \to \mathcal{F}\boldsymbol{fib}$ to be the functor that takes $f: X_1 \to X_2$ over $B\mathcal{F}$ from $p_1: X_1 \to B\mathcal{F}$ to $p_2: X_2 \to B\mathcal{F}$, to the $\mathcal{F}$-pairwise map $(\theta(f), f)$, from $p_1^*(p_{\mathcal{F}})$ to $p_2^*(p_{\mathcal{F}})$, where $\theta(f): X_1 \sqcap E\mathcal{F} \to X_2 \sqcap E\mathcal{F}$ is defined by $\theta(f)(x_1, \ell) = (f(x_1), \ell)$, with $p_1(x_1) = p_{\mathcal{F}}(\ell)$. It will be convenient to sometimes write $X_i \sqcap E\mathcal{F}$ as $\theta(X_i)$, for $i = 1$and 2, so then $\theta(f): \theta(X_1) \to \theta(X_2)$.

Theorem 2.1. If $(\mathcal{F}, F)$ is such that F and $\mathcal{F}(F)$ are in $\mathcal{W}$ then there is a functor $\phi: \mathcal{F}\boldsymbol{fib} \to \boldsymbol{Dfib}_{B\mathcal{F}}$ that is "homotopy inverse" to θ in the sense that:

(i) θ is left adjoint to ϕ,

(ii) there is a natural transformation $d = d_X: X \to \phi\theta X$ consisting of $FHEs$ d_X, one for each object $p: X \to B\mathcal{F}$ of $\boldsymbol{Dfib}_{B\mathcal{F}}$,

(iii) there is a natural transformation $e = e_Z: \theta\phi Z \to Z$ consisting of $\mathcal{F}PHEs$ e_Z, one for each object $r: Z \to C$ of $\mathcal{F}\boldsymbol{fib}$,

(iv) for all pairs of objects $p_i: X_i \to B\mathcal{F}$ of $\boldsymbol{Dfib}_{B\mathcal{F}}$ with $i = 1$ and 2, the rule $\theta_{12}(f) = (\theta(f), f)$ determines a homotopy equivalence
$\theta_{12}: Map_{B\mathcal{F}}(X_1, X_2) \to \mathcal{F}Pws(\theta X_1, \theta X_2)$, where $f \in Map_{B\mathcal{F}}(X_1, X_2)$, and

(v) for all pairs of objects $r_i: Z_i \to C_i$ of $\mathcal{F}\boldsymbol{fib}$ with $i = 1$ and 2, ϕ defines a homotopy equivalence $\phi_{12}: \mathcal{F}Pws(Z_1, Z_2) \to Map_{B\mathcal{F}}(\phi Z_1, \phi Z_2)$.

The proof is removed to sections 6 and 7 below.

Given an $\mathcal{F}$-fibration $r: Z \to C$, we will use $k: C \to B\mathcal{F}$ and $q(k): \bar{C} \to B\mathcal{F}$ to denote the corresponding classifying map and "classifying fibration" (= k made into a fibration as described in the introduction) respectively.

Corollary 2.2. $\mathcal{E}_{\mathcal{F}}(Z) \cong \mathcal{E}_{B\mathcal{F}}(\bar{C})$

Proof. There are bijections
$\pi_0(\mathcal{F}Pws(\bar{C} \sqcap E\mathcal{F}, \bar{C} \sqcap E\mathcal{F}) \to \pi_0\mathcal{F}Pws(C \sqcap E\mathcal{F}, C \sqcap E\mathcal{F})) \to \pi_0(\mathcal{F}Pws(Z, Z))$ by 1.9 and 1.11b respectively, that clearly restrict to group isomorphisms $\mathcal{E}^{\mathcal{F}}(\bar{C} \sqcap E\mathcal{F}) \cong \mathcal{E}^{\mathcal{F}}(C \sqcap \mathcal{E}\mathcal{F}) \cong \mathcal{E}^{\mathcal{F}}(Z)$.

If h is a self-FHE of $\bar{C}$ then pulling back over $q(k) \circ h = q(k)$ we obtain $\bar{C} \sqcap (\bar{C} \sqcap E\mathcal{F}) = \bar{C} \sqcap E\mathcal{F}$; using this in conjunction with 1.9 it follows that $(\theta(h), h)$ is a self-$\mathcal{F}PHE$ for $q(k)^{*}p_{\mathcal{F}}$, and the homotopy equivalence $\theta: Map_{B\mathcal{F}}(\bar{C}, \bar{C}) \to \mathcal{F}Pws(\bar{C} \sqcap E\mathcal{F}, \bar{C} \sqcap E\mathcal{F})$ of 2.1(d) determines a function $\mathcal{E}_{B\mathcal{F}}(\bar{C}) \to \mathcal{E}^{F}(\bar{C} \sqcap E\mathcal{F}), [h] \to [(\theta(h), h)]$. It is immediate that this function is a bijection; the functoriality of θ implies that it is also an isomorphism.

Example 2.3. $\mathcal{E}^{\mathcal{F}}(E\mathcal{F}) = 0$

Proof. We notice that $p_{\mathcal{F}}$ has classifying map $1_{B\mathcal{F}}$, that since $1_{B\mathcal{F}}$ is a fibration $q(1_{B\mathcal{F}})$ is FHE to $1_{B\mathcal{F}}$ [Do, th. 6.1] and so
$\mathcal{E}^{\mathcal{F}}(E\mathcal{F}) \cong \mathcal{E}_{B\mathcal{F}}(\overline{B\mathcal{F}}) \cong \mathcal{E}_{B\mathcal{F}}(B\mathcal{F}) = 0$.

Example 2.4. Considering the path fibration
$q_{B\mathcal{F}}: P(B\mathcal{F}) \to B\mathcal{F}, \mathcal{E}_{B\mathcal{F}}(P(B\mathcal{F})) \cong \pi_0(\mathcal{F}(F))$.

Proof. The $\mathcal{F}$-fibration $F \to *$ has classifying map $k: * \to B\mathcal{F}$ and $q(k) = q_{B\mathcal{F}}$ so $\mathcal{E}_{B\mathcal{F}}(P(B\mathcal{F})) \cong \mathcal{E}^{\mathcal{F}}(F) = \pi_0(\mathcal{F}(F))$.

Corollary 2.5. The fibration $q(k)$ has fibres homotopy equivalent to $P = Prin_F Z$ and hence has a classifying map $c: B\mathcal{F} \to P\mathcal{H}(P)$, and

$$\mathcal{E}^{\mathcal{F}}(Z) \cong \pi_1(Map(B\mathcal{F}, B\mathcal{H}(P)), c).$$

Proof. The fibration $(prin_F(p_{\mathcal{F}}))^{*}q(k)$ possesses a contractible base space $Prin_F(E\mathcal{F})$ (1.12) so its fibres (homeomorphic to the fibres of $q(k)$) have the homotopy type of its total space $\bar{C} \sqcap (Prin_F(E\mathcal{F}))$, of $C \sqcap (Prin_F(E\mathcal{F})$ (1.9), of $Prin_F(C \sqcap E\mathcal{F})$ (1.4), and hence of P. The result now follows from 2.2 because $\mathcal{E}_{B\mathcal{F}}(\bar{C}) \cong \pi_1(Map(B\mathcal{F}, B\mathcal{H}(P)), k)$ [BHMP, th.3.3].

3 Groups of self-equivalences I

We start this section with some easy short exact sequences which determine groups $\mathcal{E}_B(X)$ up to extension, with emphasis on the case where X is a sphere. Then, in conjunction with corollary 2.2, these enable us to either determine $\mathcal{E}^{\mathcal{F}}(Z)$, or determine it up to extension, especially in cases where Z is an $\mathcal{F}$-space over a sphere.

Proposition 3.1. (i) If $p: X \to B$ is a Dold fibration then there is a short exact sequence of groups and homomorphisms:

$$\pi_1(Map(X, B), p)/p_{\#}\pi_1(\mathcal{H}(X)), 1_X) \to \mathcal{E}_B(X) \to \{[f] \in \pi_0(\mathcal{H}(X)) | pf \simeq p\},$$

where $p_{\#}: \pi_1(\mathcal{H}(X), 1_X) \to \pi_1(Map(X,B),p)$ is induced by the Dold fibration $\hat{p}: \mathcal{H}(X) \to Map(X,B)$ that is defined by composition with p.

(ii) Let $X = S^n$. If, regarding $[p]$ as an element of $\pi_n(B)$,

(a) $2[p] \neq 0$ then $\mathcal{E}_B(S^n) \cong \pi_1(Map(S^n,B),p)/p_{\#}\pi_1(\mathcal{H}(S^n),1)$; if

(b) $2[p] = 0$ there is a short exact sequence:
$\pi_1(Map(S^n,B),p)/p_{\#}\pi_1(\mathcal{H}(S^n),1) \to \mathcal{E}_B(S^n) \to Z_2$.

Proof. (i) The Dold fibration $\hat{p}$ has distinguished fibre $\mathcal{H}_B(X)$ and exact sequence:

$$\pi_1(\mathcal{H}(X),1) \underset{p_{\#}}{\to} \pi_1(Map(X,B),p) \underset{\partial}{\to} \mathcal{E}_B(S^n) \to \pi_0(\mathcal{H}(X)) \underset{p_{\#}}{\to} \pi_0(Map(X,B)).$$

The verification of (i) is now immediate, except that we must show ∂ to be a homomorphism. Taking $PMap\,(X,B)$ and $\Omega Map(X,B)$ as the spaces of paths and loops in $Map(X,B)$ starting at p and based at p respectively, $\mathcal{H}(X) \sqcap PMap\,(X,B)$ the pullback space obtained using $\hat{p}$ and the path fibration $PMap\,(X,B) \to Map(X,B)$, $i: \Omega Map(X,B) \to \mathcal{H}(X) \sqcap PMap\,(X,B), i(\ell) = (1_X, \ell)$, and $j: \mathcal{H}_B(X) \to \mathcal{H}(X) \sqcap (PMap\,(X,B), j(f) = (f, c_p)$, where c_p is the constant loop value p, we recall that j is a homotopy equivalence since it is the inclusion of a fibre in the total space of a fibration ($= \hat{p}$ pulled back over $PMap\,(X,B)$) with contractible base space, and ∂ is π_0 of the map $j^{-1} \circ i: \Omega Map(X,B) \to \mathcal{H}_B(X)$, where j^{-1} is a homotopy inverse of j.

Let us assume that there are paths from $(1_X, \ell^\alpha)$ to (f^α, c_p) in $\mathcal{H}(X) \sqcap PMap\,(X,B)$, denoted by $(f_t^\alpha, \ell_t^\alpha) \in \mathcal{H}(X) \sqcap PMap\,(X,B)$, where t ranges over I, for $\alpha = 1$ and $\alpha = 2$. We define a path $\ell_t^1 \cdot f_t^2: I \to Map(X,B)$ by $(\ell_t^1 \cdot f_t^2)(s) = \ell_t^1(s) \circ f_t^2$ where $s \in I$, noticing that $(\ell_t^1 \cdot f_t^2)(0) = p \circ f_t^2$ and $\ell_t^1 \cdot f_t^2(1) = p \circ f_t^1 \circ f_t^2$. Then there is also a path $(f_t^1 \circ f_t^2, (\ell_t^1 \cdot f_t^2) + \ell_t^2)$ from $(1_X, \ell^1 + \ell^2)$ to $(f^1 \circ f^2, c_q)$, where $\ell^1 + \ell^2$ denotes the path obtained by attaching the beginning of ℓ^1 to the end of ℓ^2. Hence $\partial[\ell^\alpha] = [f^\alpha]$ for $\alpha = 1$ and 2 implies that $\partial[\ell^1 + \ell^2] = [f^1 \circ f^2]$.

(ii) Let $\mu \in \mathcal{H}(S^n)$ denote the "coinverse" involved in the co-H-space structure on S^n, so $\pi_0(\mathcal{H}(S^n)) = \{[1],[\mu]\}$. Then $\{[f] \in \pi_0(\mathcal{H}(S^n)) | pf \simeq p\}$ is Z_2 if $p\mu \simeq p$, i.e. if $2[p] = [0]$; otherwise it is zero.

Corollary 3.2. If $p: S^n \to B (n > 2)$ is a Dold fibration and either B is 2-connected or B is a 1-connected H-space with unit then

(i) if $2[p] \neq 0$ then $\mathcal{E}_B(S^n) \cong \pi_{n+1}(B)/\langle p \circ \eta \rangle$ where η denotes the generator of $\pi_{n+1}(S^n)$ and $\langle\,\rangle$ means "subgroup generated by",

(ii) if $2[p] = 0$ there is a short exact sequence

$$\pi_{n+1}(B)/\langle p \circ \eta \rangle \to \mathcal{E}_B(S^n) \to Z_2$$

Proof. Using the fibrations $\mathcal{H}(S^n) \to S^n$ and $Map(S^n, B) \to B$ that evaluate at a given point of S^n and the result [Th, p. 31] that the path-components of an H-group (in this case $Map_0(S^n,S^n)$ or $Map_0(S^n,B)$) have the same homotopy type, we have

$$\pi_1(\mathcal{H}(S^n),1) \cong \pi_1(\mathcal{H}_0(S^n),1) \cong \pi_1(Map_0(S^n,S^n),*) \cong \pi_{n+1}(S^n).$$

When B is 2-connected

$$\begin{aligned}&\pi_1(Map(S^n,B),p)\\ &\qquad \cong \pi_1(Map_0(S^n,B),p) \cong \pi_1(Map_0(S^n,B),*) \cong \pi_{n+1}(B)\end{aligned}$$

and when B is a 1-connected H-space with operation $\cdot$ and unit, $Map(S^n,B) \to B$ has a section $s: B \to Map(S^n,B)$ determined by $s(b)(x) = b \cdot p(x)$, $b \in B$, $x \in S^n$ so again $\pi_1(Map(S^n,B),p) \cong \pi_1(Map_0(S^n,B),*)$ and the above argument proves the latter group isomorphic to $\pi_{n+1}(B)$. Hence $p_{\#}: \pi_0(\mathcal{H}(X),1_X) \to \pi_1(Map(X,B),p)$ can be identified with the homomorphism $\pi_{n+1}(S^n) \to \pi_{n+1}(B)$ induced by p; the result follows from 3.1 (ii).

Theorem 3.3. Let us assume that $F \in \mathcal{W}$ and $\mathcal{F}(F) \in \mathcal{W}$.

(i) Let $r: Z \to C$ be an $\mathcal{F}$-fibration with classifying map $k: C \to B\mathcal{F}$. Then there is a short exact sequence of groups

$$\pi_1(Map(C, B\mathcal{F}), k)/k_{\#}(\pi_1(\mathcal{H}(C), 1_C)) \to \mathcal{E}^{\mathcal{F}}(Z) \to \{[f] \in \pi_0(\mathcal{H}(C)) | kf \simeq k\},$$

where $k_{\#}$ is induced by the map $\mathcal{H}(C) \to Map(C, B\mathcal{F})$ determined by composition with k.

(ii) Let $C = S^n$, for some $n > 2$. If

(a) $2[k] \neq 0$ then $\mathcal{E}^{\mathcal{F}}(Z) \cong \pi_1(Map(S^n, B\mathcal{F}, k)/k_{\#}(\pi_1(\mathcal{H}(S^n), 1))$, if

(b) $2[k] = 0$ then there is a short exact sequence.
$\pi_1(Map(S^n, B\mathcal{F}), k)/k_{\#}(\pi_1(\mathcal{H}(S^n), 1)) \to \mathcal{E}^{\mathcal{F}}(Z) \to Z_2$

Proof. The existence of a canonical homotopy equivalence $S^n \to \overline{S^n}$ enables us to apply 3.1 to the fibration $q(k)$; the result follows via 2.2.

4 Groups of self-equivalences II

In this section we leave the generality of $\mathcal{F}$-fibrations and focus on particular theories.

4.1 Principal G-fibrations

Let G and $\mathcal{G}$ be as described in 1.1. We define a *principal G-fibration* to consist of a right G-space Z and a map $r: Z \to C$ that is numerally $L\mathcal{G}FHT$ in the sense of 1.5. Thus if $C \in \mathcal{W}$ it follows from 1.5 that a principal G-fibration over C is precisely a $\mathcal{G}$-fibration in the sense of 1.7. There is a principal G-fibration $p_G: EG \to BG(= p_{\mathcal{G}}: E\mathcal{G} \to B\mathcal{G})$ that is universal (1.11) amongst principal G-fibrations over spaces in $\mathcal{W}$ [Bol].

Parallel theories of principal G-fibrations satisfying the $\mathcal{G}CHP$ can be obtained, but using different categories $\mathcal{G}$ (see ex. 6.2(i) and (ii), the paragraph following those examples and also cor. 9.4, all of [May]).

Taking $\mathcal{F} = \mathcal{G}$ then 2.1 - 2.5 apply to principal G-fibrations, so the categories $\mathcal{G}$***fib*** and ***Dfib***$_{BG}$ are homotopy equivalent and groups $\mathcal{E}^{\mathcal{G}}(Z)$ can be studied via 2.2 - 2.5. In the case of 2.5, the homeomorphism $\mathcal{G}(G) \to G$ described in 1.1 easily extends to a homeomorphism $Prin_G Z \cong Z$, so that result takes the following form.

(4.1.1) If $G \in \mathcal{W}$ and $r: Z \to C$ is a principal G-fibration then
$\mathcal{E}^{\mathcal{G}}(Z) \cong \pi_1 (Map(BG, B\mathcal{H}(Z)), c)$.

Example 4.1.2. If $G \in \mathcal{W}$ is a grouplike topological monoid then $\mathcal{E}_{BG}(EG) \cong \pi_0(G)$. In particular if X is a pointed CW-complex and $q_X: PX \to X$ denotes the path fibration over X then $\mathcal{E}_X(PX) \cong \pi_1(X)$.

Proof. We first consider the special case. Defining PX to consist of the Moore paths in X starting at $*$, i.e. all maps $f: [o, e_f] \to X$ where e_f is a non-negative real number and $f(0) = *$, then q_X evaluates at e_f, i.e. it takes f to $f(e_f)$. It follows from the numerable contractibility of X (see the hypothesis of [Do, th. 6.3]) that q_X is $L\mathcal{G}FHT$ relative to the action of the monoid $G = \Omega X$ of Moore paths that also end at $*$. Now PX is contractible so q_X is a universal principal ΩX-fibration and $B\Omega X = X$. Applying 2.4 to the category of principal ΩX-fibrations we have $\mathcal{E}_X(PX) \cong \pi_O(\Omega X) \cong \pi_1(X)$.

There is a map $PBG \to EG$ over BG if there is a section to the induced fibration $(q_{BG})^*(p_G)$ and this induced fibration has a contractible base space PBG, so the obstructions to constructing such a section are located in cohomology groups that are zero [Wh., Chapter VI, section 5] and the lifting $PBG \to EG$ exists. Now $G \in \mathcal{W}$ ensures that $EG \cong Prin_G EG$ is contractible (1.12), so the lifting is a homotopy equivalence and also a FHE [Do, th. 6.1]. Hence $\mathcal{E}_{BG}(EG) \cong \mathcal{E}_{BG}(PBG)$, as explained above $\mathcal{E}_{BG}(PBG) \cong \pi_0(\Omega BG) \cong \pi_1(BG)$ and it follows from the exact sequence of $p_G\colon EG \to BG$ that $\pi_1(BG) \cong \pi_0(G)$.

Our next result determines $\mathcal{E}^{\mathcal{G}}(Z)$ in a broad range of situations.

Theorem 4.1.3. Let $G \in \mathcal{W}$ be a 1-connected grouplike topological monoid and $r\colon Z \to S^n (n > 2)$ be a principal G-fibration with characteristic element $\chi \in \pi_{n-1}(G)$.

If (i) $2\chi \neq 0$ then $\mathcal{E}^{\mathcal{G}}(Z) \cong (\pi_n(G)/\langle \chi \circ \eta \rangle)$, where η is the generator of $\pi_n(S^{n-1})$.

If (ii) $2\chi = 0$ then there is a short exact sequence

$$(\pi_n(G)/\langle \chi \circ \eta \rangle) \to \mathcal{E}^{\mathcal{G}}(Z) \to Z_2.$$

Proof. It follows from the exact sequence of $p_G\colon EG \to BG$ that $\pi_{n-1}(G) \cong \pi_n(B_G)$, and it is standard that χ corresponds to $[k]$ under this isomorphism, where $k\colon S^n \to BG$ is a classifying map for r. Now $k_{\#}\colon \pi_1(\mathcal{H}(S^n), 1) \to \pi_1(Map(S^n, BG), k)$ can be identified, as in 3.2, with the homomorphism $\pi_{n+1}(S^n) \to \pi_{n+1}(BG)$ induced by k and, since $\pi_{n+1}(S^n) \cong \pi_n(S^{n-1})$ and $\pi_{n+1}(B_G) \cong \pi_n(G)$, with the homomorphism $\pi_n(S^{n-1}) \to \pi_n(G)$ defined by composition with χ. Hence $\pi_1(Map(S^n, BG), k)/k_{\#}(\pi_1(\mathcal{H}(S^n), 1))$ is isomorphic to $\pi_n(G)/\langle \chi \circ \eta \rangle$. The result follows from 3.3 (ii).

Remark 4.1.4. If we wish to delete the $\pi_1(G) = 0$ assumption in 4.1.3 then $\pi_n(G)$ must be replaced by $\pi_n(G)/[k, \pi_1(G)]$, where $[\ ,\ ]$ denotes the Samelson product. The appropriate modification of the proof is similar to parts of the proof of [Ts, th. 2.2].

4.2 Principal G-bundles

Let G be a topological group, $r_i\colon Z_i \to C_i$ ($i = 1$ and 2) be principal G-bundles and $f\colon Z_1 \to Z_2$ be a G-map. Now C_1 is, to within a homeomorphism, the quotient space Z_1/G so it follows that there is a unique map $g\colon C_1 \to C_2$ such that $r_2 f = g r_1$, hence there is a natural bijective correspondence between the set $GMap(Z_1, Z_2)$ of G-maps $f\colon Z_1 \to Z_2$ and the set of G-pairwise maps (f, g) from r_1 to r_2. *Hence we see that the use of $\mathcal{F}$-pairwise maps in our general theory is a natural generalization of the use of G-maps for principal G-bundles.*

We now describe the homotopy equivalence of categories discussed in the first section of this paper. Assume $G \in \mathcal{W}$ and taking $p_G\colon EG \to BG$ to denote a universal principal G-bundle we see that the functor $\phi = \phi(p_G)\colon \boldsymbol{FrGSp} \to \boldsymbol{Dfib}_{BG}$ is a homotopy equivalence of categories in the sense described in theorem 2.1, but with BG and $GMap(\ ,\ )$ replacing $B\mathcal{F}$ and $\mathcal{F}Pws(\ ,\ )$ respectively. This is not quite a special case of 2.1 as we are working with $\mathcal{F}$-bundles rather than $\mathcal{F}$-fibrations, but the proof is parallel. However principal bundles are examples of principal fibrations so, for topological groups G, we have $\mathcal{E}^G(Z) = \mathcal{E}^{\mathcal{G}}(Z)$ and hence 2.2 - 2.5 (with $\mathcal{F} = \mathcal{G}$) and 4.1.1 - 4.1.3 all apply to $\mathcal{E}^G(Z)$.

The group $\mathcal{E}^G(Z)$ has been investigated in the case where C is a sphere [Ts] or C is a suspension [OT]. Theorem 4.1.3 should be compared to [Ts, th. 2.1] which embeds that same group in an equivalent 5-term exact sequence and then uses this sequence to compute quite a number of specific groups $\mathcal{E}^G(Z)$. Our formulation is on the one hand the more general, in the sense that it applies to principal G-fibrations where G is a grouplike topological monoid, rather than to principal G-bundles where G is a compact topological group, and on the other hand less general in that we add the condition $\pi_1(G)$ to the [Ts, th. 2.1] assumption that $\pi_0(G) = 0$ (but see our remark 4.1.4). Our reason for imposing this restrictive condition is just that 4.1.3 then takes on a rather simple form.

4.3 Hurewicz fibrations and Dold fibrations

Let F and $\mathcal{H}$ be as described in 1.2. If we follow 1.6 $\mathcal{H}$-fibrations $r: Z \to C$ are just Hurewicz fibrations (see [May ex. 6.6(ii) and cor. 9.5 (ii)]), if 1.7 Dold fibrations, in each case with fibres homotopy equivalent to F. On either approach there exists universal $\mathcal{H}$-fibration $p_{\mathcal{H}}: E\mathcal{H} \to B\mathcal{H} = B\mathcal{H}(F)$ (often written $p_\infty: E_\infty \to B_\infty$). Then $\mathcal{H}\boldsymbol{fib}$ (either sense) and $\boldsymbol{Dfib}_{B\mathcal{H}(F)}$ are homotopy equivalent categories, as described in 2.1, and the group $\mathcal{E}^{\mathcal{H}}(Z)$ may be studied via 2.2 - 2.5 and theorem 3.3.

One reason for being interested in $\mathcal{E}^{\mathcal{H}}(Z)$ is that it sometimes coincides with $\mathcal{E}(Z)$; hence it follows from 2.2 that $\mathcal{E}(Z)$ is then isomorphic to the group of self-$FHEs$ $\mathcal{E}_{B\mathcal{H}(F)}(\bar{C})$. A detailed analysis of this topic is beyond the scope of this paper, but we illustrate the relationship by means of an example.

First we must introduce a based version $\mathcal{E}^{\mathcal{H}}_{\bullet}(Z)$ of $\mathcal{E}^{\mathcal{H}}(Z)$ by assuming that Z and C have base points and $r: Z \to C$ is a base point preserving map, and then modifying the definition of $\mathcal{E}^{\mathcal{H}}(Z)$ by requiring that every map in sight preserves base points.

Lemma 4.3.1. Let $r: Z \to C$ be a base point preserving map between spaces with non-degenerate base points. If further r is a Hurewicz fibration with fibres homotopy equivalent to F, and the spaces F and C simply connected then the groups $\mathcal{E}^{\mathcal{H}}(Z)$ and $\mathcal{E}^{\mathcal{H}}_{\bullet}(Z)$ are isomorphic.

Proof. This is a direct generalization of the proof for the easy result that if X is simply connected and has a non-degenerate base point then the sets of free and based homotopy classes of maps $X \to X$ coincide; details are left to the reader.

Example 4.3.2. (i) Let Z be a path-connected space with a non-degenerate base point and just two non-zero homotopy groups, i.e. π_1 and π_2 (π_2 being finitely generated) in dimensions n_1 and n_2, respectively, where $1 < n_1 < n_2$. Taking a Postnikov factorization of Z, $K(\pi_2, n_2) \to Z \underset{r}{\to} K(\pi_1, n_1)$, where the fibration r has classifying map $k: K(\pi_1, n_1) \to B\mathcal{H}(K(\pi_2, n_2))$,

then $\mathcal{E}(Z) \cong \mathcal{E}_{B\mathcal{H}(K(\pi_2,n_2))}(\overline{K(\pi_1, n_1)})$.

(ii) If $c: B\mathcal{H}(K(\pi_2, n_2)) \to B\mathcal{H}(Prin_F Z)$ is a classifying map for $q(k)$ then $\mathcal{E}(Z) \cong \pi_1(Map(B\mathcal{H}(K($

Proof. (i) It follows from [Ka(D), th. 2.2] that $\mathcal{E}(Z) \cong \mathcal{E}^{\mathcal{H}}_{\bullet}(Z)$ and the result follows from 4.3.1 and 2.2 (as explained in 2.2 π_2 is finitely generated implies that $\mathcal{H}(K(\pi_2, n_2)) \in \mathcal{W}$ and hence 2.2 applies).

(ii) This is immediate from (i) and 2.5.

4.4 Sectioned fibrations

In recent years there has been a development of the topic of sectioned fibrations (= ex-fibrations) and of the associated ex-homotopy theory. Much of this interest has been related to the work of I.M. James; two basic references are [Ja1] and [Ja2].

Let $(F, *)$ and $\mathcal{H}_0$ be as described in 1.3. The classifying space $B\mathcal{H}_0 = B\mathcal{H}_0(F)$ is the classifying space for $\mathcal{H}_0$-fibrations in the sense of either 1.6 or 1.7. In the latter case such a fibration is just a map $r: Z \to C$ with a section $s: C \to Z$ which, at least in cases with $C \in \mathcal{W}$ is numerably $L\mathcal{H}_0FHT$ (see 1.5). For the 1.6 sense see [May, ex. 6.9 (ii) and cor. 9.8].

It follows from 2.1 that the categories $\mathcal{H}_0\boldsymbol{Fib}$ of corresponding *sectioned fibrations* over spaces in $\mathcal{W}$ and *section preserving pairwise maps* = $\mathcal{H}_0$-pairwise maps) is homotopy equivalent to the category $\boldsymbol{Dfib}_{B\mathcal{H}_0(F)}$. The group $\mathcal{E}^{\mathcal{H}_0}(Z)$ may be studied via 2.2 - 2.5 and the results of section 3.

5 Generalized fibre bundles

Before attempting to prove theorem 2.2 one has to decide how the functor ϕ used there might be defined, taking into account that in the principal G-bundle case this is done in terms of the fibre bundle construction
$q_Z: Y \times_G Z \to B$.

Let G be a topological group, $q: Y \to B$ be a principal G-bundle and Z be a right G-space; there is an obvious induced right action of G on $Y \times Z$, which generates the equivalence relation that determines the quotient set $Y \times_G Z$. Now the proof that this relation is an equivalence relation uses the fact that G has inverses to show that it is a symmetric relation, so it is not clear that $Y \times_G Z$ can be defined in the case where G is a grouplike topological monoid and r is a principal G-fibration.

The key to resolving this difficulty lies in noticing that if $y \in Y$ and $z \in Z$ then the equivalence class $[y, z] = \{(yg, zg) | g \in G\}$ is in fact a function, *for a function is by definition a set of ordered pairs*, with domain yG and range zG. If the projection $Z \to Z/G = C$ is denoted r, $q(y) = b \in B$ and $r(z) = c \in Z/G$ then $yG = q^{-1}(b), zG = r^{-1}(c)$ and $[y, z]$ is a G-*Map* from $q^{-1}(b)$ to $r^{-1}(c)$. In fact it may be shown that each such G-map coincides with such an equivalence class. Hence $Y \times_G Z$ has underlying set $\bigcup_{b\in B, c\in C} GMap(Y_b, Z_c)$.

Theorem 5.1. (from [Be, appendix B]). If B is a Hausdorff space, $q: Y \to B$ is a principal G-bundle and Z is a right G-space then $Y \times_G Z$, regarded as a set of G-maps, can be given a type of compact-open topology that agrees with the usual quotient topology there.

This result, whilst not essential to our main line of argument, does motivate our reasoning. We therefore define the mapping space topology but do not give full details of the proof.

Outline proof. If Z is any space then $Z^\sim$ will be defined to have as underlying set the disjoint union of Z and a point ∞, and topology specified by the condition that any subset C of $Z^\sim$ is closed if either $C = Z^\sim$ of if C is closed in Z. Then we can define a function $j: Y \times_G Z \to Map(Y, Z^\sim)$ by $j(f)(y) = f(y)$ if $y \in dom(f)$ and $= \infty$ otherwise, where $f \in GMap(q^{-1}(b), r^{-1}(c))$ for some $b \in B$ and $c \in C$. We define $q_Z(f) = b$ and $r_Y: Y \times_G Z \to C$ by $r_Y(f) = c$. Topologizing $Map(Y, Z^\sim)$ with the ordinary compact-open topology we can topologize $Y \times_G Z$ with the initial (= strong) topology in ***CgTop*** relative to the outgoing functions j, q_Z and r_Y.

The usual and mapping space topologies on $Y \times_G Z$ can be seen to agree, in the special case where q is the projection $Y = B \times G \to B$, as follows. $Y \times_G Z$ in the quotient sense is then homeomorphic to $B \times (G \times_G Y) \cong B \times Y$; $Y \times_G Z$ with the mapping space topology may be shown (using a result similar to [BB, th 2.1]) to then be homeomorphic to $B \times GMap(G, Y) \cong B \times Y$.

Now an arbitrary principal G-bundle $q: Y \to B$ is locally trivial so the above special case tells us that, in general, the two topologies on $Y \times_G Z$ agree locally; it follows by the usual local to global routine that they agree globally.

The important point about this mapping space formulation of $Y \times_G Z$, is that it is not dependent on G having inverses, in fact will be shown elsewhere that it allows the development of a systematic theory of fibre bundles with structure monoid. For now we merely notice that there is a possibility of defining ϕ in the $\mathcal{F}$-fibration situation, i.e. where $q: Y \to B$ and $r: Z \to C$ are $\mathcal{F}$-fibrations, by replacing $Y \times_G Z$ by a "fibred mapping space" $Y \square_{\mathcal{F}} Z$ with underlying set $\bigcup_{b\in B, c\in C} \mathcal{F}(q^{-1}(b), r^{-1}(c))$. Fortunately the topology of such spaces has been investigated elsewhere, the next section includes a review of some of their properties.

6 Fibred product and fibred mapping spaces

In this section we set up the framework for the proof of theorem 2.1. The main result, proposition 6.5, should be viewed as a "fibred" version of the following set of easy statements concerning the category ***CgTop*** .

Let Y be a given space. There is a functor θ: ***CgTop*** $\to$ ***CgTop*** defined by $\theta(X) = X \times Y$ and $\theta(f) = f \times 1_Y$ where X is any space and f any map, and a functor ϕ: ***CgTop*** $\to$ ***CgTop*** defined by $\phi(Z) = Map(Y, Z)$ where Z is any space and $\phi(g)\colon Map(Y, Z_1) \to Map(Y, Z_2)$ is defined by $\phi(g)(h) = g \circ h$, where $g \in Map(Z_1, Z_2)$ and $h \in Map(Y, Z_1)$. Then

(i) θ is left adjoint to ϕ,

(ii) there is a natural transformation $d_X\colon X \to \phi\theta(X) = Map(Y, X \times Y)$ defined by $d_X(x)(y) = (x, y)$, where $x \in X$ and $y \in Y$,

(iii) there is a natural transformation $e_Z\colon \theta\phi(Z) = Map(Y, Z) \times Y \to Z$,
$e_Z(f, y) = f(y)$, where $f \in Map(Y, Z)$ and $y \in Y$.

(iv) for all choices of spaces X_1, X_2 the function
$\theta_{12}\colon Map(X_1, X_2) \to Map(X_1 \times Y, X_2 \times Y), \theta_{12}(f) = \theta(f)$, where
$f \in Map(X_1, X_2)$, is continuous and

(v) for all choices of spaces Z_1 and Z_2 the function
$\phi_{12}\colon Map(Z_1, Z_2) \to Map(Map(Y, Z_1), Map(Y, Z_2)), \phi_{12}(g) = \phi(g)$, where $g \in Map(Z_1, Z_2)$, is continuous.

In proposition 6.5 product spaces are replaced by fibred product spaces, mapping spaces by fibred mapping spaces and the verification, which in the above case depends on the ordinary exponential law, is dealt with using a fibred exponential law. Furthermore the result is presented in the language of $\mathcal{F}$-fibrations.

We review some point set topological properties of fibred mapping spaces: more information on them is given in [BHP1] and [BHP2].

Let $(\mathcal{F}, F)$ be a category of enriched spaces, and $q\colon Y \to B$ and $r\colon Z \to C$ be $\mathcal{F}$-spaces. We form the set:

$$Y \Box_{\mathcal{F}} Z = \bigcup_{\substack{b \in B \\ c \in C}} \mathcal{F}(q^{-1}(b), r^{-1}(c)),$$

where $\mathcal{F}(q^{-1}(b), r^{-1}(c))$ denotes the set of morphisms in $\mathcal{F}$ from $q^{-1}(b)$ to $r^{-1}(c)$, and define $q \swarrow_{\mathcal{F}} r\colon Y \Box_{\mathcal{F}} Z \to B$ and $q \searrow_{\mathcal{F}} r\colon Y \Box_{\mathcal{F}} Z \to C$ to be the obvious projections. Referring back to section 5 the reader will notice that these projections correspond to q_Z and r_Y respectively. Defining $j\colon Y \Box_{\mathcal{F}} Z \to Map(Y, Z^{\sim})$ by $j(f)(y) = f(y)$ if $y \in dom(f)$ and $= \infty$ otherwise, $Y \Box_{\mathcal{F}} Z$ will be given the initial (= strong) topology in ***CgTop*** relative to the functions $q \swarrow_{\mathcal{F}} r, q \searrow_{\mathcal{F}} r$ and j.

Proposition 6.1. (Fibred exponential law). Let B be a Hausdorff space, $p\colon X \to B$ a map, and $q\colon Y \to B$ and $r\colon Z \to C$ be $\mathcal{F}$-spaces. Then there is a bijective correspondence between maps $f\colon X \to Y \Box_{\mathcal{F}} Z$ over B, i.e. such that $(q \swarrow_{\mathcal{F}} r) \circ f = p$, and $\mathcal{F}$-pairwise maps (g, h) of p^*q to r, where $g\colon X \sqcap Y \to Z$ is defined by $g(x, y) = f(x)(y)$ for all x and y with $p(x) = q(y)$, and $h = (q \searrow_{\mathcal{F}} r) \circ f$. Further the rule $f \to (g, h)$ is a homeomorphism $\epsilon = \epsilon(X, Z)\colon Map_B(X, Y \Box_{\mathcal{F}} Z) \to \mathcal{F}Pws(X \sqcap Y, Z)$.

For example if $B = C = * =$ a singleton space, then p^*q is the projection $pr_X\colon X \times Y \to X$, Y and Z are objects of $\mathcal{F}$ and $Y \Box_{\mathcal{F}} Z = \mathcal{F}(Y, Z)$. Then 6.1 describes a bijective correspondence between maps $f\colon X \to \mathcal{F}(Y, Z)$ and $\mathcal{F}$-pairwise maps (g, h) from the $\mathcal{F}$-space pr_X to the $\mathcal{F}$-space $Z \to *$. Now $h\colon X \to *$ is the same constant map in every case so we can describe the relationshp completely by $f(x)(y) = g(x, y)$, for $x \in X$ and $y \in Y$.

Proof of 6.1. The bijection is justified in [BHP2, lem.1.1]; it can be shown to be a homeomorphism by using exponential laws to verify that functions k from spaces into $Map_B(X, Y \Box_{\mathcal{F}} Z)$ are continuous if and only if the corresponding $\epsilon \circ k$ are continuous.

Corollary 6.2.(i) (Fibred evaluation map). There is an $\mathcal{F}$-pairwise map $\langle e_Z, q \searrow_{\mathcal{F}} r\rangle$ from $(q \swarrow_{\mathcal{F}} r)^* q$ to r, where $e_Z: (Y\square_{\mathcal{F}}Z) \sqcap Y \to Z$ is defined by $e_Z(s,y) = s(y)$.

(ii) There is a map $d_X: X \to Y\square_{\mathcal{F}}(X \sqcap Y)$ over B; i.e., such that we have $(q \swarrow_{\mathcal{F}} (p^* q))d_X = p$, defined by $d_X(x)(y) = (x,y)$, where $p(x) = q(y)$

For example if $B = C = *$ then $e_Z: \mathcal{F}(Y,Z) \times Y \to Z$ is the evaluation map in a more familiar sense. In the same situation $Y\square_{\mathcal{F}}(X \times Y)$ can be identified with $X \times \mathcal{F}(Y,Y)$. At the underlying set level this is clear from the definitions of the mapping spaces involved; at the topological level it can be checked by using exponential laws to verify that functions into one of these spaces are continuous if and only if the corresponding functions into the other are continuous. Then $d_X: X \to X \times \mathcal{F}(Y,Y)$ is given by $d_X(x) = (x, 1_Y), x \in X$.

Proof. These maps are obtained by applying 6.1 to the identity on $Y\square_{\mathcal{F}}Z$ viewed as a map over B and to the pairwise map $\langle 1_{X\sqcap Y}, 1_X\rangle$, respectively.

Corollary 6.3. Given $\mathcal{F}$-spaces $q: Y \to B, r_i: Z_i \to C_i$ for $i = 1$ and 2 and an $\mathcal{F}$-pairwise map $\langle g,h\rangle$ from r_1 to r_2 there is an induced map over $B, \phi(\langle g,h\rangle): Y\square_{\mathcal{F}}Z_1 \to Y\square_{\mathcal{F}}Z_2$ over $B, \phi(\langle g,h\rangle)(s) = (g|r_1^{-1}(c)) \circ s$ where $s \in \mathcal{F}(q^{-1}(b), r_1^{-1}(c))$ and $g|r_1^{-1}(c) \in \mathcal{F}(r_1^{-1}(c), r_2^{-1}(h(c)))$ for some $b \in B$ and $c \in C$.

For example if $B = C_1 = C_2 = *$ then Y, Z_1 and Z_2 are objects of $\mathcal{F}$, and an $\mathcal{F}$-pairwise map $\langle g,h\rangle$ from r_1 to r_2 consists of a morphism $g: Z_1 \to Z_2$ in $\mathcal{F}$ and the constant map $h: * \to *$. We have $Y\square_{\mathcal{F}}Z_i = \mathcal{F}(Y,Z_i)$ $(i = 1$ and $2)$ and $(\phi(\langle g,h\rangle): \mathcal{F}(Y,Z_1) \to \mathcal{F}(Y,Z_2)$ is the map induced by composition with g in the usual way.

Proof. Taking $\epsilon = \epsilon(Y\square_{\mathcal{F}}Z_1, Z_2), \phi(\langle g,h\rangle)$ is ϵ^{-1} of the composite of the $\mathcal{F}$-pairwise maps $\langle g,h\rangle$ and $\langle e(Z_1), q \searrow_{\mathcal{F}} r_1\rangle$.

The category of spaces and maps over B will be denoted by $\boldsymbol{Sp_B}$; that of $\mathcal{F}$-spaces and $\mathcal{F}$-pairwise maps by $\boldsymbol{\mathcal{F}sp}$. We notice that $\boldsymbol{Dfib_B}$ and $\boldsymbol{\mathcal{F}fib}$ are full subcategories of $\boldsymbol{Sp_B}$ and $\boldsymbol{\mathcal{F}sp}$ respectively. Extending a definition from section 2: if $q: Y \to B$ is an $\mathcal{F}$-space there is a functor $\theta = \theta(q): \boldsymbol{Sp_B} \to \boldsymbol{\mathcal{F}sp}$ defined by $\theta(q)(p): \theta(X) \to X$ is $p^*q: X \sqcap Y \to X$, where $p: X \to B$ is any object of $\boldsymbol{Sp_B}$, and morphisms are sent to morphisms by $\theta(f) = \langle f \times 1_Y, f\rangle$.

Proposition 6.4. If B is a Hausdorff space the $\mathcal{F}$-space $q: Y \to B$ determines a covariant functor $\phi: \boldsymbol{\mathcal{F}sp} \to \boldsymbol{Sp_B}$ as follows: $\phi(r): \phi(Z) \to B$ is $q \swarrow_{\mathcal{F}} r: Y\square_{\mathcal{F}}Z \to B$ and $\phi(\langle g,h\rangle)$ is as defined in 6.3, where $r: Z \to C$ is an $\mathcal{F}$-space and $\langle g,h\rangle$ an $\mathcal{F}$-pairwise map.

Proof. This follows from 6.3.

Proposition 6.5. If B is a Hausdorff space and $q: Y \to B$ is an $\mathcal{F}$-space then the functors $\theta = \theta(q)$ and $\phi = \phi(q)$, as described in this section, are related by:

(i) θ is left adjoint to ϕ,

(ii) there are a natural transformation $d = d_X: X \to \phi\theta X$, for all spaces X over B,

(iii) the $\mathcal{F}$-pairwise maps $\langle e_Z, q \searrow_{\mathcal{F}} r\rangle$ are a natural tranformation, $\theta\phi Z \to Z$ for all $\mathcal{F}$-spaces $r: Z \to C$ over any space C,

(iv) for all spaces $p_i: X_i \to B$ over B, $i = 1$ and 2,

$$\theta_{12}: Map_B(X_1, X_2) \to \mathcal{F}Pws(\theta X_1, \theta X_2),\ \theta_{12}(f) = \langle\theta(f), f\rangle$$

is continuous, and

(v) for all $\mathcal{F}$-spaces $r_i: Z_i \to C_i, i = 1$ and 2,

$$\phi_{12}: \mathcal{F}Pws(Z_1, Z_2) \to Map_B(\phi Z_1, \phi Z_2),\ \phi_{12}(\langle g,h\rangle) = \phi(\langle g,h\rangle)$$

is continuous.

Proof. (i) is immediate from 6.1, whilst (ii) and (iii) are consequences of an easy standard result about adjoint functors [Mac, p.80], i.e. that the families of morphisms $X \to \phi\theta X$ and $\theta\phi Z \to Z$ corresponding to the families of identity morphisms $\theta X \to \theta X$ and $\phi Z \to \phi Z$, respectively, are natural transformations.

(iv) The map $d = d(X_2)$ induces a map

$$d_{\#}: Map_B(X_1, X_2) \to Map_B(X_1, Y \square_{\mathcal{F}} (X_2 \sqcap Y)), d_{\#}(f) = d \circ f,$$

$f \in Map_B(X_1, X_2)$, and $\theta_{12} = \epsilon \circ d_{\#}$, where ϵ is the homeomorphism

$$\epsilon\,(X_1, X_2 \sqcap Y): Map_B(X_1, Y \square_{\mathcal{F}} (X_2 \sqcap Y)) \to \mathcal{F}Pws(X_1 \sqcap Y, X_2 \sqcap Y)$$

(v) The map $e = e(Z_1)$ induces a map

$$e^{\#}: \mathcal{F}Pws(Z_1, Z_2) \to \mathcal{F}Pws(Y \square_{\mathcal{F}} Z_1) \sqcap Y, Z_2),\ e^{\#}(g) = g \circ e,$$

$g \in \mathcal{F}Pws(Z_1, Z_2)$, and $\phi_{12} = \epsilon^{-1} \circ e^{\#}$, where ϵ is the homeomorphism

$$\epsilon(Y \square_{\mathcal{F}} Z_1, Z_2): Map_B(Y \square_{\mathcal{F}} Z_1, Y \square_{\mathcal{F}} Z_2) \to \mathcal{F}Pws((Y \square_{\mathcal{F}} Z_1) \sqcap Y, Z_2).$$

7 Proof of theorem 2.1

We first require two more results concerning fibred mapping spaces. Let $q: Y \to B$ and $r: Z \to C$ be $\mathcal{F}$-spaces.

(7.1) If $b \in B$ then the fibre of $q \swarrow_{\mathcal{F}} r: Y \square_{\mathcal{F}} Z \to B$ over b is $Prin_{p^{-1}(b)} Z$.

Proof. This identification of spaces is obvious at the level of underlying sets. It can be shown to be a homeomorphism by checking, using exponential laws, that any function from any space into one of these spaces is continuous if and only if the corresponding function into the other space is also continuous.

(7.2) If the $\mathcal{F}$-space $q: Y \to B$ is also an $\mathcal{F}$-fibration then $q \square_{\mathcal{F}} r: Y \square_{\mathcal{F}} Z \to B$ is a Dold fibration.

The result of [Bo1] verifying this when we work in the 1.7 sense of $\mathcal{F}$-spaces satisfying the $\mathcal{F}WCHP$ obviously also includes the 1.6 case of $\mathcal{F}$-spaces satisfying the $\mathcal{F}CHP$.

Proof of theorem 2.1. (i) Our argument is built on the basis of proposition 6.5, taking $q: Y \to B$ to be $p_{\mathcal{F}}: E\mathcal{F} \to B\mathcal{F}$. If $p: X \to B\mathcal{F}$ is a Dold fibration with fibres in $\mathcal{W}$ then X, the base space for $p^*(p_{\mathcal{F}})$, is also in $\mathcal{W}$ [Sc, th. 2], so $\theta(p_{\mathcal{F}})(p) = p^*(p_{\mathcal{F}})$ is an object of $\mathcal{F}$***fib***.

If $r: Z \to C$ is an $\mathcal{F}$-space then $p_{\mathcal{F}} \swarrow_{\mathcal{F}} r$ is a Dold fibration (7.2). Given that $C \in \mathcal{W}$ we know that $\mathcal{F}(F) \in \mathcal{W}$ and, applying [Sc, th.2] to $prin_F r$, that $Prin_F Z \in \mathcal{W}$. Now the fibres of $(p_{\mathcal{F}}) \swarrow_{\mathcal{F}} r$ have the homotopy type of $Prin_F Z$ (7.1) and so are in $\mathcal{W}$, hence $\phi(r)$ is an object of $\boldsymbol{Dfib}_{B\mathcal{F}}$. The proof of 6.5(a) now covers that for 2.1(a).

(ii) Given a Dold fibration $p: X \to B\mathcal{F}$ we notice that if $b \in B\mathcal{F}$

$$d|p^{-1}(b): p^{-1}(b) \to ((p_{\mathcal{F}}) \swarrow_{\mathcal{F}} (p^*(p_{\mathcal{F}})))^{-1}(b)$$

is the composite:

$$p^{-1}(b) \underset{i}{\to} X \sqcap Prin_{p_{\mathcal{F}}^{-1}(b)} E\mathcal{F} \underset{\xi}{\to} Prin_{p_{\mathcal{F}}^{-1}(b)}(X \sqcap E\mathcal{F}),$$

where i is the inclusion of the fibre in the total space of the Dold fibration $(prin_{p_{\mathcal{F}}^{-1}(b)}p_{\mathcal{F}})^{*}p$ with $i(w) = (w, 1: p_{\mathcal{F}}^{-1}(b) \to p_{\mathcal{F}}^{-1}(b))$ for $w \in p^{-1}(b)$, and ξ the homeomorphism described in 1.4. Now this last fibration has contractible base space (it has the homotopy type of $Prin_{\mathcal{F}}(E\mathcal{F})$ and see 1.12), hence i is a homotopy equivalence [Va, p. 267] and so is $d|p^{-1}(b)$. Further, the map $(p_{\mathcal{F}}) \swarrow_{\mathcal{F}} (p^{*}(p_{\mathcal{F}}))$ is a Dold fibration (see 1.8 and 7.2) and d is therefore a map over $B\mathcal{F}$ between the total spaces of Dold fibrations; it follows by [Do, th. 6.3] that d is an FHE.

(iii) We first consider the case where the induced $\mathcal{F}$-fibration r used is of the form $p^{*}(p_{\mathcal{F}})$ for some Dold fibration $p: X \to B\mathcal{F}$. Now $\langle \theta(d_X), d_X \rangle$ is an $\mathcal{F}$-pairwise map so it follows (also using 6.5c) that there is a commutative diagram:

$$\begin{array}{ccccc} X \sqcap E\mathcal{F} & \xrightarrow{\theta(d)} & [E\mathcal{F} \square_{\mathcal{F}} (X \sqcap E\mathcal{F})] \sqcap E\mathcal{F} & \xrightarrow{e = e(X \sqcap E\mathcal{F})} & X \sqcap E\mathcal{F} \\ \downarrow {\scriptstyle p^{*}(p_{\mathcal{F}})} & & \downarrow {\scriptstyle [p_{\mathcal{F}} \swarrow_{\mathcal{F}} (p^{*}(p_{\mathcal{F}}))]^{*} p_{\mathcal{F}}} & & \downarrow {\scriptstyle p^{*}(p_{\mathcal{F}})} \\ X & \xrightarrow{d = d_X} & E\mathcal{F} \square_{\mathcal{F}} (X \sqcap E\mathcal{F}) & \xrightarrow{(p_{\mathcal{F}}) \searrow_{\mathcal{F}} (p^{*}(p_{\mathcal{F}}))} & X \end{array}$$

If $(x, w) \in X \sqcap E\mathcal{F}$ then $(e\theta(d))(x, w) = e(d(x), w) = d(x)(w) = (x, w)$, and so $e \circ \theta(d) = 1: X \sqcap E\mathcal{F} \to X \sqcap E\mathcal{F}$; also $([(p_{\mathcal{F}}) \searrow_{\mathcal{F}} (p^{*}(p_{\mathcal{F}}))] \circ d)(x)$ is, by the definition of $\searrow_{\mathcal{F}}$, the image under $p^{*}(p_{\mathcal{F}})$ of the range of $d(x)$ and $d(x)(w) = (x, w)$, so $([(p_{\mathcal{F}}) \searrow_{\mathcal{F}} (p^{*}(p_{\mathcal{F}}))] \circ d)(x) = (p^{*}(p_{\mathcal{F}}))(x, w) = x$ and hence $[(p_{\mathcal{F}}) \searrow_{\mathcal{F}} (p^{*}(p_{\mathcal{F}}))] \circ d = 1_X$. We know $[(p_{\mathcal{F}}) \swarrow_{\mathcal{F}} (p^{*}(p_{\mathcal{F}}))] \circ d = p$ (6.5b) so

$$p^{*}(p_{\mathcal{F}}) = [[(p_{\mathcal{F}}) \swarrow_{\mathcal{F}} (p^{*}(p_{\mathcal{F}}))] \circ d]^{*}(p_{\mathcal{F}}) = (d)^{*}[(p_{\mathcal{F}}) \swarrow_{\mathcal{F}} (p^{*}(p_{\mathcal{F}}))]^{*}(p_{\mathcal{F}}),$$

i.e. $p^{*}(p_{\mathcal{F}})$ is the pullback of $[(p_{\mathcal{F}}) \swarrow_{\mathcal{F}} (p^{*}(p_{\mathcal{F}}))]^{*}(p_{\mathcal{F}})$ over the homotopy equivalence d (see (ii)), hence the left hand rectangle is a pullback diagram and so by 1.9 $\langle \theta(d), d \rangle$ is an $\mathcal{F}PHE$. It follows that its left inverse, the right hand rectangle or more precisely the $\mathcal{F}$-pairwise map $\langle e, [p_{\mathcal{F}} \searrow_{\mathcal{F}} (p^{*}(p_{\mathcal{F}}))] \rangle$, is also an $\mathcal{F}PHE$.

We now move on to the case of an $\mathcal{F}$-fibration $k^{*}(p_{\mathcal{F}}): X \sqcap E\mathcal{F} \to X$ induced via a map $k: X \to B\mathcal{F}$. Factoring k as $q(k) \circ h(k)$, where $h(k): X \to \bar{X}$ is a homotopy equivalence, we have $k^{*}(p_{\mathcal{F}}) = (q(k) \circ h(k))^{*}(p_{\mathcal{F}}) = h(k)^{*}(q(k)^{*}(p_{\mathcal{F}}))$, hence it follows from 1.9 that $k^{*}(p_{\mathcal{F}})$ is $\mathcal{F}PHE$ to $q(k)^{*}(p_{\mathcal{F}})$ and then, by the naturality of θ and ϕ, that (c) holds for all $\mathcal{F}$-fibrations of the form $k^{*}(p_{\mathcal{F}})$. Now all $\mathcal{F}$-fibrations over spaces in $\mathcal{W}$ are $\mathcal{F}FHE$ to $\mathcal{F}$-fibrations of the form $k^{*}(p_{\mathcal{F}})$ (see 1.11b), so (c) holds in general.

(iv) and (v). These follow from (ii) and (iii) for $d_{\#}$ and $e_{\#}$ in the proofs of 6.5(iv) and 6.5(v) must be homotopy equivalences.

REFERENCES

Be R. Bencivenga, On the groups of automorphisms of principal and fibre bundles, Ph.D. thesis, Memorial University of Newfoundland, March 1982.

Bo1 P. Booth, Classifying spaces for a general theory of fibrations (to appear).

Bo2 P. Booth, Maps between classifying spaces and the classification of fibrations (to appear).

BB P. Booth and R. Brown, On the application of fibred mapping spaces to exponential laws for bundles, ex-spaces and other categories of maps, Gen. Top. and its Applications 8 (1978) 165-179.

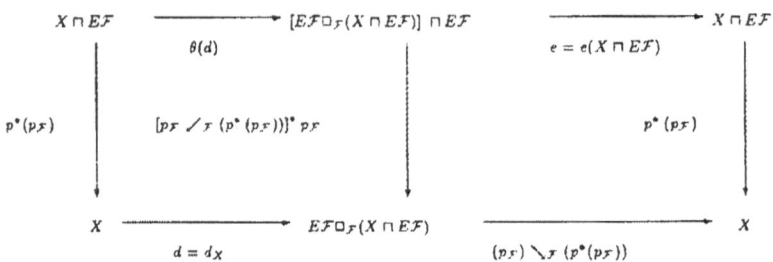

BHP1 P. Booth, P. Heath and R. Piccinini, Fibre preserving maps and functional spaces, Lecture notes in Math. 673 (Springer-Verlag, Berlin, 1978) 158-167.

BHP2 P. Booth, P. Heath and R. Piccinini, Characterizing universal fibrations, Lecture notes in Math. 673 (Springer-Verlag, Berlin 1978) 168-184.

BHMP P. Booth, P. Heath, C. Morgan, and R. Piccinini, H-spaces of self-equivalences of fibrations and bundles, Proc. Lond. Math. Soc. (3) 49 (1984) 111-127.

Bor A. Borel, Topics in the homology theory of fibre bundles, Lecture notes in Math. 36, Springer-Verlag, Berlin 1967.

Do A. Dold, Partitions of unity in the theory of fibrations, Ann. of Math. 78 (1963) 223-255.

DDK E. Dror, W.G. Dwyer and D.M. Kan, Equivariant maps which are self-homotopy equivalences, Proc. Amer. Math. Soc. 80 (1980) 670-672.

DK W.G. Dwyer and D.M. Kan, Reducing equivariant homotopy theory to the theory of fibrations, Contemporary Mathematics, Vol 37 (1985) 35-49.

Fu1 M. Fuchs, Borel fibrations and G-spaces, Manuscripta Math. 58 (1987) 377-380.

Fu2 M. Fuchs, Equivariant maps up to homotopy and Borel spaces. Pub. Math. Univ. Aut. de Barcelona (1984) 79-102.

Ja1 I.M. James, Ex-homotopy theory, Illinois J. Math. 15 (1971), 324-337.

Ja2 I.M. James, Alternative homotopy theories, L'Enseignement Mathématique, XXIII, fasc. 3-4 (1977) 221-237.

Ka(D) D. Kahn, Induced maps for Postnikov systems, Trans. Amer. Math. Soc. 107 (1963) 432-450.

Ka(P) P.J. Kahn, Some function spaces of CW-type Proc. Amer. Math. Soc. 90 (1984) 599-607.

Mac S. MacLane, Categories for the working mathematician. Berlin: Springer-Verlag, Berlin, 1971.

May J.P. May, Classifying spaces and fibrations, Memoirs Amer. Math. Soc. 155 (1975).

Mi1 J. Milnor, Construction of universal bundles II, Ann. of Math. 63 (1956) 430-436.

Mi2 J. Milnor, On spaces having the homotopy type of a CW- complex. Trans. Amer. Math. Soc. 90 (1959) 272-280.

OT H. Oshima and K. Tsukiyama, On the group of equivariant self equivalences of free actions, Publ. RIMS, Kyoto Univ. 22 (1986) 905-923.

Sc R. Schön, Fibrations over a CWh-base, Proc. Amer. Math. Soc. 62 (1977) 165-166.

SV R. Schwaenzl and R.M. Vogt, Coherence in homotopy group actions, Transformation groups, Lecture notes in Math. 1217 (Springer-Verlag, Berlin, 1986) 364-390.

Sh W. Shih, On the group $\mathcal{E}[X]$ of homotopy equivalence maps. Bull. Amer. Math. Soc. 70 (1964) 361-365.

Sp E. Spanier, Algebraic Topology, McGraw-Hill, New York, 1966.

St N. Steenrod, A convenient category of topological spaces, Michigan Math. J. 14 (1967) 133-152.

Th R. Thom, L'homologie des espaces fonctionnels, Colloque. de Topologie Algébrique, Louvain (1956).

Ts K. Tsukiyama, Equivariant self equivalences of principal fibre bundles, Proc. Camb. Phil. Soc. 98 (1985) 87-92.

Va K. Varadarajan, On fibrations and category, Math. Zeitschr. 88 (1965) 267-273.

Vo R.M. Vogt, Convenient categories of topological spaces for homotopy theory, Arch. Math. 22 (1971) 545-555.

Wh G.W. Whitehead, Elements of homotopy theory, Springer-Verlag, Berlin-Heidelberg-New York, (1978).

On the Groups $\mathcal{E}(X \times Y)$ and $\mathcal{E}_B^B(X \times_B Y)$ †.

P.I. Booth and P.R. Heath
Department of Mathematics and Statistics
Memorial University of Newfoundland
St. John's, Newfoundland, Canada A1C 5S7

The main aim of this paper is to study the group $\mathcal{E}(X \times Y)$, under composition, of based homotopy classes of based self homotopy equivalences of the product $X \times Y$ of the based topological spaces X and Y. Let $[X,Y]^*$ denote the set of basepoint preserving homotopy classes of based maps from X to Y, and $H(X)$ the topological monoid of free self homotopy equivalences of X, with the identity 1_X as base point. In our main theorem we will give simple conditions under which there is a split short exact sequence

$$1 \to [X,H(Y)]^* \xrightarrow{\iota} \mathcal{E}(X \times Y) \xrightarrow{\Phi} \mathcal{E}(X) \times \mathcal{E}(Y) \to 1,$$

of groups and homomorphisms; it will then follow that $\mathcal{E}(X \times Y)$ is a semidirect product of $\mathcal{E}(X) \times \mathcal{E}(Y)$ and $[X,H(Y)]^*$ (theorem 2.7).

This result clarifies, unifies, and generalizes results of [AY], [N], [OSS] and [Y], the most general of these being [AY] and [Y]. J.W. Rutter [R] has pointed out that the analogue of our result given in [AY] may require that a further condition be imposed: he suggests (using our notation) that X might be either an H-space or a co-H-space. We avoid this type of condition by replacing the (a) portion of the hypothesis of [AY] (augmented in this way) by something more general, namely that $[Y,H(X)]^* = 0$ (but see corollary 2.8, and compare it with [AY] and [R]). We also introduce another type of condition, the *induced equivalence property* (IEP), which is a necessary and sufficient condition for the map Φ above to be well defined (see 2.3(i)). This important point is dealt with elsewhere (except where it is not considered at all) by imposing more specific conditions. For further details, including an explanation of how our 2.7 generalizes the

---

† This paper is in final form and no version of it will be submitted elsewhere.

main theorems of [Y], the reader should see the preamble to Example 2.10

One of our motives for writing this paper was to produce a theory that could be generalized to the situation where X and Y are "over B under B" spaces i.e. ex-spaces. We make an announcement concerning such a theory for the group $\mathcal{E}_B^B(X \times_B Y)$, of ex-homotopy classes of ex-fibre homotopy equivalences of a fibred product (pullback) of ex-fibrations.

The outline of the paper is as follows: in Section 1 we give an overview of the notation and also the preliminaries, in section 2, the exposition and results about $\mathcal{E}(X \times Y)$, while Section 3 contains the announcement mentioned above.

We would like to express our appreciation to Chris Morgan for many helpful discussions and contributions.

Section 1. Notation and preliminaries.

We work in the context of the category of *compactly generated spaces* [V;theorem 5.1], i.e. spaces having the final (= weak) topology relative to all incoming maps from compact Hausdorff spaces. Any space can be *cg-ified*, i.e. made into a compactly generated space, by retopologizing it with this final topology. All mapping spaces will be assumed to be compactly generated, that is they will carry the cg-ification of the compact-open topology.

We recall the exponential law for mapping spaces: if X, Y and Z are spaces and M(Y,Z) denotes the space of maps of Y into Z then there is a bijective correspondence between the set of maps $f:X \to M(Y,Z)$ and the set of maps $\tilde{f}:X \times Y \to Z$ defined by $\tilde{f}(x,y) = f(x)(y)$, for all $x \in X$, $y \in Y$ [V;theorem 3.6]. For a *Top* version under appropriate conditions see [Sp;p6].

Let A and B be given spaces. An *over B under A space* X, more precisely $(p,s) = (p:X \to B, s:A \to X)$, will consist of maps $p:X \to B$ and $s:A \to X$. If (p,s) and $(q,t) = (q:Y \to B, t:A \to Y)$ are over B under A spaces then an *over B under A map* $X \to Y$ will be a map $f:X \to Y$ such that $qf = p$ and $fs = t$. The space of such maps will be denoted by $M_B^A(X,Y)$. An *over B under A homotopy*, denoted by $\simeq_B^A$, will consist of a homotopy $F:X \times I \to Y$ such that, for each $u \in I$, $F|X \times \{u\}:X \times \{u\} \to Y$ is an over B under A map (we identify $X \times \{u\}$ with X in the obvious way). The set of such homotopy classes of over B under A maps from X to Y will be denoted by $[X,Y]_B^A$. Clearly there is an associated concept of *over B under A homotopy equivalence.* We will use $H_B^A(X)$ to denote the *topological monoid, under composition, of over B under A homotopy equivalences of* X *into itself* and $\mathcal{E}_B^A(X) = \pi_0(H_B^A(X))$ the associated *group of over B under A homotopy classes of over B under A self homotopy equivalences of X.* Note that the basepoint of $H_B^A(X)$ is 1_X.

In order to conform more closely with standard notation whenever $A = \emptyset$ (denoting the empty space) occurs we will supress it, we will also supress $*$ (denoting a point) when it occurs as a subscript, but not when it occurs as a superscript. Thus for example we write $[X,Y]_*^*$ as $[X,Y]^*$, $H_*^*(X)$ as $H^*(X)$ and $H_*^{\emptyset}(X)$ as $H(X)$. Again in the interest of conformity to standard notation, we will be slightly inconsistent with our own convention and denote by $\mathcal{E}(X)$ the group $\pi_0(H^*(X))$ of based homotopy classes of based self homotopy equivalences of X. Since we do not consider groups of free homotopy equivalence classes in any sense in this paper, this should not cause any confusion.

We will need the following in Section 3: if we take $A = B$ and require that $ps = 1_B$, then over B under B spaces, maps, homotopies and homotopy equivalences will be called *ex-spaces*, *ex-maps*, *ex-homotopies* and *ex-fibre homotopy equivalences* (ex-FHEs) respectively. Then $H_B^B(X)$ will denote the topological monoid of self ex-FHEs of X and $\mathcal{E}_B^B(X)$ the group $\pi_0(H_B^B(X))$.

Let $pr_X:X \times Y \to X$ and $pr_Y:X \times Y \to Y$ be defined by $pr_X(x,y) = x$, and $pr_Y(x,y) = y$ where $(x,y) \in X \times Y$, then pr_X and pr_Y are Hurewicz fibrations. Further we define maps $in_X:X \to X \times Y$ and $in_Y:Y \to X \times Y$ by $in_X(x) = (x,y_0)$ and $in_Y(y) = (x_0,y)$ where x_0 and y_0 are the basepoints of X and Y respectively, and where $x \in X$ and $y \in Y$. If x_0 and y_0 are closed non-degenerate base points, then the maps in_X and in_Y are cofibrations (e.g. in_X can be identified with the inclusion $X \times \{y_0\} \subset X \times Y$, see [Br, 7.3.2]). If $h:X \times Y \to X \times Y$ then we will use the notation h_{IJ} for the composite $(pr_I)h(in_J)$ where $I,J \in \{X,Y\}$. We will sometimes write a map h, as above, as an ordered pair (f,g) where $f:X \times Y \to X$, and $g:X \times Y \to Y$. By the universal property of products we must have $f = pr_X h$, and $g = pr_Y h$. In this case we observe, for example, that $(f,g)_{XX} = f(in_X)$ and $(f,g)_{YY} = f(in_Y)$.

Section 2. On the Group $\mathcal{E}(X \times Y)$.

We assume throughout the section that *X and Y are path connected, equipped with closed non-degenerate base points, and that unless otherwise stated all maps are basepoint preserving.*

There is a function $\phi:[X \times Y, X \times Y]^* \to [X,X]^* \times [Y,Y]^*$ given by $\phi([h]) = ([h_{XX}, h_{YY}])$ or in the alternate form $\phi([(f,g)]) = [f(in_X), g(in_Y)]$. Note also that there are obvious inclusions $\mathcal{E}(X \times Y) \subset [X \times Y, X \times Y]^*$, and $\mathcal{E}(X) \times \mathcal{E}(Y) \subset [X,X]^* \times [Y,Y]^*$. We consider two questions:

I when does ϕ restrict to a function $\Phi:\mathcal{E}(X \times Y) \to \mathcal{E}(X) \times \mathcal{E}(Y)$, and II in the case that I holds, when is Φ a homomorphism?

Condition I does not hold in all cases. For example let X be a topological space that is not contractible, take Y = X and let $h:X \times Y \to X \times Y$ to be the switch map given by $h(x,y) = (y,x)$ for $x \in X$ and $y \in Y$. Then h is a homotopy equivalence so $[h] \in \mathcal{E}(X \times Y)$. Now $\phi([h])$, being the class of a pair of constant maps, belongs to $[X,X]^* \times [Y,Y]^*$, but not to $\mathcal{E}(X) \times \mathcal{E}(Y)$.

Definition 2.1. Let X and Y be topological spaces. We say that X and Y have the *induced equivalence property* (IEP) if whenever $h:X \times Y \to X \times Y$ is a homotopy equivalence, then the composites $h_{XX}:X \to X$ and $h_{YY}:Y \to Y$ are homotopy equivalences.

It is not hard to show for path connected spaces with non-degenerate base points, that the IEP is independent of the basepoints chosen. As shown above not every pair has the IEP.

Examples 2.2. (i) The spaces X and Y will be said to be *homotopically disjoint* if for each integer $i > 0$, at least one of $\pi_i(X)$ and $\pi_i(Y)$ is zero. If the homotopically disjoint spaces X and Y have the homotopy type of CW-complexes, then they satisfy the IEP.

(ii) If X is an n-connected CW-complex and Y is a CW-complex such that $\dim Y \leq n$, then X and Y have the IEP. In particular, we have:

(iii) If $X = S^n$ and $Y = S^m$ with $m \neq n$, then X and Y have the IEP.

Proof. Part (i) follows from a theorem of J.H.C. Whitehead, because h_{XX} and h_{YY} are weak homotopy equivalences (see for example [Sp;7.6.24])

(ii) This proof (although it is not stated there in these terms) is the major portion of the proof of theorem B in [Y].

A direct proof of (iii) is as follows: there are isomorphisms $(h_{XX})_*:H_n(X) \to H_n(X)$ in homology and so, by the naturality of the Hurewicz isomorphism, $(h_{XX})_*:\pi_n(X) \to \pi_n(X)$ is an isomorphism. If g is an inverse of this last isomorphism then a homotopy inverse of h_{XX} is given by $g(1_X)$. The argument for h_{YY} is similar.

We recall the following concept: a space will be said to be *numerably contractible* if it admits a numberable cover $\mathcal{C}$ such that the inclusion map $U \to B$ is null homotopic for each $U \in \mathcal{C}$. Such spaces include CW-complexes, more generally locally contractible paracompact spaces, classifying spaces B_G and also spaces homotopy equivalent to any numerably contractible space (see [D; theorem 6.3]).

Proposition 2.3. (i) The spaces X and Y have the IEP if and only if the function $\phi:[X \times Y, X \times Y]^* \to [X,X]^* \times [Y,Y]^*$ restricts to a function $\Phi:\mathcal{E}(X \times Y) \to \mathcal{E}(X) \times \mathcal{E}(Y)$.

(ii) If X and Y have the IEP, Y is numerably contractible and $[Y,H(X)]^* = 0$, then Φ is a homomorphism of groups.

We will take the next few pages to establish this proposition.

There is a natural multiplication on $[Y,H(X)]^*$, given by *pointwise composition* $(f_*g)(x,y) = f(y)g(y)(x)$, where $x \in X$, $y \in Y$ and $[f]$ and $[g] \in [Y,H(X)]^*$.

Proposition 2.4. If Y is numerably contractible and X and Y have the IEP then $[Y,H(X)]^*$ is a group under the above multiplication, and there is an exact sequence of groups and homomorphisms

$$1 \to [Y,H(X)]^* \xrightarrow{\lambda} \mathcal{E}_Y^*(X \times Y) \xrightarrow{\mu} \mathcal{E}(X) \to 1,$$

where $\lambda([f]) = [(\bar{f},pr_Y)]$, and $\mu([h]) = [h_{XX}]$; moreover this sequence splits via the homomorphism ϵ that takes the class $[g] \in \mathcal{E}(X)$ to the class $[g \times 1_Y] \in \mathcal{E}_Y^*(X \times Y)$.

Proof: Consider the map $w:H_Y^*(X \times Y) \to H^*(X)$ given by $w(h) = h_{XX}$. Then the fibre of w over 1_X is $H_Y^X(X \times Y)$. It is clear that $g \to (g \times 1)$ is a section to w and that the various maps are monoid homomorphisms. We show (i) that w is a fibration that gives rise to an exact sequence

$$1 \to \mathcal{E}_Y^X(X \times Y) \xrightarrow{\rho} \mathcal{E}_Y^*(X \times Y) \xrightarrow{\mu} \mathcal{E}(X) \to 1,$$

and then (ii) replace $\mathcal{E}_Y^X(X \times Y)$ by $[Y,H(X)]^*$.

(i) In the course of the following argument we sometimes identify X with $X \times \{y_0\}$. Consider a map g, and a homotopy G such that the left hand diagram commutes.

$$\begin{array}{ccc} A \times \{0\} & \xrightarrow{g} & H_Y^*(X \times Y) \\ \downarrow & & \downarrow w \\ A \times I & \xrightarrow{G} & H^*(X) \end{array} \qquad \begin{array}{ccc} (A\times 0\times X\times Y)\cup(A\times I\times X\times y_0) & \xrightarrow{\bar{g}\cup(\bar{G},c)} & X \times Y \\ \downarrow & & \downarrow pr_Y \\ A \times I \times X \times Y & \xrightarrow{pr_Y} & Y \end{array}$$

The existence of such a g and G is equivalent via the exponential law to the existence of maps making the right hand diagram commutative, where $c:A \times I \times X \times \{y_0\} \to Y$ denotes the constant map with value y_0. Now the inclusion of $A \times X \times \{y_0\}$ into $A \times X \times Y$ is a cofibration [Br;7.3.2] so by [St; Theorem 4] the right hand diagram has a relative lift $\tilde{F}:A \times I \times X \times Y \to X \times Y$, maintaining the commutativity of that diagram. Again by the exponential law we obtain a map $F:A \times I \to M(X \times Y,X \times Y)$. We show that the image of F is in $H_Y^*(X \times Y)$. Note first that $\tilde{F}$ extends $(\bar{G},c)$ and $\bar{G}(a,u,x_0,y_0) = x_0$, for each $u \in I$ and $a \in A$, so that $F:A \times I \to M_Y^*(X \times Y,X \times Y)$. Also for each $u \in I$ and $a \in A$, $F(a,u)$ is homotopic to a homotopy equivalence $F(a,0)$ and so must itself be a homotopy equivalence. By [Br, 7.3.2] the inclusion $\{(x_0,y_0)\} \to X \times Y$ is a cofibration and so [BK;theorem 4.5] $F(a,u)$ is an over Y under (x_0,y_0) homotopy equivalence, i.e. $F:A \times I \to H_Y^*(X \times Y)$ as required. Note that $wF = G$ and $F(\ ,0) = g$.

(ii) If $f:Y \to H(X)$ is a based map then $(\bar{f},pr_Y):X \times Y \to X \times Y$ is a map over the numerably contractible space Y; it restricts to homotopy equivalences between fibres, i.e. over points of Y, and so by [D; theorem 6.3] it is itself a homotopy equivalence. Now $(\bar{f},pr_Y)$ is also under X (since $f(y_0) = 1_X$) so (again by [BK]) it is an under X over Y homotopy equivalence, and $[(\bar{f},pr_Y)] \in \mathcal{E}_Y^X(X \times Y)$. Hence the function $\omega:[Y,H(X)]^* \to \mathcal{E}_Y^X(X \times Y)$ given by $\omega([f]) = [(\bar{f},pr_Y)]$ for $[f] \in [Y,H(X)]^*$, is well defined; viewing $[Y,H(X)]^*$ as a monoid under pointwise composition it is easily verified that this function is an isomorphism of monoids, and so $[Y,H(X)]^*$ is a group. □.

Corollary 2.5. Let Y be numerably contractible, X and Y have the IEP, $[Y,H(X)]^* = 0$, and $h \in H^*(X \times Y)$, then $[pr_X h] = [h_{XX} pr_X]$ in $[X \times Y,X]^*$.

Proof Since $[Y,H(X)]^* = 0$, the homomorphism μ in 2.4 is an isomorphism with the section ε as inverse. It follows that for any $[k] \in \mathcal{E}_Y^*(X \times Y)$, $[k] = \varepsilon\mu([k])$. In particular when $[k] = [(pr_X h,pr_Y)]$ for $[h] \in \mathcal{E}(X \times Y)$, then we have that $[(pr_X h,pr_Y)] = [(h_{XX} pr_X,pr_Y)]$ and so there is a based homotopy $F:pr_X h \simeq^* h_{XX} pr_X$. □.

We need the following definition for the proof of proposition 2.3.

Definition 2.6. If $d:M \to P$ and $k:N \to Q$ are basepoint preserving maps then a *pairwise map* $\langle f,g\rangle$ from d to k consists of basepoint preserving maps $f:M \to N$ and $g:P \to Q$ such that $kf = gd$; a pairwise map from $d \times 1_I:M \times I \to P \times I$ to k that preserves basepoints for all $u \in I$ will be called a *pairwise homotopy*. Clearly there is a corresponding concept of *pairwise homotopy equivalence*.

It is clear that if $\langle f,g\rangle$ is a pairwise homotopy equivalence then both f and g are ordinary (based) homotopy equivalences.

Let $\mathcal{E}(p)$ denote *the group, under composition, of pairwise homotopy classes of pairwise homotopy equivalences of* $p:X \to B$ *into itself*. Note that $\mathcal{E}(p)$ is quite different from $\mathcal{E}^*_B(X)$.

Proof of Proposition 2.3. The first part is obvious. For the second define homomorphisms $\eta:\mathcal{E}(pr_X) \to \mathcal{E}(X \times Y)$ and $\Psi:\mathcal{E}(pr_X) \to \mathcal{E}(X) \times \mathcal{E}(Y)$ by $\eta([\langle h,g\rangle]) = [h]$, and $\Psi([\langle h,g\rangle]) = ([g],[h|\{x_0\} \times Y])$. Now $[h|\{x_0\} \times Y] = [h_{YY}]$ and $g = g(pr_X)in_X = pr_Xh(in_X) = h_{XX}$ so $\Psi = \Phi\eta$. We will show that η is an isomorphism of groups, by showing that the function $\zeta:\mathcal{E}(X \times Y) \to \mathcal{E}(pr_X)$ given by $\zeta([h]) = [\langle(h_{XX}pr_X,pr_Yh),h_{XX}\rangle]$ is the inverse of η; the proposition then follows.

We need to show that ζ is well defined, i.e. that $\zeta([h]) \in \mathcal{E}(pr_X)$, in other words that $\langle(h_{XX}pr_X,pr_Yh),h_{XX}\rangle$ is a pairwise homotopy equivalence. Given $[h] \in \mathcal{E}(X \times Y)$, we first show that if $[k]$ is the inverse of $[h]$ in $\mathcal{E}(X \times Y)$, then $[k_{XX}]$ is an inverse of $[h_{XX}]$ in $\mathcal{E}(X)$. By 2.5 $k_{XX}h_{XX} = k_{XX}h_{XX}pr_Xin_X \simeq^* k_{XX}pr_Xh(in_X) \simeq^* pr_Xkh(in_X) \simeq^* 1_X$. Similarly we have that $h_{XX}k_{XX} \simeq^* 1_X$, and so the first part is proven.

We now have to verify that the following composite is pairwise homotopic to the identity $\langle 1_{X\times Y},1_X\rangle$.

$$\begin{array}{ccccc} X \times Y & \xrightarrow{(h_{XX}pr_X,\,pr_Yh)} & X \times Y & \xrightarrow{(k_{XX}pr_X,\,pr_Yk)} & X \times Y \\ \downarrow{\scriptstyle pr_X} & & \downarrow{\scriptstyle pr_X} & & \downarrow{\scriptstyle pr_X} \\ X & \xrightarrow[h_{XX}]{} & X & \xrightarrow[k_{XX}]{} & X. \end{array}$$

As $(k_{XX}pr_X,pr_Yk)(h_{XX}pr_X,pr_Yh) = (k_{XX}h_{XX}pr_X,pr_Yk(h_{XX}pr_X,pr_Yh)$, and $k_{XX}h_{XX} \simeq^* 1_X$, we have the required pairwise homotopy if $pr_Yk(h_{XX}pr_X,pr_Yh) \simeq^* pr_Y$. Now by 2.5 we have

$$pr_Yk(h_{XX}pr_X,pr_Yh) \simeq^* pr_Yk(pr_Xh,pr_Yh) = pr_Ykh \simeq^* pr_Y1_{X\times Y} = pr_Y,$$

and hence ζ is well defined.

Using a based homotopy $F: pr_X h \simeq^* h_{XX} pr_X$ which exists by 2.5, it is clear that $\eta\zeta = 1$; to prove that $\zeta\eta = 1$ we notice firstly that, as above, if $\langle h,g\rangle \in \mathcal{E}(pr_X)$ then $g = h_{XX}$, and secondly that the homotopy $X \times Y \times I \to X \times Y$ that takes (x,y,u) to $(F(x,y_0,u), pr_Y h(x,y))$, and the homotopy from $X \times I \to X$ that takes (x,u) to $F(x,y_0,u)$, where $x \in X$, $y \in Y$ and $u \in I$, together constitute a pairwise homotopy from $\langle (h_{XX}pr_X, pr_Y h), h_{XX}\rangle$ to $\langle h,k\rangle$. In particular, the verification that this pairwise homotopy starts at $\langle h,k\rangle$ goes as follows.
$F(x,y,0) = pr_X h(x,y)$ (by the definition of F) $= k(x)$ (because $\langle h,k\rangle$ is a pairwise map) and so $F(x,y_0,0) = k(x)$; also
$(F(x,y_0,0), pr_Y h(x,y)) = (k(x), pr_Y h(x,y)) = (pr_X h(x,y), pr_Y h(x,y)) = h(x,y)$ □.

We now give our main result on $\mathcal{E}(X \times Y)$, together with corollaries and examples; the proofs are delayed until after the examples.

Theorem 2.7. If X and Y are numerably contractible path connected spaces with closed nondegenerate basepoints (e.g. X and Y are CW-complexes), have the IEP and satisfy the condition $[Y,H(X)]^* = 0$, then there is a short exact sequence of groups and homomorphisms

$$1 \to [X,H(Y)]^* \xrightarrow{\iota} \mathcal{E}(X \times Y) \xrightarrow{\Phi} \mathcal{E}(X) \times \mathcal{E}(Y) \to 1.$$

which splits by a homomorphism $\sigma: \mathcal{E}(X) \times \mathcal{E}(Y) \to \mathcal{E}(X \times Y)$ given by $\sigma(([f],[g])) = [f \times g]$, where $[f] \in \mathcal{E}(X)$, and $[g] \in \mathcal{E}(Y)$. Thus *$\mathcal{E}(X \times Y)$ is the semidirect product of the groups $\mathcal{E}(X) \times \mathcal{E}(Y)$ and* $[X,H(Y)]^*$ determined by the section σ in the usual way (see [MB;corollary p463]).

The next result is an adjusted version of the theorem of [AY]; it incorporates the comment of [R] and the necessary (see Proposition 2.3 (i)) additional IEP condition.

Corollary 2.8. Let X and Y be numerably contractible path-connected spaces with closed non-degenerate base points, X also being either an H-space or a co-H-space. If X and Y have the IEP and satisfy $[X \wedge Y,X]^* = [Y,X]^* = 0$, then theorem 2.7 applies.

The term H-space is to be understood in the sense that involves associativity, identity, and inverse conditions, the term co-H-space in the sense that involves the corresponding "co" concepts.

Corollary 2.9. (c.f. [AY;corollary]) Let X and Y be numerably contractible path connected spaces with closed non-degenerate base points, have the IEP and satisfy the conditions $[Y,H(X)]^* = 0$ and $[X,H(Y)]^* = 0$, then there is an isomorphism $\mathcal{E}(X) \times \mathcal{E}(Y) \cong \mathcal{E}(X \times Y)$ given by $([f],[g]) \to [f \times g]$.

As mentioned in the introduction our use of the IEP and the $[Y,H(X)]^* = 0$ condition clarifies, unifies, and generalizes several similar published results. More specifically example 2.10(i) below is [Y; theorem 6] whilst 2.10(ii) which restricts 2.7 to the case where $X = K(\pi,n)$ generalizes the corollary on p.466 of [Y]; 2.10(iii) is Theorem 7 of [Y], 2.10(iv) (see theorem 7.9 of [S]) is a development of a particular case of (ii) or (iii); 2.10(v) is an example of corollary 2.9 and a case where [S; example 3.6] greatly simplifies; and 2.10(vi) seems to be new. In what follows $\circledS$ will denote the semi-direct product as described in 2.7.

Examples 2.10(i) Let $n > 0$. If X and Y are CW-complexes, $\pi_i(X) = 0$ for all $i > n$ and Y is n-connected then theorem 2.7 applies to $\mathcal{E}(X \times Y)$.

(ii) Let $n > 0$, $X = K(\pi,n)$ and Y be a CW-complex, satisfying $H^n(Y,\pi) = 0$, and $\pi_n(Y) = 0$, then

$$\mathcal{E}(K(\pi,n)\times Y) \cong (\mathrm{Aut}(\pi) \times \mathcal{E}(Y)) \circledS [K(\pi,n),H(Y)]^*.$$

(iii) If $n > 0$, X is an n-connected CW-complex, T us a /cw-complex with dimension $\leq n$ and $[Y,H(X)]^* = 0$, then theorem 2.7 applies.

(iv) If $n > 1$ then $\mathcal{E}(S^1 \times S^n) = (Z_2)^3$.

(v) $\mathcal{E}(S^5 \times S^6) \cong (Z_2)^2$.

(vi) In the cases $m \geq 6$ and $n = 4$, $m \geq 7$ and $n = 5$, and $m = 7,8$ or ≥ 14 and $n = 12$, then

$$\mathcal{E}(S^m \times S^n) \cong (Z_2)^2 \circledS \pi_m(H(S^n)).$$

Proof of 2.7. We have seen (proposition 2.3) that the function Φ exists and is a homomorphism under our assumptions. Define ι by $\iota([f]) = [(pr_X,\vec{f})]$ for $[f] \in [X,H(Y)]^*$, i.e. if $(x,y) \in X \times Y$ then $\iota([f])$ is the homotopy class of the map which takes (x,y) to $(x,f(x)(y))$. That ι is a homomorphism follows from the fact that $(x,(f*g)(x)(y)) = (x,f(x)g(x)(y))$.

To show that ι is injective we factor it as the composite $[X,H(Y)]^* \xrightarrow{\omega} \mathcal{E}_X^Y(X \times Y) \xrightarrow{\rho} \mathcal{E}_X^*(X \times Y) \xrightarrow{\nu} \mathcal{E}(X \times Y)$. The isomorphism ω and the monomorphism ρ are obtained by swapping the roles of X and Y in the proof of 2.4. Note that it is here that we use the numerable contractibility if X. Define $\nu([g]) = [g]$ for $[g] \in \mathcal{E}_X^*(X \times Y)$, by simply forgetting that g is over X. To see that ν is injective we supose that $g,h \in H_X^*(X \times Y)$ with $g \simeq^* h$, then $g = (pr_X, pr_Y g) \simeq_X^* (pr_X, pr_Y h) = h$ as required.

It remains to show exactness at $\mathcal{E}(X \times Y)$. Let $[f] \in [X,H(Y)]^*$, then $\Phi\iota([f]) = \Phi([(pr_X,\tilde{f})] = ([pr_X in_X, \tilde{f}(in_Y)]) = ([1_X],[1_Y])$, so $\text{Im}\ \iota \subseteq \text{Ker}\ \Phi$ Let $[h] = [(pr_X h, pr_Y h)] \in \text{Ker}\ \Phi$, then $[h_{XX}] = [1_X]$, and $[h_{YY}] = [1_Y]$. Using a homotopy $F: pr_X h \simeq^* h_{XX} pr_X$ (corollary 2.5), we see that $[(pr_X h, pr_Y h)] = [(pr_X, pr_Y h)]$. Since in_Y is a cofibration and $pr_Y h(in_Y) = h_{YY} \simeq^* 1_Y$ there is a map $\bar{g} \simeq^* pr_Y h$ with $\bar{g}(in_Y) = 1_Y$ i.e. $[(pr_X,\bar{g})] \in \mathcal{E}_X^Y(X \times Y)$, and if g denotes the right adjoint of $\bar{g}$ then $[g] \in [X,H(Y)]^*$. Finally $\iota([g]) = \nu\rho\omega([g]) = \nu\rho([pr_X,\tilde{g}])$ $= \nu([(pr_X, pr_Y h)] = [h]$, as required. □.

Proof of 2.8. Let $c:X \to X$ and $1:X \to X$ denote the constant map with value the base point of X and the identity of X respectively. Then $M^*(X,X;c)$ and $M^*(X,X;1)$ will denote the path components of $M^*(X,X)$ containing c and 1 respectively, these two maps being taken as base points. Applying the result of [T,p.31] that the path components of an H-space, such as $M^*(X,X)$, all have the same homotopy type we see that $[Y,M^*(X,X;c)]^* \cong [Y,M^*(X,X;1)]^* = [Y,H(X)]^*$. Now $[Y,M^*(X,X;c)]^* = [Y,M^*(X,X)]^* = [Y\#X,X]^* = 0$.

The fibration $H(X) \to X$ that evaluates at the base point of X has distinguished fibre $H^*(X)$, hence there is an exact sequence $\ldots \to [Y,H^*(X)]^* \to [Y,H(X)]^* \to [Y,X]^*$; and it follows that $[Y,H(X)]^* = 0$. □.

Proof of 2.9. This is an immediate consequence of 2.7.

Proof of 2.10. (i) X and Y are homotopically disjoint so the IEP is satisfied (2.2(i)). To prove $[Y,H(X)]^* = 0$ by the exponential law it is sufficient to show that there is a single under X homotopy class of maps $X \times Y \to X$, i.e. relative to $in_X: X \to X \times Y$ and $1:X \to X$. Now the obstructions to the existence of this map are located in the groups $H^i(X \times Y, X \times \{y_0\}; \pi_i(X)) \cong H^i(Y;\pi_i(X))$ [HW;10.6.10c] $= 0$ (from the data) for $i \geq 1$.

(ii) $\pi_n(Y) = 0$ ensures that $K(\pi,n)$ and Y are homotopically disjoint so the IEP is satisfied (see 2.2(i)). The path components of $H(K(\pi,n))$ are copies of $K(\pi,n)$ (see [M;proposition 25.2]) so

$[Y,H(K(\pi,n))]^* = [Y,K(\pi,n)]^* = H^n(Y,\pi) = 0.$

(iii) This follows from 2.2 (ii).

In the remainder of this proof we use 2.2 (iii) and the following fundamental exact sequence (based on [W, theorem 3.2], for details see also for example, [K;p.899 and 900]):

$$\cdots \to \pi_{n+1}(S^m) \xrightarrow[P_{n+1}]{} \pi_{m+n}(S^m) \to \pi_n(H(S^m)) \to \pi_n(S^m) \xrightarrow[P_n]{} \pi_{m+n-1}(S^m) \to \cdots .$$

where $P_n(\alpha) = -[\iota_m,\alpha]$, with $\alpha \in \pi_n(S^m)$, ι_m the homotopy class of the identity on S^m and [,] the Whitehead product brackets.

(iv) It follows from the above sequence that for $n > 1$ $\pi_n(H(S^1)) = 0$, for $n > 2$, $\pi_1(H(S^n)) \cong Z_2$, whilst $\pi_1(H(S^2)) \cong Z_2$ since for $P_2:\pi_2(S^2)) \to \pi_3(S^2)$, $P_2(\iota_2) = -[\iota_2,\iota_2] = \pm 2 \in Z = \pi_3(S^2)$ [K;proof of 3.9]. Now $\mathrm{Aut}(Z_2) = 1$ so

$$\mathcal{E}(S^1\times S^n) \cong \mathcal{E}(S^1) \times \mathcal{E}(S^n) \times Z_2 \cong (Z_2)^3.$$

(v) We have $\pi_5(H(S^6)) = 0$ [K;lemma 3.11]. The homomorphisms P_6 and P_7 in the exact sequence

$$\cdots \to \pi_7(S^5) \xrightarrow[P_7]{} \pi_{11}(S^5) \to \pi_6(H(S^5)) \to \pi_6(S^5) \xrightarrow[P_6]{} \pi_{10}(S^5) \to \cdots$$

are determined by $P_6(\eta_5) \neq 0$ and $P_7((\eta_5)^2) \neq 0$ [HLS;lemma 5.1], so that $\pi_6(H(S^5)) = 0$ and $\mathcal{E}(S^5 \times S^6) \cong \mathcal{E}(S^5) \times \mathcal{E}(S^6) \cong (Z_2)^2$.

(vi) Inserting the information that under the listed conditions, $\pi_{m+n}(S^m) = \pi_n(S^m) = 0$ into the exact sequence given above we see that $[Y,H(X)]^* = \pi_n(H(S^m)) = 0$. □

3. On the Group $\mathcal{E}_B^B(X \times_B Y)$.

In this section we make the anouncement about $\mathcal{E}_B^B(X \times_B Y)$ which we mentioned in the introduction. First some notation.

Let $p:X \to B$ and $q:Y \to B$ be maps and let

$$X \times_B Y = \{(x,y) \in X \times Y : p(x) = q(y)\}$$

denote the pullback or fibred product space of X and Y. Thus we have a commutative diagram

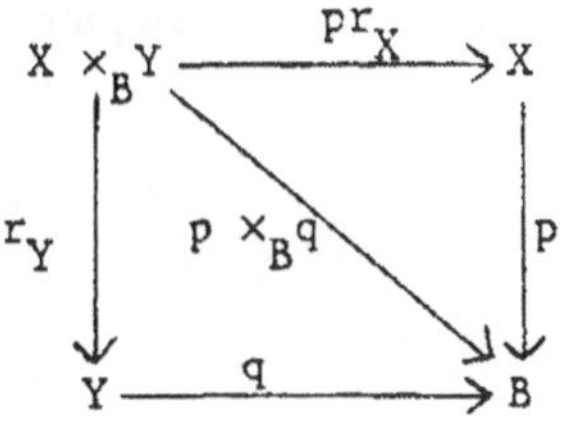

with the usual universal properties. Here $p \times_B q : X \times_B Y \to B$ is the map given by $(p \times_B q)(x,y) = p(x)$ and, generalizing definitions from section 1 $pr_X(x,y) = x$, and $pr_Y(x,y) = y$ where $(x,y) \in X \times_B Y$. If p and q are Hurewicz fibrations then so also are pr_X, pr_Y and $p \times_B q$, the last map being the composite of p and pr_X.

Let (p,s) and (q,t) be ex-spaces, then the map $p \times_B q$ has a section $(s,t): B \to X \times_B Y$, given by $(s,t)(b) = (s(b),t(b))$, for $b \in B$. So then $(p \times_B q,(s,t))$ is a ex-space, called the *fibred product of the ex-spaces* (p,s) *and* (q,t).

We generalize the maps in_X and in_Y of section 1, defining $in_X : X \to X \times_B Y$ and $in_Y : Y \to X \times_B Y$ by $in_X(x) = (x, tp(x))$ and $in_Y(y) = (sq(y),y)$, where $x \in X$ and $y \in Y$.

An *ex-fibration* (p,s) is an ex-space for which p is a Hurewicz fibration. If in addition s is a closed cofibration we say that (p,s) is a *cofibred ex-fibration*.

Given ex-spaces (p,s) and (q,t), then $\mathcal{E}_B^B(X \times_B Y)$ is the group of the ex-space $(p \times_B q,(s,t))$. We consider the question of determining $\mathcal{E}_B^B(X \times_B Y)$ in terms of $\mathcal{E}_B^B(X)$ and $\mathcal{E}_B^B(Y)$.

Let us assume that the fibres of the cofibred ex-fibrations (p,s) and (q,t) have the based homotopy types of the based spaces D and F repectively, where B is numerably contractible and D and F have the IEP. It can be shown using [D;theorem 6.3], and [H;theorem 2.2] that if h is a self ex-FHE of $(p \times_B q,(s,t))$ then $h_{XX} = pr_X h(in_X)$ and $h_{YY} = pr_Y h(in_Y)$ are self FHEs of (p,s) and (q,t) respectively. So in this situation there is a function $\Phi: \mathcal{E}_B^B(X \times_B Y) \to \mathcal{E}_B^B(X) \times \mathcal{E}_B^B(Y)$ given by $\Phi([h]) = ([h_{XX}, h_{YY})]$.

In fact the proofs of Section 2 can be modified, generalized and supplemented to obtain the following theorem.

Theorem 3.1. Let B be a numerably contractible space, $(p: X \to B, s: B \to X)$ and $(q: Y \to B, t: B \to Y)$ be cofibred ex-fibrations whose path connected fibres have the IEP with respect to each other, and $\mathcal{E}_Y^X(X \times_B Y) = 0$. Then there is a short exact sequence of groups and homomorphisms

$$1 \to \mathcal{E}_X^Y(X \times_B Y) \to \mathcal{E}_B^B(X \times_B Y) \underset{\Phi}{\to} \mathcal{E}_B^B(X) \times \mathcal{E}_B^B(Y) \to 1$$

which splits by a homomorphim σ: $\mathcal{E}_B^B(X) \times \mathcal{E}_B^B(Y) \to \mathcal{E}_B^B(X \times_B Y)$ given by $\sigma(([f],[g])) = [f \times g]$, where $[f] \in \mathcal{E}_B^B(X)$, and $[g] \in \mathcal{E}_B^B(Y)$. Thus

$\mathcal{E}_B^B(X \times_B Y)$ is the semi-direct product of the groups $\mathcal{E}_B^B(X) \times \mathcal{E}_B^B(Y)$ and the group $\mathcal{E}_X^Y(X \times_B Y)$ determined by σ in the usual way.

Here $\mathcal{E}_Y^X(X \times_B Y)$, and $\mathcal{E}_X^Y(X \times_B Y)$ are the groups associated with the over and under spaces (pr_Y, in_X) and (pr_X, in_Y) respectively. In order to apply Theorem 3.1 we need a technique for computing these groups: in part (ii) of the proof of proposition 2.4 we were able, using the numerable contractibility of Y, to identify $\mathcal{E}_Y^X(X \times_B Y)$ with $[Y,H(X)]*$ via the ordinary exponential law of spaces. This identification facilitates the computation of the former group. In the general situation the matter is not quite so simple. One method that sometimes works is to reformulate these groups in terms of fibred mapping spaces (see [B], [BBI] and [BBII]). This technique involves setting up isomorphisms between each of these groups and the groups of path components of the spaces of sections to suitably chosen fibred mapping space projections. A successful application of these techniques can be made in the following situation: let $(q:Y \to S^4, t:S^4 \to Y)$ be an ex-bundle, and

$$((pr_{S^4}):S^1 \times S^4 \to S^4, (c,1):S^4 \to S^1 \times S^4)$$

be a trivial ex-fibration, where $c:S^1 \to S^4$ denotes the constant map with value $*$. Then $Y \times_{S^4} (S^1 \times S^4) \cong S^1 \times Y$, and it can then be shown that:

$$\mathcal{E}_{S^4}^{S^4}(S^1 \times S^4) \cong Z_2, \quad \mathcal{E}_Y^{S^1 \times S^4}(S^1 \times Y) = 0 \quad \text{and} \quad \mathcal{E}_{S^1 \times S^4}^Y(S^1 \times Y) = Z_2$$

(the last two by obstruction theory) and hence:

$$\mathcal{E}_{S^4}^{S^4}(S^1 \times Y) \cong \mathcal{E}_{S^4}^{S^4}(Y) \times (Z_2)^2$$

Details of this and other examples will appear elsewhere.

References

[AY] Ando Y., Yamaguchi K., On homotopy self-equivalences of the Product A × B. Proc. Japan Acad., 58, Ser. A No.7 (1982) 323-325.

[B] Booth P.I., The section problem and the lifting problem, Math. Z. 121, (1971) 273-287.

[BBI] Booth, P.I., Brown R., Spaces of partial maps, fibred mapping spaces and the compact-open topology. Gen. Topology and its Applications. 8 (1978) 181-195.

[BBII] ________________ , On the application of fibred mapping spaces to exponential laws for bundles, ex-spaces and other categories of maps. Gen Topology and its Applications, 8 (1978) 165-179.

[BK] Berrick A.J., Kamps K.H. Comparison theorems in homotopy theory via operations on homotopy sets. Math. Nachr. 121 (1985) 25-32.

[Br] Brown R., Elements of modern topology. McGraw-Hill, Maidenhead, 1968.

[D] Dold A., Partitions of unity in the theory of fibrations. Annals of Math., Vol 78, No. 2, (1964) 223-255.

[H] Heath P.R., Homotopy equivalence of a cofibre-fibre composite. Can. J. Math., Vol. XXIV, No. 6, (1977), 1152-1156.

[HW] Hilton P.J., Wylie S., Homology theory, Cambridge University Press, 1960.

[HLS] Hsiang W.C., Levine J., Szczarba R.H., On the normal bundle of a homotopy sphere embedded in Euclidean space, Topology Vol 3, 173-181. (1965).

[K] Koh S.S., Note on the homotopy properties of the components of the mapping space X^{S^p}, Proc Amer. Math. Soc., 11 (1960), 896-904.

[M] May J.P. Simplicial objects in algebraic topology. Van Nostrand Co., Inc., Princeton, 1967.

[MB] MacLane S., Birkhoff G., Algebra. The MacMillan Company, New York 1967.

[N] Nomura Y., Homotopy equivalences in a principal fibre space. Math. Z. 92 (1966) 380-388.

[OSS] Oka S., Sawashita N., Sugawara M., On the group of self-equivalences of a mapping cone. Hiroshima Math. J. 4 (1974) 9-28.

[R] Rutter J. W. Review of [AY], Zentralblatt 537 (1985) review # 55010, p. 285.

[S] Sawashita N., On the group of self-equivalences of the product of spheres, Hiroshima Math. J. 5 (1975) 69-86.

[Sp] Spanier E., Algebraic topology, McGraw-Hill, New York, 1966.

[St] Strøm A., Note on cofibrations, Math. Scand. 22 (1968), 11-14.

[T] Thom R., L'homologie des espaces fonctionnels, Colloque de topologie algébrique, Louvain 1956, 29-39. Georges Thone, Liège; Masson and Cie, Paris, 1957.

[V] Vogt R., Convenient categories of topological spaces for homotopy theory, Arch. Math., 22, (1971) 545-555.

[W] Whitehead G.W., On products in homotopy groups, Annals of Math. 47 (1946) 460-475.

[Y] Yamanoshita T., On the spaces of self-homotopy equivalences of certain CW-complexes. J Math. Soc. Japan Vol 37, No 3, (1985) p. 455-470.

HOMOTOPIE DES ESPACES D'EQUIVALENCES

G.Didierjean

L'étude de l'homotopie de E(X),l'espace des équivalences d'homotopie d'un espace X,se fait essentiellement à partir d'une décomposition,soit en squelette,soit en tour de Postnikov de l'espace X.Cette dernière a l'avantage de nous donner une "unicité" de la construction effectuée.

Dans cet exposé, on montre comment la connaissance explicite de la décomposition en tour de Postnikov d'un espace X permet le calcul des groupes d'homotopie de son espace d'équivalence.

Dans une première partie,l'augmentation du nombre de groupes d'homotopie non nuls de l'espace X en dessous de sa dimension,nous amène naturellement à considérer les espaces d'équivalences fibrées.On donne dans la seconde partie la suite spectrale des équivalences fibrées.Enfin dans la dernière partie,on explicite l'algorithme de calcul obtenu sur un exemple.

§1 Historique de ce procédé de calcul

Dans ce qui suit,on note E(X) l' espace des équivalences d'homotopie de X,$\mathcal{E}(X)$ le groupe d'homotopie des équivalences d'homotopie de X et $\mathcal{E}^{\#}(X)$ le sous-groupe du groupe $\mathcal{E}(X)$ formé des équivalences qui induisent l'identité sur les groupes d'homotopie de l'espace X.On décompose l'espace X en tour de Postnikov notée (X^n) avec ses invariants (ξ^n), $\xi^n \in H^{n+1}(X^n,\Pi_n(X))$,cohomologie éventuellement prise à coefficients locaux, si $\Pi_1(X)$ est non nul.La notation adoptée est celle de Shih-Cartan ,c'est à dire $\Pi_i(X^n) = \Pi_i(X)$ si $i<n$, $\Pi_i(X^n)=0$ sinon.

La première remarque fondamentale est celle d **Arkowitz et Curjel**, seuls doivent être pris en compte les groupes d'homotopie en dessous de la dimension N de X:

This paper is in final form and no version of it will be submitted for publication elsewhere.

Proposition (1.1) (Arkowitz et Curjel [1]): *Soit* X *un* C.W *complexe de dimension* N,*alors :* $\mathcal{E}(X) \approx \mathcal{E}(X^n)$ si $n > N = \dim X$

On a les résultats suivants:

Un groupe d'homotopie non nul en dessous de la dimension:

Proposition (1.2) (J.P. May [7]) : *Soit X un espace simplicial de même type d'homotopie qu'un espace d'Eilenberg-Mac Lane* $K(\Pi,n)$ *avec* Π *un groupe abélien alors*

$$\mathrm{Aut}(K(\Pi,n)) \cong K(\Pi,n) \times \mathrm{Aut}\,\Pi \quad \textit{(produit semi-direct)}$$

en tant que groupe simplicial .

Deux groupes d'homotopies non nuls en dessous de la dimension:

Proposition (1.3) (Shih [9]- Tsukiyama [11]): *Soit X un C.W. complexe n'ayant que deux groupes d'homotopie non nuls* $\Pi_s = \Pi_s(X)$ $s = i,j$,*avec* $i < j$,*d'invariant de Postnikov* $\xi \in H^{j+1}(\Pi_i,i,\Pi_j)$.

Pour $1<i<j$, *Shih démontre l'exactitude de la suite :*

$$0 \to H^j(\Pi_i,i,\Pi_j) \to \mathcal{E}(X) \to (\mathrm{Aut}\Pi_i \times \mathrm{Aut}\Pi_j)_\xi \to 0$$

avec les groupes $\mathrm{Aut}\,\Pi_i$ *et* $\mathrm{Aut}\,\Pi_j$ *opèrant sur* $H^{j+1}(\Pi_i,i,\Pi_j)$ *et où on note* $(\mathrm{Aut}\Pi_i \times \mathrm{Aut}\,\Pi_j)_\xi$ *le stabilisateur de* ξ *pour l'opération de* $\mathrm{Aut}\Pi_i \times \mathrm{Aut}\,\Pi_j$ sur $H^{j+1}(\Pi_i,i,\Pi_j)$

Pour $i = 1$,*Tsukiyama montre que* $\mathcal{E}^{\#}(X) = H_j(\Pi_i, i, \Pi_j)$ *(cohomologie à coefficients locaux)*

Trois groupes d'homotopie non nuls en dessous de la dimension:

Proposition (1.4) ([4]):*Soit X un C.W. complexe de dimension N n'ayant que trois groupes d'homotopie non nuls en dessous de sa dimension N, soit* $\Pi_s(X) = \Pi_s$,$s = i,j,k$, *avec* $1<i<j<k\leq N=\dim X$, *on a la suite exacte:*

$$H^{j-1}(\Pi_i,i,\Pi_j) \to H^k(X^k,\Pi_k) \to \mathcal{E}^{\#}(X) \to H^j(\Pi_i,i,\Pi_j) \to H^{k+1}(X^k,\Pi_k)$$

Remarque: Soit X un C.W complexe de dimension 4,1-connexe ,on a la suite exacte:

$$\mathrm{Hom}(\Pi_2,\Pi_3) \to K^4(X,\Pi_4) \to \mathcal{E}^{\#}(X) \to \mathrm{Ext}(\Pi_2,\Pi_3) \to H^5(X^4,\Pi_4)$$

Pour démontrer ceci on applique la proposition précédente .

Dans la décomposition de l'espace X on considère la fibration $X^{k+1} \to X^j \to X^i = K(\Pi_j, i)$, on veut comparer l'espace d'équivalences $E(X^{k+1})$ à $E(X^j)$. Pour cette raison on est amené à se placer dans le cadre simplicial où cette décomposition est "unique" et "fonctorielle".
La proposition précédente se déduit alors de la proposition (1.5):.

Proposition (1.5) : *Soit X un ensemble simplicial de Kan, (X^n) sa tour de Postnikov ,pour $p<q$ on a une fibration de Kan* $E(X^q) \to E(X^p)$ *de fibre* $E_X{}^p(X^q)$ *au dessus de l'identité de* X^p *et pour $p<q<r$ on a la fibration des fibres:* $E_X{}^q(X^r) \to E_X{}^p(X^r) \to E_X{}^p(X^q)$.

Pour passer à plus de trois groupes d'homotopie en dessous de la dimension de l'espace X,il faut calculer les groupes Π_i ($E_X{}^p(X^q)$).Ceci peut se faire à l'aide de la suite spectrale du paragraphe suivant.

§ 2 Suite spectrale des équivalences d'homotopies fibrées

On considère un fibré de Kan $X \to B$ de fibre F,et $E_B(X)$ désigne l'ensemble simplicial des équivalences d'homotopie fibrées au dessus de l'identité de B,espace pointé par l'identité de X.Pour simplifier les écritures,on supposera dans tout ce qui suit,que l'espace X est simplement connexe.Soit $\mathcal{E}_B(X) = \Pi_0(E_B(X))$ et $\mathcal{E}^{\#}_B(X)$ est le sous-groupe de $\mathcal{E}_B(X)$ formé des équivalences d'homotopie qui induisent l'identité sur les groupes d'homotopie de X.Tout fibré simplicial possèdant un sous fibré minimal,rétract par déformation,on se place dans le cadre simplicial minimal.

Dans un premier temps,on suppose que la fibre F n'a qu'un groupe d'homotopie non nul Π,abélien.L'espace X a donc même type d'homotopie que l'espace simplicial $K(\Pi,n) \times_\tau B$ avec $\tau : B \to \mathrm{Aut}(K(\Pi,n))$.Dans ce cas,l'espace $E_B(X)$ est le groupe simplicial des automorphismes de X au dessus de l'identité de la base B,$\mathrm{Aut}_B(K(\Pi,n)\times_\tau B)$.De plus:

Proposition (2.1) :*Soit* $K(\Pi,n) \times_\tau B \to B$ *un fibré principal, on a la suite exacte de groupes simpliciaux*

$$0 \to \mathrm{Hom}(B,K(\Pi,n)) \to \mathrm{Aut}_B(K(\Pi,n)\times_\tau B) \to \mathrm{Aut}\,\Pi_\xi \to 0$$

Cette suite exacte provient du fait que tout automorphisme ϕ de X au dessus de l'identité de B s'écrit $\phi(w,b) = (\alpha w + \beta(b), b)$ avec $\alpha \in \mathrm{Aut}\,\Pi_\xi$ et $\beta(b) \in K(\Pi,n)$.

A présent,de façon générale, soit $F \to X \to B$ un fibré de Kan , minimal dont la base et la fibre sont connexes . A ce fibré est associé sa tour de Postnikov fibrée [7], ($F^n \to X^n \to B$) telle que $X^{n+1} \to X^n$ soit un fibré principal de fibre $K(\Pi_n(F),n)$. Ces fibrés s'écrivent $K(\Pi_n(F),n) \times_\tau X^n \to X^n$ avec $\tau : X^n \to K(\Pi_n(F),n)$. On note à present $\xi^n \in H^{n+1}(X^n,\Pi_n(F))$, le $n^{\text{ième}}$ invariant de Postnikov fibré.

Avec ces hypothèses et notations,on a ,comme dans le cas non fibré,une suite de fibrations:

$$E_B(X^q) \to E_B(X^{q-1}) \to \dots \to E_B(X^2) \to E_B(X^1) = E_B(B) = \{ id_B \}$$

On sait (Federer [7],pour la décomposition en squelettes;Shih [10] ,pour la décomposition en tour de Postnikov;puis Bousfield-Kan [2] qui a formalisé le procédé) qu'une telle succession de fibrations conduit à une suite spectrale.

Dans le cas qui nous occupe, cette suite spectrale est non abélienne,c'est à dire:

$$\underset{\textit{groupe abélien}}{E_r^{p-r,\,-p+r-1}} \xrightarrow[\text{morphisme}]{d_r} \underset{\textit{groupe}}{E_r^{p,-p}} \xrightarrow[\text{opération}]{d_r} \underset{\textit{ensemble}}{E_r^{p+r,-p-r+1}}$$

Pour effectuer des calculs en degré zero, il nous faut $p+q = 1$ et $p+q = -1$. La suite spectrale fait alors intervenir des groupes non abéliens en degré 0 et des ensembles en degré 1. Une étude précise de ce type de suite spectrale se trouve faite par A.Legrand [6].Ici,il nous reste à identifier les termes initiaux , on a:

Théorème (2.2) [4]: *Soit* $X \to B$ *un fibré de Kan ,de fibre F et de base connexes et X 1-connexe.Au monoïde des équivalences fibrées de X au dessus de l'identité de B ,on associe une suite spectrale non abélienne ,limitée de degré total* $p+q \leq 1$ *et dont les termes initiaux vérifient :*

(1) pour $p+q < 0$ *et* $2p+q \geq 0$

$$E_1^{p,q} = H^{2p+q}(X, \Pi_p(F))$$

(2) pour $p+q = 0$ *et* $p \geq 1$ *on a la suite exacte :*

$$0 \to H^p(X^p, \Pi_p(F)) \to E_1^{p,-p} \to \operatorname{Aut} \Pi_p(F)_{\xi p} \to 0$$

(3) pour $p+q = 1$

$$E_1^{p,1-p} = H^{p+1}(X^p, \Pi_p(F)) / \operatorname{Aut} \Pi_p(F)$$

Lorsque l'espace X est de dimension finie ou si les groupes d' homotopie de la fibre sont nuls à partir d'un certain rang , cette suite spectrale converge vers le bigradué :

$$\mathfrak{F}^p \Pi_*(E_B(X)) / \mathfrak{F}^{p+1} \Pi_*(E_B(X))$$

$$\text{où} \quad \mathfrak{F}^p \Pi_*(E_B(X)) = \operatorname{Ker} [\, \Pi_*(E_B(X)) \to \Pi_*(E_B(X^p)) \,]$$

Remarques sur la démonstration:

La partie (1) provient des fibrations successives,(2) est obtenu en généralisant la proposition (1.2) à un fibré principal $K(\Pi,n) \times_\tau B \to B$ et (3) est donnée par l'obstruction.De plus on démontre que $\mathcal{E}_B(X) \cong \mathcal{E}_B(X^n)$ si $n > \dim(X)$ et $H^i(X, \Pi_s(F)) \cong H^i(X^p, \Pi_s(F))$ si $1 \leq i \leq p-1$.

La différentielle d_1 est définie pour $p+q \leq 0$ par l'opérateur bord
$\delta^{p+q} : \Pi_{-(p+q)}(E_X{}^p(X^{p+1})) \to \Pi_{-(p+q+1)}(E_X{}^{p+1}(X^{p+2}))$.

§3 Applications,algorithme de calcul [4].

Quelques conséquences du théorème:

Si B = {pt} ,on retrouve la suite spectrale des équivalences construite par Shih.

Soit $X \to B$ un fibré de fibre F.Si F n'a que deux groupes d'homotopie non nuls, $\Pi_i = \Pi_i(F)$,i = k,m avec k < m ,la suite spectrale n'a que deux colonnes non nulles.On obtient par la méthode classique,une suite exacte de Gysin:

Proposition (3.1): *Soit* $X \to B$ *un fibré dont la base et la fibre sont connexes ,et l' espace X simplement connexe.On suppose de plus que la fibre n'a que deux groupes d'homotopie non nuls,* $\Pi_i(F) = \Pi_i$ *pour* i = k,m *avec* k<m *.On a alors:*

* *pour* $i > m$ $\Pi_i(E_B(X)) = 0$

* *pour* $k < i \leq m$ $\Pi_i(E_B(X)) = H^{m-i}(X,\Pi_m)$

* *pour* $i < k$ *on a la suite exacte :*

$$H^{m-k}(X,\Pi_m) \to \Pi_k(E_B(X)) \to \Pi_k \to \dots \to \Pi_i(E_B(X)) \to H^{k-i}(X,\Pi_k) \to H^{m-i+1}(X,\Pi_m)$$
$$\to \Pi_{i-1}(E_B(X)) \to \dots \to \Pi_1(E_B(X)) \to H^{k-1}(X,\Pi_k)$$
$$\to E_1^{m,-m} \to \mathcal{E}_B(X) \to E_1^{k,-k} \to E_1^{m,-m+1}$$

avec

-*l'application de* $H^{k-i}(X,\Pi_k)$ *dans* $H^{m-i+1}(X,\Pi_m)$ *est la différentielle* d_{m-k} *de la suite spectrale des équivalences d'homotopie fibrées.*

-*les groupes* $E^{m,-m}$ *et* $E^{k,-k}$ *sont donnés par les suites exactes:*

$$0 \to H^m(X^m,\Pi_m) \to E_1^{m,-m} \to [\,\mathrm{Aut}\,\Pi_m]_{\xi^m} \to 1$$

$$0 \to H^k(X^k,\Pi_k) \to E_1^{k,-k} \to [\mathrm{Aut}\,\Pi_k]_{\xi^k} \to 1$$

$$-E^{m,-m+1} = H^{m+1}(X^m, \Pi_m)/\,\mathrm{Aut}\,(\Pi_m)$$

Exemples d'utilisation,algorithme:

Soit $\Pi_i = \Pi_i(X)$ et l'on note $K^s(X,\Pi) = \mathrm{Ker}\ u \cdot v$ [9],où
$v: H^s(X,\Pi) \to \mathrm{Hom}(H_s(X),\Pi))$ est le morphisme des coefficients universels et
$u: \mathrm{Hom}(H_s(X,\Pi)) \to \mathrm{Hom}(\Pi_s(X),\Pi)$,celui induit par le morphisme d'Hurewicz.

Dans la proposition (1.4) on voit apparaitre un algorithme de calcul des groupes $\Pi_i(E(X))$.En effet,on considère, à partir de la décomposition en tour de Postnikov de l'espace X,des fibrations successives $X^k \to X^j$ n'ayant dans la fibre que deux groupes d'homotopie non nuls.Ainsi,par exemple,pour un espace n'ayant que quatre groupes d'homotopie non nuls en dessous de sa dimension,on a:

Proposition (3.2): *Soit X un C.W. complexe 1-connexe ayant 4 groupes d'homotopie non nuls entre les crans 2 et N = dim(X) .On note* $\Pi_s = \Pi_s(X)$ $s = i,j,k,m$ *avec* $1<i<j<k<m\leq N$, *le groupe* $\mathcal{E}^{\#}(X)$ *se calcule à partir des suites exactes suivantes:*

$$\ldots \to H^{j-1}(X,\Pi_j) \to \mathcal{E}^{\#}_{X^{j+1}}(X^{m+1}) \to \mathcal{E}^{\#}(X) \to H^j(\Pi_i,i,\Pi_j)$$

$$\|$$

$$\ldots \to H^{k-1}(X,\Pi_k) \to K^m(X,\Pi_m) \to \mathcal{E}^{\#}_{X^{j+1}}(X^{m+1}) \to K^k(X,\Pi_k) \to H^{m+1}(X^m,\Pi_m)$$

où $K^s(X,\Pi) = \mathrm{Ker}\, u \bullet v$ *a été défini précédement.*

Proposition (3.3): *L'espace X vérifiant toujours les hypothèses précédentes , le groupe* $\mathcal{E}(X)$ *se calcule à partir des suites exactes suivantes:*

$$0 \to H^j(\Pi_i, i, \Pi_j) \to \mathcal{E}(X^{j+1}) \to (\mathrm{Aut}\,\Pi_i \times \mathrm{Aut}\,\Pi_j)_\xi \to 0$$

$$\|$$

$$\ldots \to H^{j-1}(X,\Pi_j) \to \mathcal{E}_{X^{j+1}}(X^{m+1}) \to \mathcal{E}(X) \to \mathcal{E}(X^{j+1})$$

$$\|$$

$$\ldots \to E_1^{m,-m} \to \mathcal{E}_{X^{j+1}}(X^{m+1}) \to E_1^{k,-k} \to E_1^{m,-m+1}$$

Etapes de la démonstration:

1 ière étape :

$$X^{i+1} \approx K(\Pi_i,i) \quad \text{et} \quad \Pi_n(E(X^{i+1})) = \begin{cases} \mathrm{Aut}\,\Pi_i & \text{si } n=0 \\ \Pi_i & \text{si } n=i \\ 0 & \text{sinon} \end{cases}$$

2ième étape:

Soit X^{j+1} l'espace de Postnikov tel que $\Pi_s(X^{j+1}) \approx \Pi_s$ pour $s = i,j$.

si $n \neq 0,i,i-1$ $\Pi_n(E(X^{j+1})) \approx H^{j-n}(X,\Pi_j)$

si $n = i,i-1$ $0 \to H^{j-i}(X,\Pi_j) \to \Pi_i(E(X^{j+1})) \to \Pi_i \to H^{j-i+1}(X,\Pi_j) \to \Pi_{i-1}(E(X^{j+1})) \to 0$

si $n = 0$ ces groupes ont déja été calculés par Shih et par Tsukiyama, on retrouve:

$$0 \to H^k(\Pi_i,i,\Pi_j) \to \mathcal{E}(X^{j+1}) \to (\mathrm{Aut}\,\Pi_i \times \mathrm{Aut}\,\Pi_j)_\xi \to 0$$

3ième étape:

La fibre du fibré $X^{m+1} \to X^{j+1}$ a deux groupes d'homotopie non nuls Π_k, Π_m et de plus on a $\mathcal{E}(X^{m+1}) \approx \mathcal{E}(X)$,d'où le résultat.

Un exemple:

Prenons $X = SU(3)$,dim(X) = 8,les groupes d'homotopies de cet espace sont:

$\Pi_1(X) = \Pi_2(X) = \Pi_4(X) = \Pi_7(X) = 0$

$\Pi_3(X) = \Pi_5(X) = \mathbb{Z}$

$\Pi_6(X) = \mathbb{Z}/6\mathbb{Z}$

$\Pi_8(X) = \mathbb{Z}/12\mathbb{Z}$

$\Pi_9(X) = \mathbb{Z}/3\mathbb{Z}$

Cet espace a donc 4 groupes d'homotopie non nuls en dessous de sa dimension.De plus, l'espace X est fibré sur S^7 de fibre S^3 la suite spectrale de Serre dégénère et l'anneau de cohomologie dans un anneau unitaire R est l'algèbre extérieure sur R , $\Lambda_R[x_3,x_5]$,ayant pour générateurs x_3 et x_5,ces éléments sont les générateurs respectifs des modules $H^3(X,R)$ et $H^5(X,R)$.

Proposition (3.4): $\mathcal{E}_{\#}(SU(3)) = \mathbb{Z}/12\mathbb{Z}$

En effet,en appliquant les deux propositions précédentes on a le diagramme:

$$\begin{array}{ccccccccc} & & \ldots \to H^4(X,\mathbb{Z}) & \to & \mathcal{E}^{\#}_{X^6}(X^9) & \to & \mathcal{E}^{\#}(X) & \to & H^5(\mathbb{Z},3,\mathbb{Z}) \\ & & & & || & & & & \\ H^5(X,\mathbb{Z}/6\mathbb{Z}) & \xrightarrow{d} & K^8(X,\mathbb{Z}/12\mathbb{Z}) & \to & \mathcal{E}^{\#}_{X^6}(X^9) & \to & K^6(X,\mathbb{Z}/6\mathbb{Z}) & \to & H^9(X^8,\mathbb{Z}/12\mathbb{Z}) \end{array}$$

La proposition découle alors des résultats suivants : $d = 0$, $K^8(X,\mathbb{Z}/12\mathbb{Z}) = \mathbb{Z}/12\mathbb{Z}$ et $K^6(X, \mathbb{Z}/6\mathbb{Z}) = 0$, $H^4(X,\mathbb{Z}) = 0$, $H^5(\mathbb{Z},3,\mathbb{Z}) = 0$.

De même on retrouve les résultats d'Oka,Sawashita,Sugawara [8]:

Proposition (3.5):

$$0 \to \mathbb{Z}/12\mathbb{Z} \to \mathcal{E}(SU(3)) \to \mathbb{Z}/2\mathbb{Z} \oplus \mathbb{Z}/2\mathbb{Z} \to 0$$

En poursuivant l'explicitation des diagrammes emboités de l'algorithme de calcul,on a:

Proposition (3.6): *La suite suivante est exacte:*

$$0 \to \mathbb{Z} \to \mathbb{Z} \to \mathbb{Z}/3\mathbb{Z} \to \Pi_1(E(SU(3))) \to \mathbb{Z}/6\mathbb{Z} \to 0$$

où le morphisme de $\mathbb{Z}$ *dans* $\mathbb{Z}/3\mathbb{Z}$ *est induit par la différentielle de la suite spectrale de la fibration* $E_{X^9}(X^{10}) \to E_{X^8}(X^{10}) \to E_{X^8}(X^9)$.

De même:

Proposition (3.7):

$$\Pi_2(E(SU(3)) = \mathbb{Z} \oplus \Pi_2(E_{X^6}(X^{11}))$$

avec $0 \to \Pi_3(E_{X^6}(X^{11})) \to \Pi_3(E_{X^6}(X^9)) \to \mathbb{Z}/30\mathbb{Z} \to \Pi_2(E_{X^6}(X^{11})) \to 0$

et $0 \to \mathbb{Z}/12\mathbb{Z} \to \Pi_3 (E_{X^6}(X^9)) \to \mathbb{Z}/6\mathbb{Z} \to 0$

Remarque: La difficulté dans de tels calculs réside,outre l'explicitation de de la différentielle,dans la conaissance explicite des invariants de Postnikov de l'espace X dont on veut calculer les groupes d'homotopie de son espace d'équivalence.

BIBLIOGRAPHIE

[1] M.Arkowitz et C.R.Curjel,*Groups of homotopies classes*,Lecture Notes in Math.4, Springer Verlag,(1964) chap.IV.

[2]A.K. Bousfield et P.M.Kan,*Homotopy limits completions and localisations*, Lecture Notes in Math.304, Springer Verlag,(1972),chap.IX.

[3] G.Didierjean,*Homotopie de l'espace de l'espace des équivalences fibrés*,Ann.Inst.Fourier 35 n°3,(1985) 33-47.

[4] ------------,*Homotopie de l'espace des équivalences d'homotopies*, Trans.Amer.Math.Soc (à paraitre),(1988).

[5] H. Federer,*A study of function spaces by spectral sequence*,Trans.Amer.Math.Soc. 82, (1956) 340-361.

[6] A.Legrand,*Homotopie des espaces de sections*,Lecture Notes in Math.941, Springer Verlag (1981).

[7] J.P.May,*Simplicial object in algebraic topology*,Van Nostrand,(1967).

[8] S.Oka,N.Sawshita et M.Sugawara ,*On the group of self-equivalences of a mapping cone*,Hiroshima Math.J.4,(1974) 9-28.

[9] W.Shih,*On the group $\mathcal{E}(X)$ of equivalences maps*, Bull.Amer.Math.Soc.492,(1964) 361-365.

[10] ---------,*Classes d'applications d'un espace dans un groupe topologique*,Séminaire E.N.S Cartan ,(1962/1963).

[11] K.Tsukiyama,*Self homotopy equivalences of a space with two non vanishing homotopy group*,Proc.Amer.Math.soc.71 n°1,(1980) 134-138.

UNIVERSITE LOUIS PASTEUR

Institut de Recherche Mathématique Avancée

7 rue R.Descartes,F67084 Strasbourg.

THE SPACE OF SELF MAPS ON THE 2-SPHERE

VAGN LUNDSGAARD HANSEN

Abstract. In this paper we review contributions to the homotopy theory of manifolds of maps between closed orientable surfaces, and in particular those results which provide a full homotopy type of a component. As a main case, we describe the complete homotopy type of the space of orientation preserving self homotopy equivalences on the 2-sphere (the component containing the maps of degree 1) in terms of well known spaces in topology. As a new result, we prove that the component in the space of self maps on the 2-sphere containing the maps of degree k admits a unique k-fold covering space, and that this covering space has the homotopy type of the space of orientation preserving self homotopy equivalences.

1980 Mathematics subject classifications: Primary 55P15, 58D15.

Keywords: Homotopy type, component, space of maps between surfaces, self maps on the 2-sphere.

Topology of mapping spaces is an important subject in algebraic topology. It should suffice to point out that one of the basic problems in algebraic topology is the homotopy classification of maps, which is equivalent to the enumeration of the (path-)components in the underlying spaces of maps. Also the central role played by loop spaces could be mentioned.

Less well known, maybe, is the role played by topology of mapping spaces in global analysis, and thus in the study of nonlinear phenomena. As a first approximation one could define global analysis as analysis in nonlinear spaces, in particular in manifolds of maps. As analysis in finite dimensional spaces requires knowledge of the topology of euclidean spaces, so analysis in manifolds of maps requires foundational knowledge of the topology of mapping spaces. By general results on infinite dimensional manifolds, the homeomorphism type of such manifolds is completely determined by the homotopy type [**10**]. Since almost any mapping space can be given the structure of a Hilbert manifold [**1**], the classification problem for mapping spaces is reduced to homotopy theory. However, not very many complete homotopy types of mapping spaces are known. In this paper we review contributions to the homotopy theory of manifolds of maps between closed orientable surfaces, and in particular those results which provide a full homotopy type of a component. As a main case, we describe the complete homotopy type of the space of orientation preserving self homotopy equivalences on the 2-sphere (the component containing the maps of degree 1) in terms of well known spaces in topology. As a new result, we prove that the component in the space of self maps on the 2-sphere containing the maps of degree k admits a unique k-fold covering space, and that this covering space has the homotopy type of the space of orientation preserving self homotopy equivalences.

This work was supported by the Danish Natural Science Research Council.

This paper is in final form and no version of it will be submitted for publication elsewhere.

1. Local structure of the components.

Let T and S denote closed orientable surfaces, equipped with base points, when necessary. Denote by $M(T,S)$ and $F(T,S)$, the space of free, respectively based, (continuous) maps of T into S with the uniform topology induced from a metric on S.

The (path-)components in $F(T,S)$ are enumerated by the based homotopy classes of based maps $\pi(T,S)$ of T into S. Similarly, the components in $M(T,S)$ are enumerated by the free homotopy classes of free maps $[T,S]$ of T into S. Each component in $M(T,S)$ is determined by a based map, since the obvious map $\pi(T,S) \to [T,S]$ is always surjective. For a based map $f: T \to S$, let $M(T,S;f)$, respectively $F(T,S;f)$, denote the component in $M(T,S)$, respectively $F(T,S)$, which contains the map f. When the surface S is the 2-sphere, each of the spaces of maps $M(T,S)$ and $F(T,S)$ has a countable infinite number of components, classified by the degrees of maps; denote by $M_k(T,S)$ and $F_k(T,S)$ the components containing the maps of degree k for $k \in \mathbb{Z}$ an arbitrary integer.

A sufficiently small neighbourhood of a (based) continuous map $f: T \to S$ in $M(T,S)$ can be identified with a neighbourhood of zero in the space of (based) vector fields along f by bending vector fields onto S along geodesics in a riemannian structure on S. The space of vector fields along f has an obvious Banach space structure, and therefore $M(T,S;f)$ and similarly $F(T,S;f)$ are infinite dimensional Banach manifolds: in particular, they are ANR's.

2. Homotopy type of the components.

If the genus of the surface S is ≥ 1, then S is an Eilenberg-MacLane space with the fundamental group as the only nonvanishing homotopy group. As a special case of a general theorem of Gottlieb ([3], Lemma 2), see also ([7],Theorem 2), we get then

THEOREM 2.1. *Let S be a surface of genus ≥ 1. Then the component $M(T,S;f)$ in $M(T,S)$ defined by a based map $f: T \to S$ has the centralizer for the image of the fundamental group of T in S as its fundamental group, and all other homotopy groups vanish.*

Now recall that two Eilenberg-MacLane spaces are homotopy equivalent if and omly if they have the only nonvanishing homotopy group in the same dimension and the groups are isomorphic.

If S has genus 1, i.e. S is a torus, then any centralizer in the fundamental group of S is the full group, since this group is abelian. Hence any component in $M(T,S)$ has S as its homotopy type.

The fundamental group of a surface S of genus ≥ 2 is highly non abelian. In particular only three types of centralizers are possible in the fundamental group of S: the full group, the trivial group, or an infinite cyclic group. Correspondingly [5], a component in $M(T,S)$ can only have one of the following homotopy types: the surface S, a point, or the circle. Examples: the component in $M(S,S)$ containing the constant maps has S as its homotopy type, the identity component in $M(S,S)$ is contractible ([2], Theorem III.2), and the maps onto a closed geodesic in S has the homotopy type of the circle.

For spaces of maps of a surface T of genus $g \geq 0$ into the 2-sphere S, the situation is more complicated. In ([4], Theorem 1), the fundamental group of the component

$M_k(T,S)$ containing the maps of degree k was determined up to an extension by the short exact sequence

$$0 \to \mathbf{Z}_{2|k|} \to \pi_1(M_k(T,S)) \to \mathbf{Z}^{2g} \to 0,$$

where $\mathbf{Z}^{2g}$ denotes the free abelian group on $2g$ generators, and $\mathbf{Z}_{2|k|}$ denotes the cyclic group of order $2|k|$, the integers $\mathbf{Z}$ for $k=0$. The extension was subsequently determined by Larmore and Thomas [**13**]. The higher homotopy groups of $M_k(T,S)$ have been computed in terms of homotopy groups of spheres in [**8**].

The complete homotopy type of a component in a space of maps into the 2-sphere has so far only been determined in the special case of orientation preserving self homotopy equivalences on the 2-sphere, [**8**]. In §4 we shall present this description. In §5 we prove a new theorem on the structure of the components in the space of self maps on the 2-sphere, which at least give partial complete homotopy types of the components corresponding to nonzero degrees of maps.

3. Factors in the homotopy of the space of self maps on the 2-sphere.

In the remaining part of the paper, the surface S is the 2-sphere. We shall first define a map which splits the homotopy of the components in the space of self maps on S and in this connection describe an exact sequence constructed by G.W. Whitehead.

For each degree k, evaluation at the base point in S defines a Hurewicz fibration

$$p_k : M_k(S,S) \to S \quad \text{with fibre} \quad F_k(S,S).$$

By considering $SO(3)$ as the group of orientation preserving isometries on S, we get similarly the fibration

$$ev : SO(3) \to S \quad \text{with fibre} \quad SO(2) = S^1.$$

Finally, we shall also need the trivial fibration

$$pr_k : SO(3) \times F_k(S,S) \to SO(3),$$

defined by the projection map pr_k onto the first factor.

For each degree k we can define a map

$$\Phi_k : SO(3) \times F_k(S,S) \to M_k(S,S),$$

by associating to the isometry $A \in SO(3)$ and the based map $f \in F_k(S,S)$ the free map $A \circ f \in M_k(S,S)$. It is easy to check that Φ_k defines a map between fibrations,

$$\begin{array}{ccc} F_k(S,S) & \xrightarrow{1} & F_k(S,S) \\ \downarrow & & \downarrow \\ SO(3) \times F_k(S,S) & \xrightarrow[\Phi_k]{} & M_k(S,S) \\ {\scriptstyle pr_k}\downarrow & & \downarrow{\scriptstyle p_k} \\ SO(3) & \xrightarrow[ev]{} & S \end{array}$$

which induces the identity map on fibres.

Consider the induced map between homotopy sequences for the map between fibrations defined by Φ_k . Since the map $ev : SO(3) \to S$ is a fibration with the circle as fibre, it induces an isomorphism between homotopy groups in dimensions $i \geq 3$. By a simple application of the 5-lemma it follows that the map Φ_k induces an isomorphism between homotopy groups in dimensions $i \geq 3$.

For the study of the map Φ_k on the homotopy level in dimensions 1 and 2, we need a formula for the boundary homomorphism in the homotopy sequence for the fibration $p_k : M_k(S,S) \to S$, which we now recall.

Let $f_k \in F_k(S,S)$ be a based map, which we use as base point in both $F_k(S,S)$ and $M_k(S,S)$. For each $i \geq 1$, there is a well known adjoint isomorphism

$$\pi_i(F_k(S,S), f_k) \simeq \pi_{i+2}(S, s_0).$$

Inserting these isomorphisms into the homotopy sequence for $p_k : M_k(S,S) \to S$ we get the exact sequence

$$\cdots \to \pi_{i+1}(S,s_0) \xrightarrow{\delta_k} \pi_{i+2}(S,s_0) \to \pi_i(M_k(S,S), f_k) \to \pi_i(S,s_0) \xrightarrow{\delta_k} \pi_{i+1}(S,s_0) \to \cdots$$

By a theorem of G.W. Whitehead [**15**], with a correction for sign in J.H.C. Whitehead [**16**], the boundary homomorphism

$$\delta_k : \pi_i(S,s_0) \to \pi_{i+1}(S,s_0)$$

is given by $\delta_k(\alpha) = -k \cdot [\iota, \alpha]$, where $[\cdot,\cdot]$ denotes Whitehead product, and $\iota \in \pi_2(S,s_0)$ is the generator defined by the identity map on S.

The Whitehead product $[\iota, \iota]$ is an element of infinite order in $\pi_3(S,s_0)$, in fact twice the generator for $\pi_3(S,s_0) \simeq \mathbf{Z}$. It is also well known that $[\iota, \alpha] = 0$ for all homotopy classes $\alpha \in \pi_i(S,s_0)$ in dimensions $i \geq 3$. Using this information on Whitehead products in the exact sequence of G.W. Whitehead, it is easy to prove that the homomorphism

$$\pi_2(F_k(S,S), f_k) \simeq \pi_4(S,s_0) \to \pi_2(M_k(S,S), f_k)$$

is an isomorphism for $k \neq 0$.

Since the second homotopy group of $SO(3)$ vanishes, it follows that the map Φ_k induces isomorphism on the homotopy level also in dimension 2, when $k \neq 0$. Altogether we have proved

PROPOSITION 3.1 ([8], PROPOSITION 2.1). *The map Φ_k induces an isomorphism between homotopy groups in all dimensions $i \geq 3$. For $k \neq 0$, it also induces an isomorphism in dimension 2.*

4. The space of self homotopy equivalences on the 2-sphere.

In this section we examine the identity component $M_1(S,S)$ in the space of self maps on the 2-sphere S. This component is exactly the space of orientation preserving self homotopy equivalences on S.

The inclusion map $SO(3) \to M_1(S,S)$ induces a map between fibrations,

$$\begin{array}{ccc} SO(2) & \longrightarrow & F_1(S,S) \\ \downarrow & & \downarrow \\ SO(3) & \longrightarrow & M_1(S,S) \\ {\scriptstyle ev}\downarrow & & \downarrow{\scriptstyle p_1} \\ S & \xrightarrow[1]{} & S \end{array}$$

Using that the Whitehead product $[\iota,\iota]$ is twice the generator in $\pi_3(S) \simeq \mathbf{Z}$, we get immediately the following theorem of Hu [**11**] from the exact sequence of G.W. Whitehead corresponding to the fibration $p_k : M_k(S,S) \to S$,

$$\pi_1(M_k(S,S)) \simeq \mathbf{Z}_{2|k|}, \quad \text{in particular,} \quad \pi_1(M_1(S,S)) \simeq \mathbf{Z}_2.$$

An easy inspection of the induced map between homotopy sequences for the above map between fibrations now reveals that the generator for $\pi_1(SO(3)) \simeq \mathbf{Z}_2$ is mapped onto the generator for $\pi_1(M_1(S,S)) \simeq \mathbf{Z}_2$ and hence that the inclusion map $SO(3) \to M_1(S,S)$ induces an isomorphism between fundamental groups.

Let $\tilde{F}_1(S,S)$ denote the universal covering space of $F_1(S,S)$. The map Φ_1 induces a map

$$\tilde{\Phi}_1 : SO(3) \times \tilde{F}_1(S,S) \to M_1(S,S).$$

From Proposition 3.1 and the above considerations follow that $\tilde{\Phi}_1$ induces an isomorphism between homotopy groups in all dimensions $i \geq 1$. A fundamental theorem of J.H.C. Whitehead now implies that $\tilde{\Phi}_1$ is a homotopy equivalence, since all spaces involved are ANR's.

All the components in F(S,S) have the same homotopy type, see e.g. ([**4**], Proposition 1), and hence we can substitute $\tilde{F}_1(S,S)$ by $\Omega = \tilde{\Omega}_0^2(S)$, which denotes the universal covering space for the component in the double loop space on S containing the constant based map. Altogether we get the following

THEOREM 4.1 ([**8**],§5). *The space of orientation preserving self homotopy equivalences on the 2-sphere has the homotopy type of* $SO(3) \times \Omega$.

The space Ω has nontrivial homotopy type. This is interesting since it was proved by Kneser [**12**] in 1926 and Smale [**14**] in 1959 that the space of orientation preserving homeomorphisms, respectively diffeomorphisms, on the 2-sphere has the homotopy type of $SO(3)$. Tacitly it seems to have been assumed that this was also the case for the space of homotopy equivalences.

5. Coverings of components in the space of self maps on the 2-sphere.

Our aim in this section is to prove the following theorem for the components in the space of self maps on the 2-sphere S.

THEOREM 5.1. *For each degree $k \neq 0$, there is a unique k-fold covering space E_k of $M_k(S,S)$. The space E_k has the homotopy type of $M_1(S,S)$.*

For the proof of this theorem the following diagram (not commutative) is essential:

$$\begin{array}{ccc}
SO(3) \times \tilde{F}_1(S,S) & & \\
\downarrow{\scriptstyle 1\times\theta} & \searrow^{\tilde{\Phi}_1} \nwarrow_{\tilde{\Psi}_1} & \\
SO(3) \times F_1(S,S) & \xrightarrow[\Phi_1]{} & M_1(S,S) \\
\downarrow{\scriptstyle 1\times\rho_k} & & \downarrow{\scriptstyle \sigma_k} \\
SO(3) \times F_k(S,S) & \xrightarrow[\Phi_k]{} & M_k(S,S)
\end{array}$$

The maps in the diagram are defined as follows:

$\theta : \tilde{F}_1(S,S) \to F_1(S,S)$ is the universal covering map,

$\rho_k : F_1(S,S) \to F_k(S,S)$ is a homotopy equivalence defined by addition of a fixed map of degree k-1, ([4],§2),

Φ_1 and Φ_k are the maps defined in §3,

$\tilde{\Phi}_1$ is the homotopy equivalence constructed in §4,

$\tilde{\Psi}_1$ is a homotopy inverse to $\tilde{\Phi}_1$.

Finally, $\sigma_k : M_1(S,S) \to M_k(S,S)$, is defined as the composite map

$$\sigma_k = \Phi_k \circ (1 \times \rho_k) \circ (1 \times \theta) \circ \tilde{\Psi}_1.$$

LEMMA 5.2. *For each $k \neq 0$, the map $\sigma_k : M_1(S,S) \to M_k(S,S)$ induces an isomorphism between homotopy groups in all dimensions $i \geq 2$. Between fundamental groups it induces multiplication by k.*

The first part of the lemma follows immediately from Proposition 3.1.

To compute σ_k on the fundamental group level, let $f_k : S \to S$ denote a fixed based map of degree $k \neq 0$. Composition with f_k defines a map between fibrations,

$$\begin{array}{ccc}
SO(2) & \longrightarrow & F_k(S,S) \\
\downarrow & & \downarrow \\
SO(3) & \xrightarrow{\bar{f}_k} & M_k(S,S) \\
{\scriptstyle ev}\downarrow & & \downarrow{\scriptstyle p_1} \\
S & \xrightarrow[1]{} & S
\end{array}$$

where $\bar{f}_k$ maps the isometry $A \in SO(3)$ into $A \circ f_k \in M_k(S,S)$.

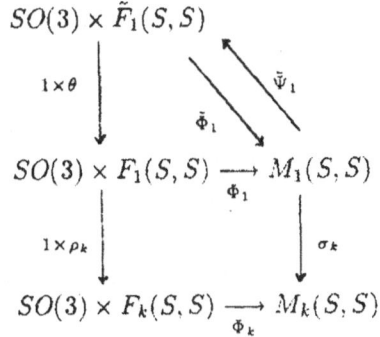

A simple analysis of the induced map between homotopy sequences, using the formula for the boundary homomorphism found by G.W. Whitehead, reveals that the generator of $\pi_1(SO(3), I) \simeq \mathbf{Z}_2$, where I denotes the identity isometry on S, is mapped onto k times the generator of $\pi_1(M_k(S,S), f_k) \simeq \mathbf{Z}_{2|k|}$. (Alternatively, one can use [**9**], Lemma 2.) On the other hand this is exactly what the map induced by σ_k does on the fundamental group level. This proves Lemma 5.2.

Now to the proof of Theorem 5.1. The fundamental group of $M_k(S,S)$ contains a unique subgroup of order 2 generated by k times the generator of $\pi_1(M_k(S,S), f_k)$. Correspondingly, there is a unique ktfold covering space E_k of $M_k(S,S)$ with this subgroup as fundamental group. By general covering space theory, there is a lift h of $\sigma_k : M_1(S,S) \to M_k(S,S)$,

$$\begin{array}{ccc} M_1(S,S) & \xrightarrow{h} & E_k \\ & {\scriptstyle \sigma_k}\searrow \quad \swarrow{\scriptstyle \bar{\sigma}_k} & \\ & M_k(S,S) & \end{array}$$

over the covering map $\bar{\sigma}_k : E_k \to M_k(S,S)$.

By construction, the map $h : M_1(S,S) \to E_k$ induces an isomorphism between homotopy groups in all dimensions $i \geq 1$, and hence it is a homotopy equivalence, since the spaces involved are ANR's. This completes the proof of Theorem 5.1.

Finally, a few words about the component $M_0(S,S)$ containing the constant maps. Since the fibration $p_0 : M_0(S,S) \to S$ has an obvious section, it is easy to prove that $M_0(S,S)$ splits as the product $S \times F_0(S,S)$ as far as homotopy groups are concerned. It is, however, an open question, first posed in [**6**], whether this splitting is true also on the space level.

Bibliography

[1] R. Geoghegan, *On spaces of homeomorphisms, embeddings and functions-I*, Topology **11** (1972), 159–177.

[2] D.H. Gottlieb, *A certain subgroup of the fundamental group*, Amer.J. Math. **87** (1965), 840–856.

[3] D.H. Gottlieb, *Covering transformations and universal fibrations*, Illinois J. Math. **13** (1969), 432–437.

[4] V.L. Hansen, *On the space of maps of a closed surface into the 2-sphere*, Math. Scand. **35** (1974), 149–158.

[5] V.L. Hansen, *On a theorem of Al'ber on spaces of maps*, J. Diff. Geom. **12** (1977), 565–566.

[6] V.L. Hansen, *Decomposability of evaluation fibrations and the brace product operation of James*, Compositio Math. **35** (1977), 83–89.

[7] V.L. Hansen, *Spaces of maps into Eilenberg-MacLane spaces*, Canadian J. Math. **XXXIII** (1981), 782–785.

[8] V.L. Hansen, *The homotopy groups of a space of maps between oriented closed surfaces*, Bull. London Math. Soc. **15** (1983), 360–364.

[9] V.L. Hansen, *On Steenrod bundles and the van Kampen theorem*, Canadian Math. Bull. **31** (1988), 241–249.

[10] D.W. Henderson, *Stable classification of infinite dimensional manifolds by homotopy type*, Invent. Math. **12** (1971), 48–56.

[11] S.T. Hu, *Concerning the homotopy groups of the components of the mapping space* Y^{S^p}, Indagationes Math. **8** (1946), 623–629.

[12] H. Kneser, *Die Deformationssätze der einfach zusammenhängenden Flächen*, Math. Z. **25** (1926), 362–372.

[13] L.L. Larmore and E. Thomas, *On the fundamental group of a space of sections*, Math. Scand. **47** (1980), 232–246.

[14] S. Smale, *Diffeomorphisms on the 2-sphere*, Proc. Amer. Math. Soc. **10** (1959), 621–626.

[15] G.W. Whitehead, *On products in homotopy groups*, Ann. Math. **47** (1946), 460–475.

[16] J.H.C. Whitehead, *On certain theorems of G.W. Whitehead*, Ann. Math. **58** (1953), 418–428.

Mathematical Institute
The Technical University of Denmark
Building 303
DK-2800 Lyngby
Denmark

FINITE PRESENTATION OF 3-MANIFOLD MAPPING CLASS GROUPS

Allen Hatcher* and Darryl McCullough*
Cornell University and the University of Oklahoma

0. Introduction

By far the greater part of work in 3-manifold theory has focused on irreducible 3-manifolds. Many problems, such as classification, need only be considered in this case. But in the study of mappings between 3-manifolds, many new phenomena arise in the reducible case. Examples appear in [H], [F-W], [K-M1], [K-M2], and [M5] (a survey is given in [M4]).

The (full) mapping class group $\mathcal{H}(M)$ of a manifold M is the group of path components (i. e. isotopy classes) of the space of homeomorphisms of M. For 2-manifolds, the mapping class groups have been heavily studied. For 3-manifolds, they have been computed in a number of special cases, for example in [R-B], [B2], [M6]. The work of Waldhausen [W] and its extensions to the nonorientable case [H1], [L] give a great deal of information. Johannson [J] used his theory of characteristic submanifolds of Haken 3-manifolds (discovered independently by Jaco and Shalen [J-S]) to prove that the mapping class groups of Haken 3-manifolds are finitely generated, and that the subgroup generated by Dehn twists about essential tori and annuli has finite index. This leads to a proof that their mapping class groups are finitely presented [W1], [G2], [M1], [M2].

In fact, the mapping class groups of irreducible orientable sufficiently large 3-manifolds (not necessarily Haken, since the boundary may be compressible) are finitely presented, and enjoy strong homological finiteness properties [M1], [M2]. To prove such results for 3-manifolds with compressible boundary, one can use the *disc complex,* for which the vertices are isotopy classes of essential properly-imbedded 2-discs in M, and a collection of vertices spans a simplex if and only if the isotopy classes can be simultaneously represented by pairwise disjoint discs. Since the vertices are isotopy classes, and the incidence condition is preserved by homeomorphisms, the mapping class group acts simplicially on the disc complex. The quotient is finite. Adapting ideas of R. Kramer [K], the disc complex can be proved to be contractible (see §5 of [M2]).

An analogous complex can be constructed for reducible 3-manifolds using isotopy classes of (smoothly) imbedded 2-spheres, essential in the sense that they do not bound 3-balls in M. However, a lemma used in the proof of contractibility of the disc complex (Lemma 5.1 of [M2]) is actually false for spheres, and the proof of contractibility of the sphere complex appears to be considerably more difficult. In the present paper we give a fairly simple proof, in §1, that the sphere complex is simply-connected, using a special technique which unfortunately does not generalize to higher homotopy groups. Then, we seek to apply the following theorem of K. Brown (Theorem 4 of [B]) to the action of $\mathcal{H}(M)$ on the sphere complex:

---

* Research of both authors supported by the National Science Foundation
This paper is in final form and no version of it will be submitted for publication elsewhere.

THEOREM: *Let G be a group which acts simplicially on a simply-connected complex so that every vertex isotropy group is finitely presented, every edge isotropy group is finitely generated, and so that the quotient has finite 2-skeleton. Then G is finitely presented.*

(In fact, Brown proves much more precise results which allow one to give an explicit presentation of G from data about the action.) In §2 are some auxiliary results, one consequence of which is the finiteness of the quotient of the sphere complex under the action of $\mathcal{H}(M)$. In §3, assuming that the universal cover of M satisfies the Poincaré Conjecture (i. e. every imbedded homotopy 3-cell is homeomorphic to the standard 3-cell), the stabilizers of the action are shown to be finitely presented; the proof relies on results from [L] and [H-L]. In the final section, an inductive argument using Brown's theorem yields our main result:

THEOREM 4.1: *Let M be a compact orientable 3-manifold whose universal cover satisfies the Poincaré Conjecture. If the mapping class group of each irreducible summand of M is finitely presented, then the mapping class group of M is finitely presented.*

The hypothesis that the universal cover of M satisfies the Poincaré Conjecture may be stronger than the assumption that M satisfies the Poincaré Conjecture, since it also excludes the possibility that M has irreducible summands covered by fake 3-spheres.

Many authors use the term *homeotopy group*, reserving the term *mapping class group* for the subgroup of orientation-preserving elements. Since this subgroup has index at most 2, and groups are finitely presented if and only if their finite-index subgroups are, Theorem 4.1 will hold using either definition once it has been proven for one of them.

Our notation for mapping class groups is fairly standard. For $A \subseteq M$, we denote by *Homeo*(M, A) the homeomorphisms that take A to A, and by *Homeo*$(M \text{ rel } A)$ those that restrict to the identity map on A; the associated mapping class groups are $\mathcal{H}(M, A)$ and $\mathcal{H}(M \text{ rel } A)$. When S is an imbedded 2-sphere in a 3-manifold (either contained in or disjoint from the boundary) a homeomorphism called a *rotation about* S can be defined by letting a noncontractible loop based at the identity in the orthogonal group SO(3) act naturally on the spheres in a product neighborhood of S; since this homeomorphism is the identity on the ends of the product it extends to M. Because product neighborhoods are unique up to isotopy, the mapping class of this rotation is well-defined, and since the loop has order 2 in $\pi_1(\mathrm{SO}(3))$, the square of a rotation is isotopic to the identity. These play a significant role in the study of mappings of reducible 3-manifolds (see [H] and §3 of [M3]). We will use them in our §3.

We thank the referee for pointing out some obscurities in our original manuscript; we hope they have all been clarified in the present version.

1. Simple-connectivity of the sphere complex

Let M be a reducible 3-manifold. Define a simplicial complex $\mathcal{S}(M)$ whose vertices are isotopy classes of imbedded 2-spheres which do not bound 3-balls in M, and whose simplices are determined by the rule that a collection $[S_0'], [S_1'], \ldots, [S_n']$ spans an n-simplex if and only if there is a submanifold $S_0 \cup S_1 \cup \ldots \cup S_n$ consisting of disjointly imbedded 2-spheres, none of which bounds a ball and no two of which are isotopic, such that S_i' isotopic to S_i for $0 \le i \le n$. We denote by $[S_0', S_1', \ldots, S_n']$ the simplex spanned by $[S_0'], [S_1'], \ldots, [S_n']$. Since we assume that M is reducible, $\mathcal{S}(M)$ is not empty.

There is another way to define a sphere complex will sometimes be useful. An n-cell will be an isotopy class of $S_0 \cup S_1 \cup \ldots \cup S_n$ as above. It has the structure of an n-simplex, with face maps obtained by passing to subcollections of the spheres. Piecing together these face maps to define attaching maps provides the structure of a cell complex. When M satisfies the Poincaré Conjecture, it follows from Lemma 3.1 below (taking $m=0$ and h equal to the identity homeomorphism of M) that if $S_0 \cup S_1 \cup \ldots \cup S_n$ and $T_0 \cup T_1 \cup \ldots \cup T_n$ are two such collections, and S_i is isotopic to T_i for $0 \le i \le n$, then $S_0 \cup S_1 \cup \ldots \cup S_n$ is isotopic to $T_0 \cup T_1 \cup \ldots \cup T_n$. Therefore when M satisfies the Poincaré Conjecture (and perhaps even when it does not), the vertices $[S_0'], [S_1'], \ldots, [S_n']$ of an n-simplex of $\mathcal{S}(M)$ determine the submanifold $S_0 \cup S_1 \cup \ldots \cup S_n$ up to isotopy, and the cell complex is isomorphic to $\mathcal{S}(M)$.

THEOREM 1.1: *$\mathcal{S}(M)$ is simply-connected.*

PROOF: We will first show that loops in $\mathcal{S}(M)$ are null-homotopic and then, by a simpler form of the argument, that $\mathcal{S}(M)$ is connected. First, we must develop a way to represent loops in $\mathcal{S}(M)$ by geometric objects.

Consider the space of imbeddings of S^2 in M whose images do not bound 3-balls, with the compact-open topology. Let J be the quotient of this space of imbeddings by the right action of the diffeomorphism group $\mathit{Diff}(S^2)$. An element $S \in J$ is regarded as a submanifold of M. Let Y be the cone $[0,1] \times J/\{0\} \times J$. A point of Y will be denoted by tS, so that $0S = 0S'$ for all S, $S' \in J$. A collection of elements $\{S_0, S_1, \ldots, S_n\}$ of J is said to be *compatible* if for each i and j, either $S_i = S_j$ or $S_i \cap S_j = \emptyset$. Let X_1 be the set of formal sums $\sum_{i=0}^{\infty} t_i S_i$ of elements of Y, where all $t_i \ge 0$, all but finitely many $t_i = 0$, $\sum_{i=0}^{\infty} t_i = 1$, and $\{S_i \mid t_i \ne 0\}$ is a compatible collection. We topologize X_1 as a subspace of the space of maps from the natural numbers into Y with the compact-open topology. There is an equivalence relation on X_1 generated by the two relations

(1) $\sum_{i=0}^{\infty} t_i S_i \sim \sum_{i=0}^{\infty} t_{\sigma(i)} S_{\sigma(i)}$ for any permutation σ of the natural numbers

(2) $0S_0 + t_1 S_1 + \sum_{i=2}^{\infty} t_i S_i \sim sS_1 + tS_1 + \sum_{i=2}^{\infty} t_i S_i$ when $s + t = t_1$.

Let X be the quotient of X_1 by this equivalence relation. Thus any unordered finite sum of nonnegative multiples of compatible 2-spheres in M, with coefficients adding up to 1, determines an element of X, and two sums $\sum t_i S_i$ and $\sum t_i' S_i'$ are close in X if, possibly after reordering and recollecting terms, both (a) t_i is close to t_i' for all i, and (b) whenever both t_i and t_i' are nonzero, S_i is close to S_i'.

A continuous map from X to $\mathcal{S}(M)$ is induced by associating to each formal sum the point $\sum t_i [S_i]$, where the sum is taken over all i such that $t_i \ne 0$.

Let $[S_0', S_1']$ be an edge in $\mathcal{S}(M)$, so that S_0' and S_1' are isotopic to disjoint spheres. Using these disjoint spheres as the $t = \frac{1}{2}$ level, we can construct an isotopy $S_0(t) \cup S_1(t)$ with $S_0(t) \cap S_1(t) = \emptyset$ for all t, so that $S_0(0) = S_0'$ and $S_1(1) = S_1'$ (as elements of J). The sums $(1-t)S_0(t) + tS_1(t)$ form a path in X, whose image in $\mathcal{S}(M)$ is the path $(1-t)[S_0'] + t[S_1']$ along the given edge $[S_0', S_1']$. Now take a sequence of edges in $\mathcal{S}(M)$ forming a loop. Choose a sphere S_i' representing each vertex of this loop and perform the preceding interpolation construction for each edge of this loop, with respect to the chosen S_i''s at its endpoints. The result is a path in X which we can write as a single

1-parameter linear combination $\Sigma_t = \sum t_i(t)S_i(t)$ where $t \in [0,1]$ is the parameter of the loop and the coefficient $t_i(t)$ is defined to be zero outside the subinterval of $[0,1]$ for which $[S_i(t)]$ is a vertex of the edge in the loop in $\mathcal{S}(M)$.

Let S_0 be a parallel copy of the base sphere $\Sigma_0 = \Sigma_1$. The goal is to construct a homotopy of paths in X, denoted by $\Sigma_{tu} = \sum t_{ij}(t,u)S_{ij}(t)$ for $u \in [0,1]$, with $\Sigma_{t0} = \Sigma_t$ and Σ_{t1} equal to $1S_0$. Then, passing to $\mathcal{S}(M)$, $\sum t_{ij}(t,u)[S_{ij}(t)]$ will represent a nullhomotopy of the given loop in $\mathcal{S}(M)$ (not necessarily in the 2-skeleton).

The first step is to arrange that all the $S_i(t)$ meet S_0 transversely. This can be done as follows. Choose product neighborhoods of the $S_i(t)$'s, varying smoothly with t, with the neighborhood of $S_i(t)$ disjoint from the neighborhood of $S_j(t)$ whenever $t_i(t)t_j(t) \neq 0$. For fixed t, most slices $S_i(t) \times \{x\}$ in this product are transverse to S_0 by Sard's theorem, and they remain transverse for nearby t as well. So by compactness of the parameter domain, finitely many disjoint parallel copies $S_{ij}(t) = S_i(t) \times \{x_{ij}\}$ of each family $S_i(t)$ may be chosen, at least one of which is transverse to S_0 for each t. Then choose the coefficients $t_{ij}(t)$ (by a partition of unity argument, for example) so that $S_{ij}(t)$ is transverse to S_0 over the support of $t_{ij}(t)$ and $\sum_j t_{ij}(t) = t_i(t)$. Since each $S_{ij}(t)$ is isotopic to the original $S_i(t)$, the new path $\sum t_{ij}(t)S_{ij}(t)$ has the same image in $\mathcal{S}(M)$ as the original Σ_t. This achieves the desired reduction to the case that all the spheres in Σ_t meet S_0 transversely. Similarly, we may assume all the spheres in Σ_t are disjoint from a chosen basepoint $p \in S_0$. We retain the original notation, with $\Sigma_t = \sum t_i(t)S_i(t)$.

Let $\mathrm{I}_i \subseteq [0,1]$ be the support of $t_i(t)$ (the closure of $\{t \mid t_i(t) > 0\}$). We may assume the functions $t_i(t)$ are piecewise linear, so I_i is a finite union of intervals. Relabeling the $S_i(t)$'s, we can take each I_i to be a single interval. Let Γ_{it} be the collection of circles of $S_i(t) \cap S_0$, and let Γ_t be the union of the Γ_{it}'s for which $t \in \mathrm{I}_i$. The elements of Γ_t are partially ordered according to the inclusion relations which hold among the disks they bound in $S_0 - \{p\}$. Choose a family of order-preserving injections $\phi_t : \Gamma_t \to (0,1)$ such that the value of ϕ_t on each circle of $S_i(t) \cap S_0$ is independent of $t \in \mathrm{I}_i$. This can be done inductively, with increasing t; as t passes an endpoint of an I_i, some circles of Γ_t are added or deleted, so first extend ϕ_t to the newly added circles, then restrict to the non-deleted circles.

Now we can construct the family Σ_{tu}. To begin, let $\gamma \in \Gamma_t$ have the smallest ϕ_t-value. For $u \leq \phi_t(\gamma)$, Σ_{tu} is independent of u, but as u increases past $\phi_t(\phi)$ surger the sphere $S_i(t) = S_{i0}(t)$ of $\Sigma_{t\phi_t(\gamma)}$ containing γ along the disk in $S_0 - \{p\}$ bounded by γ, in the usual way, producing a pair of spheres $S_{i1}(t) \cup S_{i2}(t)$, varying continuously with $t \in I_i$. We push these spheres off $S_{i0}(t)$ slightly, so that they are disjoint from $\Sigma_{t\phi_t(\gamma)}$ and can be added to it, using coefficients $t_{i1}(t,u)$ and $t_{i2}(t,u)$ which go linearly from 0 to $t_i(t)/2$ as u goes from $\phi_t(\gamma)$ to $\phi_t(\gamma) + \epsilon$, while the coefficient $t_{i0}(t,u)$ is allowed to go linearly from its initial value to zero. (If one of the resulting spheres, say $S_{i2}(t)$, bounds a ball, then it is not used, and instead the coefficient $t_{i1}(t,u)$ goes from 0 to $t_i(t)$. Since $S_{i0}(t)$ does not bound a ball, at least one of the spheres resulting from surgery does not bound a ball.) Other than these changes, let Γ_{tu} be independent of $u \in [\phi_t(\gamma), \phi_t(\gamma) + \epsilon]$.

This surgery process is repeated for circles of Γ_t with successively larger ϕ_t-values to produce the family Σ_{tu} ending with $\Sigma_{t,1-\epsilon}$, say, a sum of spheres disjoint from S_0. (Since ϕ_t is injective, we never have to perform simultaneous surgeries.) To finish the construction of Σ_{tu}, let it go linearly from $\Sigma_{t,1-\epsilon}$ to $\Sigma_{t1} = S_0$. We have shown that every loop in $\mathcal{S}(M)$ is contractible.

The proof that $\mathcal{S}(M)$ is connected is much simpler. Fix S_0 and use a sequence of surgeries to construct a path from any given vertex $[S]$ to a vertex $[S_1]$, where S_1 is disjoint from S_0; then, $[S_0, S_1]$ is a 1-simplex so $\mathcal{S}(M)$ is connected. This completes the proof of Theorem 1.1.

REMARK: The obstacle to extending this proof to show that the higher homotopy groups of $\mathcal{S}(M)$ vanish occurs in the first step, when we produced a continuous family Σ_t representing the loop in $\mathcal{S}(M)$.

2. Auxiliary results

For $1 \leq i \leq r$ let D_i be a 3-cell in the oriented compact irreducible 3-manifold P_i. Regard the oriented connected sum $M = P_1 \# P_2 \# \ldots \# P_r \# (\#_g S^2 \times S^1)$ as constructed from a punctured 3-cell B having $2g + r$ boundary components by attaching the $P_i - int(D_i)$ to r of the 2-sphere boundary components, and attaching g copies $S_j \times I$ of $S^2 \times I$ to the remaining $2g$ 2-sphere boundary components.

For some $i \leq r$, let $\widehat{M}$ be obtained from M by replacing $P_i - int(D_i)$ by a 3-ball E. Let α be an oriented arc in $\widehat{M}$ meeting E only in its endpoints. Choose an isotopy J_t of $\widehat{M}$ with the following properties:

(a) $J_0 = 1_{\widehat{M}}$
(b) $J_1|_E = 1_E$
(c) there is a regular neighborhood of $B_i \cup \alpha$ outside of which each J_t is the identity
(d) J_t moves E around α, i. e. if e is the center of E, then the trace $J_t(e)$ is a loop representing the generator of the fundamental group of the regular neighborhood of $B_i \cup \alpha$ (which is a solid torus having infinite cyclic fundamental group) determined by the orientation of α.

By a *slide homeomorphism that slides P_i around α*, we mean a homeomorphism isotopic to $h\colon M \to M$ defined by $h|_{M-(P_i-D_i)} = J_1|_{\widehat{M}-E}$ and $h|_{P_i-D_i} = 1_{P_i-D_i}$. The following lemma is essentially due to M. Scharlemann (see Appendix A of [B1]).

LEMMA 2.1: *Let Σ be a collection of pairwise disjoint essential 2-spheres in M. Then there is a composite g of slide homeomorphisms such that $g(\Sigma) \subseteq B$.*

PROOF SKETCH: An intersection curve of $\partial B \cap \Sigma$, innermost on Σ, bounds a disc D in Σ whose interior is disjoint from ∂B. If D lies in the closure of $M - B$, then there is an isotopy that pulls D into B, eliminating the intersection curve ∂D and possibly others. Eliminate all such intersections possible. Then, when D lies in B, there is a punctured 3-cell B_0 in B bounded by D and a disc in ∂B, and a loop α which starts in B_0, moves into $M - B$ and reenters B in the complement of B_0, without touching Σ (follow along the sphere that contains D) and then moves through B, intersecting Σ, to form a closed loop. Using this loop as a guide, apply slide homeomorphisms to slide the summands attached in B_0 around loops that pass through $M - B$ without intersecting Σ; the image of D after these slides can be isotoped to decrease the number of intersection curves. Repeat to eliminate all intersections. Then, each sphere in the image of Σ either lies in B, or is parallel into B, and another isotopy moves all of the image of Σ into B.

PROPOSITION 2.2: *The quotient $\mathcal{S}(M)/\mathcal{H}(M)$ is finite.*

PROOF: Let $[S_0', S_1', \ldots, S_n']$ be an n-simplex in $\mathcal{S}$. Then there is a submanifold $S_0 \cup S_1 \cup \ldots \cup S_n$ of M, with S_i isotopic to S_i'. By Lemma 2.1, there exists a homeomorphism carrying $S_0 \cup S_1 \cup \ldots \cup S_n$ into B. But there are only finitely many isotopy classes of 2-spheres in B (they are determined by the way they partition the boundary components). Therefore there are only finitely many orbits of the action of $\mathcal{H}(M)$ on $\mathcal{S}(M)$.

PROPOSITION 2.3: *Let M be a (connected) 3-manifold with finitely generated fundamental group. Suppose S is a 2-sphere boundary component of M and let $\widehat{M}$ be obtained by filling in S with a 3-ball. Then $\mathcal{H}(M,S)$ is finitely presented if and only if $\mathcal{H}(\widehat{M})$ is finitely presented.*

PROOF: Let e_0 be a point in the interior of the filled-in 3-ball E. From the restriction fibration $\mathit{Homeo}(\widehat{M}) \to \mathit{int}(M)$ (see [M]), there is an exact sequence

$$\pi_1(\widehat{M}) \longrightarrow \mathcal{H}(\widehat{M}, e_0) \longrightarrow \mathcal{H}(\widehat{M}) \longrightarrow 1.$$

Now $\mathcal{H}(\widehat{M}, e_0) \cong \mathcal{H}(\widehat{M}, E) \cong \mathcal{H}(M,S)$. The kernel of $\pi_1(\widehat{M}) \to \mathcal{H}(\widehat{M}, e_0)$ is the subgroup of traces of isotopies from $1_{\widehat{M}}$ to $1_{\widehat{M}}$. By [G1], these traces are central in $\pi_1(\widehat{M})$. According to a recent result of G. Mess [M7], the center must be finitely generated. Therefore this kernel is a finitely generated abelian group. Since $\pi_1(\widehat{M})$ is finitely generated, it is finitely presented [S]. The result now follows from the exact sequence.

3. Calculation of the stabilizers

Throughout this section, M is compact, orientable, reducible, and has no 2-sphere boundary components.

LEMMA 3.1: *Suppose that M satisfies the Poincaré Conjecture. Let $T_1, T_2, \ldots, T_n$ be a collection of pairwise disjoint pairwise nonisotopic essential imbedded 2-spheres in M and let h be a homeomorphism of M such that $h(T_i) = T_i$ for $1 \le i \le m$ and $h(T_i)$ is homotopic to T_i for $m+1 \le i \le n$. Then h is isotopic preserving $T_1, \ldots, T_m$ to a homeomorphism h' such that $h'(T_j) = T_j$ for $1 \le j \le n$.*

PROOF: An easy extension of the Lemma on p. 124 of [L] to collections of disjoint 2-spheres shows that $h(T_{m+1})$ is homotopic to T_{m+1} by a homotopy that avoids $\cup_{i=1}^m T_i$. Since in the absence of fake 3-cells homotopic 2-spheres are isotopic (Theorem III.1.3 of [L]), $h(T_{m+1})$ is isotopic to T_{m+1} in the complement of $\cup_{i=1}^m T_i$. Induction completes the proof.

Define $\mathcal{R}(M)$ to be the subgroup of $\mathcal{H}(M)$ generated by rotations about imbedded 2-spheres in M. From §3 of [M3], we have

LEMMA 3.2: (1) *Let r_Σ be a rotation about the 2-sphere $\Sigma \subseteq M$ and let F be an incompressible surface in the interior of M. Then there is a product r of rotations about 2-spheres disjoint from F so that $\langle r_\Sigma \rangle = \langle r \rangle$ in $\mathcal{H}(M)$.*
(2) *$\mathcal{R}(M)$ is a normal subgroup of $\mathcal{H}(M)$.*
(3) *$\mathcal{R}(M) \cong (\mathbb{Z}/2)^k$ for some nonnegative integer k.*

Denote by $\overline{\mathcal{H}}(M)$ the quotient $\mathcal{H}(M)/\mathcal{R}(M)$. Statement (1) of Lemma 3.2, applied when F is a 2-sphere, shows that $\mathcal{R}(M)$ acts trivially on the sphere complex $\mathcal{S}(M)$. Therefore we have immediately:

PROPOSITION 3.3: (a) *The action of $\mathcal{H}(M)$ on $\mathcal{S}(M)$ induces an action of $\overline{\mathcal{H}}(M)$ on $\mathcal{S}(M)$.*
(b) *If σ is any simplex of $\mathcal{S}(M)$, then the stabilizers of the actions in (a) appear in an exact sequence*

$$1 \longrightarrow \mathcal{R}(M) \longrightarrow \mathrm{stab}_{\mathcal{H}(M)}(\sigma) \longrightarrow \mathrm{stab}_{\overline{\mathcal{H}}(M)}(\sigma) \longrightarrow 1.$$

Suppose that $S_0, S_1, \ldots, S_n$ is a collection of pairwise disjoint essential 2-spheres in M, representing a simplex of $\mathcal{S}(M)$. Denote by $M_1, M_2, \ldots, M_m$ the components that result from cutting M along $\cup_{i=0}^n S_i$. Define $\overline{\mathcal{H}}_\partial(M_j)$ to be the subgroup generated by the elements of $\overline{\mathcal{H}}(M_j)$ that take each component of ∂M_j to itself and restrict to a degree 1 homeomorphism on each 2-sphere boundary component; this is a subgroup of finite index in $\overline{\mathcal{H}}(M_j)$. There is a well-defined homomorphism $\prod \overline{\mathcal{H}}_\partial(M_j) \to \overline{\mathcal{H}}(M)$ obtained by choosing representatives that are the identity on each 2-sphere boundary component and fitting them together to form a homeomorphism of M.

LEMMA 3.4: *Suppose that the universal cover of M satisfies the Poincaré Conjecture. Then the homomorphism $\prod \overline{\mathcal{H}}_\partial(M_j) \to \overline{\mathcal{H}}(M)$ is injective.*

PROOF: An element of the kernel can be represented by a pieced-together homeomorphism $h = h_1 \cup h_2 \cup \ldots \cup h_m$ which is isotopic to a product of rotations about 2-spheres in M. By Lemma 3.2(1), this product is isotopic to a product of rotations about spheres disjoint from $\cup_{i=0}^n S_i$; changing the h_j by these rotations (which we are free to do at any time, since we are working in $\overline{\mathcal{H}}$) we may assume that h is isotopic to the identity. To prove the lemma, we will show that h is isotopic to the identity by an isotopy that preserves $\cup_{i=0}^n S_i$.

REMARK: In contrast to the analogous situation for an incompressible surface in a Haken 3-manifold, not every isotopy is actually deformable to one that preserves $\cup_{i=0}^n S_i$ (examples are given in [M5]). What we are proving is, of course, much weaker.

Changing h by an isotopy in a neighborhood of ∂M, we may assume that h is the identity on ∂M and that the isotopy is (rel ∂M). Lemma 2.1 shows there is a punctured 3-cell P in M such that

(a) $\cup_{i=0}^n S_i \subseteq P$
(b) $M - P$ is a disjoint union of $P_i - D_i$, where D_i is a 3-cell in the irreducible 3-manifold P_i, and g copies $\Sigma_j \times (0,1)$ of $S^2 \times (0,1)$.

We will prove the following two facts:

(i) Changing h by isotopy preserving the S_i, we may assume that $h(P) = P$ and h restricted to P is the identity map.
(ii) Changing h by rotations and by isotopy preserving the S_i, we may also assume for a 3-ball $P_0 \subseteq P$ that h is isotopic to the identity (rel $P_0 \cup \partial M$).

Assuming these two facts, we are in a position to use the following result, Theorem 2 of [H-M]:

THEOREM: *Suppose none of the summands P_i has universal cover a homotopy 3-sphere nondiffeomorphic to S^3. Then the inclusion map*

$$Diff(M - int(P_0), P - int(P_0) \text{ rel } \partial(M - int(P_0))) \longrightarrow$$
$$Diff(M - int(P_0) \text{ rel } \partial(M - int(P_0)))$$

is a homotopy equivalence.

Applied to the restriction of h to $M - int(P_0)$, this theorem shows that h is isotopic preserving P to the identity map. Once P is preserved, it is easy to redefine the isotopy so that it preserves the S_i. Thus facts (i) and (ii) will complete the proof of Lemma 3.4.

To achieve (i), we consider the collection $\{T_j\}_{j=1}^s$ containing the following three kinds of 2-spheres in M:

(1) the S_i
(2) each sphere $\partial(\overline{P_i - D_i})$ that is not parallel to one of the S_i
(3) each $\Sigma_j \times \{1/2\} \subset \Sigma_j \times (0,1)$ that is not parallel to one of the S_i.

The T_j are pairwise nonisotopic and each $h(T_j)$ is homotopic to T_j, since h is isotopic to the identity. By Lemma 3.1, we may change h by an isotopy preserving $\cup S_i$ so that $h(T_j) = T_j$ for all j. By further isotopy we may assume that $h(P) = P$ and h is the identity on each boundary component of P (note that h cannot reverse the sides or reverse the orientation on any boundary component of P, since h is isotopic to the identity). It follows as on p. 126 of [L] that the restriction of h to P is isotopic to the identity, preserving $\cup S_i$, so we may assume that h is the identity on P. This proves (i).

Now choose a 3-ball $P_0 \subset P - (\partial P \cup (\cup S_i))$. We claim that the trace of the isotopy from h to the identity is trivial at a basepoint in P_0. Since $r + g \geq 2$, the fundamental group $\pi_1(M)$ is a nontrivial free product, and any inner automorphism that preserves the free factors must be the identity. Since $h(P) = P$, the induced automorphism of h preserves the free factors, and since h is isotopic to the identity, its induced automorphism is conjugation by the trace of the isotopy. Therefore the trace is trivial.

Because the trace is trivial, the results of McCarty [M] show that h is isotopic preserving P_0 to the identity, and therefore isotopic (rel P_0) to the identity or to a rotation about ∂P_0. In the latter case, we may change h by rotations about all boundary components of P (the product of these is isotopic preserving $\cup S_i$ to the rotation about ∂P_0) to obtain h isotopic to the identity (rel P_0). By a further isotopy of h near ∂M, this isotopy may be assumed to be (rel $P_0 \cup \partial M$). This establishes (ii) and completes the proof of Lemma 3.4.

4. Proof of the Main Theorem

THEOREM 4.1: *Let M be a compact orientable 3-manifold whose universal cover satisfies the Poincaré Conjecture. If the mapping class group of each irreducible summand of M is finitely presented, then the mapping class group of M is finitely presented.*

Proof: By repeated use of Proposition 2.3, it suffices to prove Theorem 4.1 for the manifold $\widehat{M}$ that results from filling in the 2-sphere boundary components of M with 3-cells. We will induct on the number of prime summands of $\widehat{M}$.

If $\widehat{M}$ is irreducible, then there is nothing to prove. If $\widehat{M} = S^2 \times S^1$, then $\mathcal{H}(\widehat{M})$ is finite [G]. So we may assume that $\widehat{M}$ has at least two prime summands which are not 3-cells.

By Theorem 1.1, the sphere complex $\mathcal{S}(\widehat{M})$ is simply-connected, and by Proposition 2.2, the quotient $\mathcal{S}(\widehat{M})/\mathcal{H}(\widehat{M})$ is finite. We will show that the stabilizer of each simplex of $\mathcal{S}(\widehat{M})$ under the action of $\mathcal{H}(\widehat{M})$ is finitely presented. By Proposition 3.3(b), this is equivalent to the assertion that the stabilizers under the action of $\overline{\mathcal{H}}(\widehat{M})$ are finitely presented.

Suppose $\sigma = [S_0', S_1', \ldots, S_n']$ is an n-simplex of $\mathcal{S}(\widehat{M})$. Then there is a submanifold $S_0 \cup S_1 \cup \ldots \cup S_n$ of $\widehat{M}$ with each S_i isotopic to S_i'. Let $M_1, M_2, \ldots, M_m$ be the components that result from cutting $\widehat{M}$ along $\cup S_i$. Clearly, the image of the injective homomorphism $\prod \overline{\mathcal{H}}_\partial(M_j) \to \overline{\mathcal{H}}(\widehat{M})$ of Lemma 3.4 lies in the stabilizer of σ.

We claim that the image has finite index in the stabilizer. Consider an element $\langle h \rangle$ in the stabilizer. Using Lemma 3.1, we may choose h within its isotopy class so that $h(\cup S_i) = \cup S_i$. Passing to a subgroup of finite index, we may assume that $h(S_i) = S_i$ for each i. Passing to a possibly smaller subgroup of finite index, we may assume that h is orientation-preserving and does not reverse the sides of any S_i; in particular, h preserves each M_j and is isotopic to the identity on each S_i. Then, $\langle h \rangle$ is in the image of $\prod \overline{\mathcal{H}}_\partial(M_j) \to \overline{\mathcal{H}}(\widehat{M})$.

Since the spheres S_i do not bound 3-balls, each $\widehat{M}_j$ has fewer prime summands than $\widehat{M}$ has. By induction, each $\mathcal{H}(\widehat{M}_j)$ and hence each $\overline{\mathcal{H}}_\partial(M_j)$ is finitely presented. This proves that the stabilizer of each simplex is finitely presented.

Applying the result of K. Brown stated in the introduction shows that $\mathcal{H}(\widehat{M})$ is finitely presented, completing the induction step and the proof of Theorem 4.1.

References

[B1] F. Bonahon, Cobordism of automorphisms of surfaces, *Ann. École Norm. Sup.* (4) 16 (1983), 237-270.

[B2] F. Bonahon, Difféotopies des espaces lenticulaires, *Topology* 22 (1983), 305-314.

[B] K. Brown, Presentations for groups acting on simply-connected complexes, *J. Pure Appl. Alg.* 32 (1984), 1-10.

F-W] J. Friedman and D. Witt, Homotopy is not isotopy for homeomorphisms of 3-manifolds, *Topology* 25 (1986), 35-44.

[G] H. Gluck, The imbedding of two-spheres in four-spheres, *Bull. Amer. Math. Soc.* 67 (1961), 586-589.

[G1] D. Gottlieb, A certain subgroup of the fundamental group, *Amer. J. Math.* 87 (1965), 840-856.

[G2] P. Grasse, Finite presentation of mapping class groups of certain 3-manifolds, *Topology Appl.* 32 (1989), 205-305.

[H1] W. Heil, On $\mathbb{P}^2$-irreducible 3-manifolds, *Bull. Amer. Math. Soc.* 75 (1969), 772-775.

[H] H. Hendriks, Applications de la théorie d'obstruction en dimension 3, *Bull. Soc. Math. France Mémoire* 53 (1977), 81-196.

[H-L] H. Hendriks and F. Laudenbach, Difféomorphismes des sommes connexes en dimension trois, *Topology* 23 (1984), 423-443.

[H-M] H. Hendriks and D. McCullough, On the diffeomorphism group of a reducible 3-manifold, *Topology Appl.* 26 (1987), 25-31.

[J-S] W. Jaco and P. Shalen, Seifert fibered spaces in 3-manifolds, *Mem. Amer. Math. Soc.* 220 (1979), 1-192.

[J] K. Johannson, *Homotopy equivalences of 3-manifolds with boundaries,* Lecture Notes in Mathematics Vol. 761, Springer-Verlag, Berlin (1979).

[K-M1] J. Kalliongis and D. McCullough, π_1-injective mappings of compact 3-manifolds, *Proc. London Math. Soc.* (3), 52 (1986), 173-192.

[K-M2] J. Kalliongis and D. McCullough, Maps inducing isomorphisms on fundamental groups of compact 3-manifolds, *J. London Math. Soc.* (2) 35 (1987), 177-192.

[K] R. Kramer, The twist group of an orientable cube-with-two-handles is not finitely generated, preprint.

[L] F. Laudenbach, Topologie de la dimension trois. Homotopie et isotopie, *Astérisque* 12 (1974), 1-152.

[M] G. S. McCarty, Homeotopy groups, *Trans. Amer. Math. Soc.* 106 (1963), 293-304.

[M1] D. McCullough, Virtual cohomological dimension of mapping class groups of 3-manifolds, *Bull. Amer. Math. Soc.* 18 (1988), 27-30.

[M2] D. McCullough, Virtually geometrically finite mapping class groups of 3-manifolds, to appear in *J. Diff. Geom.*

[M3] D. McCullough, Twist groups of compact 3-manifolds, *Topology* 24 (1985), 461-474.

[M4] D. McCullough, Mappings of reducible 3-manifolds, in *Proceeding of the Semester in Geometric and Algebraic Topology of the Banach International Mathematical Center*, ed. H. Torunczyk, Banach Center Publications, Warsaw (1986), 61-76.

[M5] D. McCullough, Homotopy groups of the space of self-homotopy-equivalences, *Trans. Amer. Math. Soc.* 264 (1981), 151-163.

[M6] D. McCullough, Automorphisms of punctured-surface bundles, in *Geometry and Topology: Manifolds, Varieties, and Knots*, ed. C. McCrory and T. Shifrin, Marcel-Dekker, New York (1987), 179-209.

[M7] G. Mess, Centers of 3-manifold groups and groups which are coarse quasiisometric to planes, preprint.

[R-B] J. H. Rubinstein and J. Birman, One-sided Heegaard splittings and homeotopy groups of some 3-manifolds, *Proc. London Math. Soc.* (3) 49 (1984), 517-536.

[S] P. Scott, Finitely generated 3-manifold groups are finitely presented, *J. London Math. Soc.* (2) 6 (1973), 437-440.

[W] F. Waldhausen, On irreducible 3-manifolds which are sufficiently large, *Annals of Math.* 87 (1968), 56-88.

[W1] F. Waldhausen, Recent results on sufficiently large 3-manifolds, in *Proceedings of Symposia in Pure Mathematics, Vol. 32*, ed. R. J. Milgram (1978), 21-38.

REPRESENTATIONS OF THE STABLE GROUP OF SELF-EQUIVALENCES*

by

Donald W. Kahn
School of Mathematics, University of Minnesota
Minneapolis, MN 55455

The group G(X) of homotopy classes of homotopy equivalences from a topological space X to itself has been the subject of many papers (see for example [1], [4], [5], or [9]). This group is a geometric version of the group of automorphisms of a group, and the concepts coincide when the space is an Eilenberg-MacLane space. The group G(X) also plays a role in understanding the degree of non-uniqueness which is present in a homotopy resolution (Postnikov Tower). If X is a 1-connected (or nilpotent) finite complex, G(X) is known to be finitely - presented (see [13] and [14]) but there are remarkably simple spaces (see [6] or [9]) for which G(X) is not even finitely - generated.

In an earlier paper [5], I studied the stable group of self - equivalences of a finite, connected complex. The method consisted of analyzing the kernel and image of the representation of the group in the group of automorphisms of the integral homology (modulo torsion). It was proved that the kernel is finite, while the image is commensur - able with an arithmetic group [2] and hence finitely - presented. In view of the great difficulty in making specific computations of the group - even for relatively simple spaces - it seems reasonable to look for other ways to determine this group, modulo some possibly finite kernel, or up to inclusion, as a subgroup of finite index, in some known group.

The present paper has several objectives. Here we write G(X) for the group of stable homotopy classes of self-equivalences of X , and we shall study these classes which lift to a wedge of spheres which has the same rational homotopy-type as X . Our first main result is

---

*This paper is in final form and no version of it will be submitted for publication elsewhere.

that these classes form a subgroup of finite index in $G(X)$. This gives a construction on $G(X)$ which is rather like rationalization or perhaps dividing out by torsion. The existence of such a construction is surprising because in the category of (non-abelian) groups there is really nothing analogous to Serre's C-theory (see [12]). An immediate corollary (to the proof) of this theorem is that the image of the natural representation of $G(X)$ in the group of graded automorphisms of $\pi_*^s(X)$, the stable homotopy of X, modulo torsion, has finite index. The second principal result of [5] may be easily recovered from this theorem.

Now, given a homology theory $h(-)$, we have a natural representation

$$p_h : G(X) \to \mathrm{Aut}(h(X))$$

where $p_h(\{f\}) = h(f) : h(X) \to h(X)$. It is reasonable to ask how faithful such a representation can be. Some easy examples show that p_h is rarely faithful (kernel zero), when h is ordinary homology. But we may then ask the next basic question: for which homology theories is $\ker p_h$ always finite, when X ranges over all finite (connected) complexes? Our second theorem asserts that p_h always has finite kernel, if and only if

$$\sigma_h : \{X,X\} \to \mathrm{End}(h(X)) ,$$

the natural representation of the ring of stable self-maps in the endomorphisms of $h(X)$, is faithful when X is a sphere.

The third theorem is more technical. It consists of an analysis of the image of those self-equivalences, which lift to a wedge of spheres as in Theorem 1, in the group $G(X)$. This is carried out in terms of a Moore - Postnikov decomposition of a fibration. We conclude with some remarks about special cases and the non-stable case.

We shall work in the category of connected complexes with basepoint, and basepoint preserving maps. Basepoints will usually be omitted from notation. Because we are studying the stable group of homotopy classes of homotopy equivalences, one may think of a space X as $(i-1)$-connected, with $\dim X << 2i-1$, whenever it is convenient. We use standard notations for the algebra of stable maps. $\{X,Y\}$ refers to the stable homotopy classes of maps $X \to Y$. We freely write the basic exact sequence for $f:X \to Y$ as

$$\cdots \to \Sigma^{-1}Cf \xrightarrow{i} X \xrightarrow{f} Y \xrightarrow{j} Cf \longrightarrow \Sigma X \to \cdots$$

If $g:X \to U$ with $g\cdot i \simeq *$ (the trivial map), there is a $\bar{g}:Y \to U$ with $\bar{g}\cdot f \simeq g$. We also recall that if X is a finite complex, whose homotopy groups are finite, then X has finite characteristic (in the sense that the identity map $1_X:X \to X$ has finite order as a stable map). This follows from the basic exact sequences of groups of homotopy classes $\{X^{(n)},X\}$.

<u>Definition</u>: If X is a finite connected complex, we write $G(X)$ for the group of stable classes of self-homotopy-equivalences of X.

It is shown in [11] and [12] that $\pi_i(S^n)$ is finite for $i \neq n$, $2n-1$, from which it is easy to see that the rational Hurewicz map is an isomorphism, for stable complexes. This implies that there is a map $F:S^{n_1}\vee\ldots\vee S^{n_k} \longrightarrow X$ inducing an isomorphism on stable homotopy groups modulo torsion. Let $G_0(X) \subseteq G(X)$ be the subgroup consisting of those classes represented by a map $f:X \to X$ so that there exists a (stable) map

$$\bar{f}:S^{n_1}\vee\ldots\vee S^{n_k} \longrightarrow S^{n_1}\vee\ldots\vee S^{n_k},$$

with $f\cdot F \simeq F\cdot\bar{f}$ (that is the maps are stably homotopic). Note that it follows easily, from Whitehead's theorem, that $\bar{f}$ is a homotopy equivalence (stable).

<u>Theorem 1</u> $G_0(X) \subseteq G(X)$ has finite index.

<u>Proof</u>: Theorem 1.1 of [5] asserts that the kernel of the representation

$$\bar{\phi}:G(X) \longrightarrow \mathrm{Aut}(H_*(X)_F) \ .$$

where $H_*(X)_F$ means homology modulo torsion, is finite. (Here $\bar{\phi}(\{f\}) = f_*$) . We are basically concerned here with the representation

$$\phi:G(X) \longrightarrow \mathrm{Aut}(\pi_*^s(X)_F) \ .$$

where $\phi(\{f\}) = f_\#$ and $\pi_*^s(X)_F$ means the stable homotopy groups of X modulo torsion. Since the stable Hurewicz map is a rational isomorphism, if we choose bases for the groups $H_*(X)$ and $\pi_*^s(X)$, modulo torsion, then $\bar{\phi}(\{f\})$ and $\phi(\{f\})$ are invertible, integral matrices, which are similar over the rational numbers. The Hurewicz map gives the similarity. It follows at once that ker ϕ is also finite.

From this, we easily conclude that

$$G_o(X) \subseteq \phi^{-1}(\phi(G_o(X)))$$

has finite index. Our theorem will be complete, if we prove that $\phi(G_o(X))$ has finite index in $\phi(G(X))$. In fact, we will show that the index of $\phi(G_o(X))$ in $\mathrm{Aut}(\pi_*^s(X)_F)$ is finite.

Now, among the subgroups of the group of automorphisms of a finitely-generated free abelian group, there is a class which is well - known to have finite index. These are the so-called congruence subgroups consisting of all these integral, invertible matrices which reduce, modulo some fixed integer $N > 1$, to the identity matrix, in other words, these subgroups are the kernels of the homomorphisms

$$G\ell(n,Z) \longrightarrow G\ell(n,Z/N) \ .$$

This means that in order to prove our theorem we must find an integer $N > 1$ so that any $a \in \mathrm{Aut}(\pi_*^s(X)_F)$, which reduces modulo N to the identity matrix, actually belongs to $\phi(G_o(X))$.

Since F is a rational homotopy equivalence, the homotopy groups of the cone on F are finite. We choose an integer N_o, which is a multiple of the characteristic of the cone on F, and which annihilates the torsion of $\pi_*^s(X)$, past the dimension of X. We then set $N = N_o^2$.

We start with a map

$$\tilde{g}: S^{n_1} \vee \ldots \vee S^{n_k} \longrightarrow S^{n_1} \vee \ldots \vee S^{n_k}$$

with $\tilde{g}_\# = a$. We will assume later that $\tilde{g}$ is trivial on torsion, that is any individual map which sends a sphere to a sphere of different dimension is trivial. Consider the following diagram

$$\begin{array}{ccc} S^{n_1} \vee \ldots \vee S^{n_k} & \xrightarrow{\tilde{g}} & S^{n_1} \vee \ldots \vee S^{n_k} \\ \downarrow F & & \downarrow F \\ X & \overset{g}{\cdots\cdots\cdots>} & X \end{array}$$

where we seek a homotopy equivalence which makes the diagram homotopy commutative. In fact, it will suffice to do this for some other map
$\tilde{g}$, just as long as $\tilde{g}_\#$ is the automorphism a on $\pi_*^s(X)_F$.

Now, because our problem is stable (so that maps may be added, etc.) we may write

$$\tilde{g} = 1 + \tilde{g}_o ,$$

where 1 is the identity on $S^{n_1} \vee \ldots \vee S^{n_k}$. Because a is the identity modulo N, $(\tilde{g}_o)_\#$ reduces modulo our N to the zero map, when taken modulo torsion.

We write $\tilde{\tilde{g}}_o$ for a map, from $S^{n_1} \vee \ldots \vee S^{n_k}$ to itself, which is $\tilde{g}_o$ modulo torsion, and which is the zero homomorphism on the torsion subgroups.

We seek $g_o: X \to X$, with $g_o \cdot F = F \cdot \tilde{\tilde{g}}_o$, and which induces the zero map on the torsion subgroup of $\pi_*^s(X)$, past the dimension of X . If we can find such a map, then I claim that $g = 1+g_0$ will be our desired element, that is $\phi(\{g\}) = a$. Note that $(g_0)_\#$ is zero on torsion, so that g induces the identity map on torsion, past the dimension of X . Because $F_\#$ may be assumed to be the identity isomorphism, modulo torsion, we see that $g = 1 + g_0$ induces the same map, modulo torsion, as $1 + \tilde{\tilde{g}}_0$ or $1 + \tilde{g}_0$.

Thus $g_\#$ is an isomorphism modulo torsion. It is then easy to check that $g_\#$ is both 1-1 and onto, past the dimension of X , so that g is a homotopy-equivalence. Because g lifts to $1+\tilde{\tilde{g}}_0$, g represents an element in $G_0(X)$. Lastly, we have that modulo torsion $g_\#$ is the same map as $(1+\tilde{\tilde{g}}_0)_\# = (1+\tilde{g}_0)_\# = \tilde{g}_\# = a$. Thus, $\phi(G_0(X))$ will contain a subgroup which has finite index in $\mathrm{Aut}(\pi_*^s(X)_F)$ provided of course that we can find g_0 .

To get g_0 , we note that any map

$$\tilde{k}:S^{n_1}V\ldots VS^{n_k} \longrightarrow S^{n_1}V\ldots VS^{n_k},$$

which is divisible by N_0 , must factor through a map

$$\tilde{\tilde{k}}:X \longrightarrow S^{n_1}V\ldots VS^{n_k} ,$$

because N_0 annihilates the characteristic of the cone CF , and we have the exact sequence of spaces

$$\ldots \longrightarrow \Sigma^{-1}CF \xrightarrow{i} S^{n_1}V\ldots VS^{n_k} \xrightarrow{F} X \longrightarrow \ldots$$

Set $k = F\cdot\tilde{\tilde{k}}:X \to X$. If in addition, we assume that $\tilde{k}$ is divisible by $N = N_0^2$, then we can take k to be divisible by N_0 and thus $k_\#$ annihilates the torsion of $\pi_*^s(X)$. In other words, if we take $\tilde{k}$ to be $\tilde{\tilde{g}}_0$, then k gives us our desired map g_0 .

Corollary 1 The image of $G(X)$ in $Aut(\pi_*^s(X)_F)$ has finite index. (With a little more work, one can conclude that the image of $G(X)$, under $\bar{\phi}$, in $Aut(H_*(X)_F)$ has finite index).

Corollary 2 $G(X)$ is finitely-presented.

Proof: A subgroup of finite index, in a finitely-presented group, is finitely-presented. Thus, $Im(\phi)$ is finitely-presented. $\ker \phi$ is finite. See [8].

Theorem 2 Given any homology theory h on our stable category of finite, connected complexes, then a necessary and sufficient condition that the representation

$$p_h : G(X) \longrightarrow Aut(h(X))$$

have finite kernel, for every X, is that the map

$$\sigma_h : \{X, X\} \longrightarrow End(h(X))$$

be faithful whenever X is a sphere.

Proof: We assume X to have been suspended sufficiently many times to be in the stable range, and select a map

$$F : S^{n_1} V \ldots V S^{n_k} \longrightarrow X$$

to be an isomorphism on stable homotopy groups modulo torsion. Recall that CF has finite characteristic. Suppose that we are given a (stable) homotopy equivalence $f : X \to X$ so that $\{f\} \in G_o(X)$ and $h(f)$ is the identity map. Then there is by assumption a map

$$\bar{f} : S^{n_1} V \ldots V S^{n_k} \longrightarrow S^{n_1} V \ldots V S^{n_k}$$

with $F \cdot \bar{f} \simeq f \cdot F$. With no loss of generality, F can be taken to be an inclusion, giving a commutative diagram.

$$\begin{array}{ccccc} S^{n_1}V\dots VS^{n_k} & \xrightarrow{F} & X & \xrightarrow{j} & CF \\ \downarrow{\scriptstyle \bar f} & & \downarrow{\scriptstyle f} & & \downarrow{\scriptstyle \bar{\bar f}} \\ S^{n_1}V\dots VS^{n_k} & \xrightarrow{F} & X & \xrightarrow{j} & CF \end{array} .$$

Because h is assumed faithful on spheres and CF has finite characteristic, $h(F)$ is a monomorphism modulo torsion. Since $h(f)$ is the identity, $h(\bar f)$ must also be the identity modulo torsion. Thus there are at most a finite number of maps $\bar f$ with $f \cdot F \simeq F \cdot \bar f$ and $h(f)$ the identity.

Now, if we can show that for any $\bar f$ there are only a finite number of possible f, the first half of the proof will be complete. To this end, suppose we have $f_1, f_2 : X \to X$ so that $f_i \cdot F \simeq F \cdot \bar f$, for $i = 1,2$. Then we have

$$(f_1 - f_2) \cdot F \simeq F \cdot (\bar f - \bar f) \simeq * .$$

the trivial map. By exactness, there is a map $e : CF \to X$ with

$$e \cdot j \simeq (f_1 - f_2) .$$

But there are only finitely many possible classes of maps e, since CF has finite characteristic. We have shown that the kernel of p_h, restricted to $G_o(X)$ is finite. Because $G_o(X)$ has finite index, the first part of the proof is complete.

For the converse, suppose $\sigma_h : \{X,X\} \to \mathrm{End}(h(X))$ is not faithful for some sphere S^n. Then, there are an infinite number of maps $g : S^n \to S^n$ with $h(g) = 0$. Let g be such a map, and let

$$q : S^n V S^n \longrightarrow S^n V S^n$$

be given by the matrix

$$\begin{bmatrix} 1 & g \\ 0 & 1 \end{bmatrix}$$

q obviously induces an isomorphism on ordinary homology groups and $h(q)$ is the identity. Since any two maps q, for different g, are not homotopic, we see that the kernel of p_h is infinite for this space. In other words, if there is a sphere for which σ_h is not faithful, there is a space with $\ker(p_h)$ infinite.

<u>Remarks</u> 1) The group of automorphisms of a finitely-generated abelian group is finitely-presented [8]. One may hope that Theorem 1, or Theorem 2 in the case when $\ker(p_h)$ is finite, can be used to construct specific generators and relations for various $G(X)$. Bounds on the number of generators would be useful.

2) Whenever $\dim H_j(X;Q) > 1$, for some j, then the image of p_H or of ϕ is a subgroup of $\mathrm{Aut}(F)$, where F is a free abelian group of rank greater than 1. Such a group contains, for example, the group $G\ell(2,Z)$ of 2×2 matrices over the integers with determinant ± 1. This group is related to the modular group $PS\ell(2,Z)$, studied long ago by Fricke and Klein. That group is a free product of a cyclic group of order 2 and a cyclic group of order 3. (See appendix B, Volume 2 of [7] as well as pp. 46-47 and pp. 100-101 of [8]). More generally, groups $G\ell(n,Z)$ - for $n \geq 2$ - contain arbitrarily large free, non-abelian subgroups. It is easy to convince oneself that the image of p_H can be very complicated.

3) For certain classes of spaces, for example when the rational cohomology is an exterior algebra, there are interesting results on the ordinary group of self-equivalences in [1]. There is also there a study of those self-equivalences which induce the identity homomorphism on homotopy groups, which in our context amounts to $\ker(p_{\pi_*})$.

For our final theorem, we study - in a specific way - the subgroup $G_0(X) \subseteq G(X)$. That is to say, we shall give a specific condition in order that a class $\{f\} \in G(X)$ actually belong to $G_0(X)$. To this end, we use Moore-Postnikov systems [10] and an obvious extension of the results in [3]. To begin, we note that if a map

$$r: S^{n_1} \vee \ldots \vee S^{n_k} \longrightarrow E$$

induces an isomorphism on all homotopy groups beyond $\max_{1 \leq j \leq k} n_j$. then any homotopy equivalence

$$f: E \longrightarrow E$$

will "lift" to a homotopy equivalence

$$\bar{f}: S^{n_1} \vee \ldots \vee S^{n_k} \longrightarrow S^{n_1} \vee \ldots \vee S^{n_k}$$

with $r \cdot \bar{f} \simeq f \cdot r$, and in fact will do so uniquely.

If the map $F: S^{n_1} \vee \ldots \vee S^{n_k} \longrightarrow X$ from Theorem 1 (an isomorphism on stable homotopy modulo torsion) is converted into a fibration then we have a Moore-Postnikov decomposition

$$S^{n_1} \vee \ldots \vee S^{n_k} \longrightarrow \ldots \longrightarrow E_m \xrightarrow{\pi_m} E_{m-1} \longrightarrow \ldots \longrightarrow X \ ,$$

where π_m is a principal filration with fibre an Eilenberg-MacLane space. The fibre of the composition $\ldots \pi_{m-1} \pi_m = t_m: E_m \longrightarrow X$ has the m homotopy-type of the fibre of the map F , which we write $\mathcal{F}$. Our ealier remarks show that a homotopy-equivalence $f: X \to X$ which lifts to homotopy equivalence of E_m , m sufficiently large, will also lift to a homotopy equivalence of $S^{n_1} \vee \ldots \vee S^{n_k}$. With these conventions, we then have

<u>Theorem 3</u>: Let $m > \dim(S^{n_1} \vee \ldots \vee S^{n_k}) = \max_{1 \leq j \leq k} n_j$. A homotopy self-equivalence $f: X \to X$ lifts to a homotopy self-equivalence $f_m: E_m \to E_m$ (that is $f \cdot \tau_m \simeq \tau_m \cdot f_m$) , if and only if there is a sequence of self-equivalences

$$f_j : E_j \longrightarrow E_j \ , \ j < m$$

and automorphisms

$$\alpha_j : \pi_j(\mathfrak{F}) \to \pi_j(\mathfrak{F}) \ , \ j \leq m \ ,$$

where $\mathfrak{F}$ is the fibre of the map F , so that

$$\alpha_j = (f_j^f)_\# \ , \ \text{if} \quad j < m \quad \text{and}$$

$$f_{j-1}^* k^{j+1} = (\alpha_j)_c k^{j+1} \ , \ \text{if} \quad j \leq m \ ,$$

with f_{j-1}^* the induced map on cohomology, $(\alpha_j)_c$ meaning α_j as a coefficient homomorphism, and $k^{j+1} \in H^{j+1}(E_{j-1};\pi_j(\mathfrak{F}))$ the k-invariant, and f_j^f means the restriction of f_j to the fibre of $\pi_j : E_j \to E_{j-1}$.

Proof: Given a map $f_{j-1} : E_{j-1} \to E_{j-1}$, then f_{j-1} lifts to a map $f_j : E_j \to E_j$ so that $f_{j-1}\pi_j = \pi_j \cdot f_j$, if and only if $(f_{j-1})^* k^{j+1} = ((f_j^f)_\#)_c k^{j+1}$, where f_j^f refers to a possible restriction of f_j to the fibre of $\pi_j : E_j \to E_{j-1}$ and the subscript c means coefficient homomorphism. See [3] for details.

If we are given f_m in our theorem we take f_j to be the induced maps and $\alpha_j = (f_j^f)_\#$ (recall that the jth homotopy group of $\mathfrak{F}$ and of the fibre of π_j are the same).

Conversely, suppose our technical condition is satisfied and we wish to find

$$f_m : E_m \to E_m \ .$$

We use the condition $f_{j-1}^* k^{j+1} = (\alpha_j)_c k^{j+1}$ for $j = m$, to construct f_m . Then f_m induces an isomorphism on the homotopy groups of the fibre of π_m , and f_{m-1} is a homotopy equivalence.

It follows at once that f_m is a homotopy equivalence which provides the desired lift.

Remarks While Theorem 3 may seem hopelessly technical, it does offer more precise information on the image of $G_o(X) \subseteq G(X)$. For example, if $\{f\} \notin G_o(X)$, then the inductive condition on k-invariants must break down at some specific stage. This means that there is a decreasing filtration on G(X) beginning with G(X) and descending to $G_o(X)$.

Note that when a k-invariant is 0 , our condition

$$f^*_{j-1}k^{j+1} = (f^f_j)_c \cdot k^{j+1}$$

becomes vacuous. Then every possible automorphism α_j is realized by some map f_j . While not every such map f_j may lift to f_m , the general philosophy should be that the vanishing of the k-invariants increases the size of $G_o(X)$, or equivalently decreases the index of $G_o(X)$.

Concluding Remarks

Most of the results in this paper would be false outside of the stable range, and to the best of my knowledge, an analysis of the representation p_h , even when h is ordinary homology, has not been carried out. For example, using Whitehead products, one may produce infinitely many self-equivalences which induce the identity on ordinary homology theory. One needs only to look at spaces of the form

$$\ldots S^n \vee S^n \vee S^{2n-1} \vee \ldots .$$

In the stable case, $G_o(X)$ appears to play to the role of dividing out by torsion. In the unstable case, it seems more sensible to go directly to the rationalization of the space (see [13] or [14]) rather than representing the self-equivalences on some homology theory. Some identification of interesting subgroups of G(X) , of geometric origin and which have finite index in the unstable case, would be very useful.

Bibliography

1. Arkowitz, M. and Curjel, C. "Groups of Homotopy Classes" Springer-Verlag Notes in Math., Vol. 4 (1964).

2. Borel, A. "Arithmetic Properties of Linear Algebraic Groups", Proc. Inter. Congress, Stockholm (1962).

3. Kahn, D.W. "Induced Maps for Postnikov Systems", Trans. A.M.S 107, No. 3 (1963) pp. 432-450.

4. Kahn, D.W. "The Group of Homotopy Equivalences", Math. Zeitschrift 84 (1964) pp. 1-8.

5. Kahn, D.W. "The Group of Stable Self-Equivalences", Topology 11 (1972) pp. 133-140.

6. Frank, D. and Kahn, D.W. "Finite Complexes with Infinitely generated Groups of Self-Equivalences", Topology 16 (1977) pp. 189-192.

7. Kuros, A. "Theory of Groups", 2nd Ed., Chelsea Co., New York 1960.

8. Magnus, Karass, Solitar, "Combinatorial Group Theory", J. Wiley, New York (1966).

9. McCullough, D. "Compact 3-manifolds with Infinitely Generated Groups of Self-Homotopy-Equivalences", Proc. A.M.S. 91 (1984) No. 4 pp. 625-629.

10. Moore, J.C. "Semi-simplicial complexes and Postnikov Systems" Inter. Symposium on Algebraic Topology", Mexico City 1956 (published 1958).

11. Serre, J.P. "Homologie Singulière des Éspaces Fibres" Annals of Math. 54 (1951) pp 425-505.

12. Serre, J.P. "Groupes d'homotopie et classes de groupes abeliens". Annals of Math. 58 (1953) pp. 258-294.

13. Sullivan, D. "Infinitesimal Computation in Topology", Publ. I.H.E.S. Paris No. 47 (1978) pp. 269-331.

14. Wilkerson, C. "Applications of Minimal Simplicial Groups" Topology 15 (1976) pp. 111-130.

HOMOTOPY EQUIVALENCES IN 2-CATEGORIES

Howard J. Marcum

The Ohio State University at Newark
University Drive
Newark, Ohio 43055 USA

A groupoid enriched category is a 2-category each of whose 2-morphisms is invertible. The example relevant to our concerns here is the track category of based topological spaces, denoted $\mathcal{T}op_*$. In $\mathcal{T}op_*$ the 2-morphisms are track classes of homotopies. In fact, commencing with the work of Gabriel-Zisman [2], various aspects of the homotopy theory familiar to $\mathcal{T}op_*$ have been developed in the setting of groupoid enriched categories.

In this paper we look specifically at the notion of homotopy equivalence in the general setting of a 2-category. Our aim is the characterization of homotopy equivalences in various lax categories. The homotopy coherence problems which arise are handled by a lemma due to Vogt [11] (see Lemma 1.5 below). As a consequence of this unified approach we can draw together several examples which previously were each considered separately by Hardie and his collaborators. In fact it was consideration of these examples which inspired this work. The author acknowledges helpful conversations with Keith Hardie and K.H. Kamps.

In [1] a somewhat different approach to some of the material is given (for example, in Section Two 2-functors and lax natural transformations can be replaced by pseudo-functors and pseudo-natural transformations). However in [1] the setting is always that of groupoid enriched categories and thus the full strength of some results is missed — consider, for instance, Theroem 2.5 below versus Corollary 2.8 below.

This paper is in final form and no version of it will be submitted for publication elsewhere.

1. Basic Definitions.

(1.1) **Definition.** Let $\mathcal{C}$ be a 2-category. A 1-morphism $f: A \to B$ is called a *homotopy equivalence* in $\mathcal{C}$ if there exists a 1-morphism $\overline{f}: B \to A$ and invertible 2-morphisms $F: \overline{f} \circ f \Rightarrow 1_A$ and $G: f \circ \overline{f} \Rightarrow 1_B$.

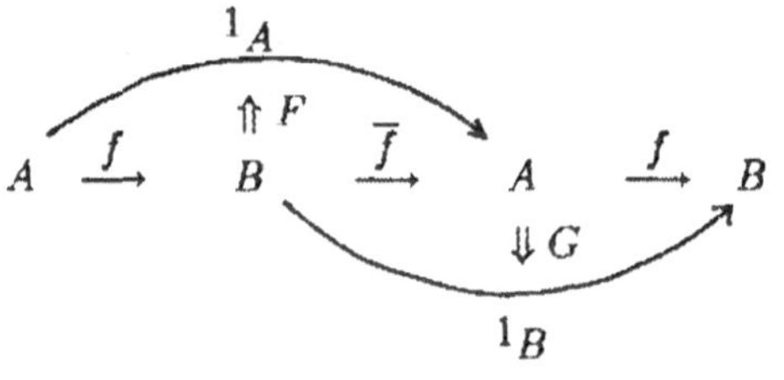

We also say that f and $\overline{f}$ are homotopy inverses.

(1.2) **Note.** This definition of homotopy equivalence does not require that each 2-morphism in $\mathcal{C}$ be invertible. Hence the notion of homotopy equivalence is meaningful when $\mathcal{C}$ is merely a 2-category as well as when $\mathcal{C}$ is a groupoid enriched category . Several previous authors (Fantham–Moore [1], Hardie–Kamps ([6], [7], [8])) have restricted themselves to the latter setting only.

(1.3) **Note.** Suppose $T: \mathcal{C} \to \mathcal{D}$ is a 2-functor between 2-categories. If $f: A \to B$ is a homotopy equivalence in $\mathcal{C}$ then $Tf: TA \to TB$ is a homotopy equivalence in $\mathcal{D}$. In particular, if $\mathcal{C}'$ is a sub-2-category of $\mathcal{C}$ and $f: A \to B$ is a homotopy equivalence in $\mathcal{C}'$ then f is also a homotopy equivalence in $\mathcal{C}$.

(1.4) **Remark.** For a 2-category $\mathcal{C}$, let $\mathrm{W} = \{f : f \text{ is a homotopy equivalence in } \mathcal{C}\}$. Then W is a family of morphisms of the underlying category of $\mathcal{C}$ (which is also denoted by $\mathcal{C}$) so one may form the category of fractions of $\mathcal{C}$ by W , denoted $\mathcal{C}[\mathrm{W}^{-1}]$. Recall that the category $\mathcal{C}[\mathrm{W}^{-1}]$ has the same objects as $\mathcal{C}$ and is provided with a canonical projection functor $P : \mathcal{C} \to \mathcal{C}[\mathrm{W}^{-1}]$ having the property that P(f) is an isomorphism in $\mathcal{C}[\mathrm{W}^{-1}]$ for each $f \in W$. This projection functor defines a notion of "homotopy" in $\mathcal{C}$, namely:

$$f \simeq g : A \to B \quad \text{if and only if} \quad P(f) = P(g) : A \to B$$

It is with respect to this relation of homotopy that the elements of W are "homotopy equivalences."

(1.5) **Vogt's Lemma.** In the 2-category $\mathcal{C}$ let $f : A \to B$ and $\overline{f} : B \to A$ be 1-morphisms which are homotopy inverses. If $F : \overline{f} \circ f \Rightarrow 1_A$ is an invertible 2-morphism then there exists an invertible 2-morphism $G : f \circ \overline{f} \Rightarrow 1_B$ satisfying

$$1_f \circ F = G \circ 1_f \quad , \quad F \circ 1_{\overline{f}} = 1_{\overline{f}} \circ G \ .$$

Proof. The proofs of this lemma given by others (e.g., [10, Korollar 1.1.10] or [8, Proposition 1.2]) in the setting of groupoid enriched categories still work in the present context. ∎

2. The 2-category $[\mathcal{A}, \mathcal{C}]$.

Let $\mathcal{A}$ and $\mathcal{C}$ be 2-categories. Following Kelly [9, §2] we review the definition of the category $[\mathcal{A}, \mathcal{C}]$. The objects of $[\mathcal{A}, \mathcal{C}]$ are 2-functors $S: \mathcal{A} \to \mathcal{C}$. A 1-morphism $\theta: S \to T$ of $[\mathcal{A}, \mathcal{C}]$ is a lax natural transformation: it assigns to each object A of $\mathcal{A}$ a 1-morphism $\theta_A: SA \to TA$ in $\mathcal{C}$ and to each 1-morphism $f: A \to B$ of $\mathcal{A}$ a 2-morphism $\theta_f: Tf \circ \theta_A \Rightarrow \theta_B \circ Sf$ in $\mathcal{C}$.

$$
\begin{array}{ccc}
A & SA \xrightarrow{\ \theta_A\ } TA \\
f\downarrow & Sf\downarrow \ \overset{\theta_f}{\Leftarrow}\ \downarrow Tf \\
B & SB \xrightarrow[\ \theta_B\ ]{} TB
\end{array}
$$

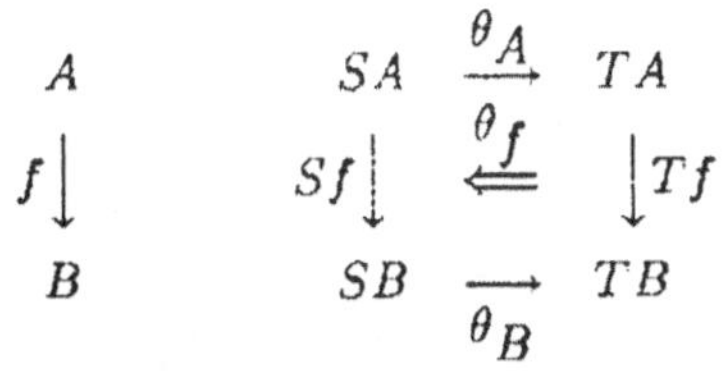

These data are to satisfy three axioms:

(1) $\theta_{1_A} = 1_{\theta_A}$

(2) $\theta_{g \circ f} = \theta_g \circ 1_{Sf} + 1_{Tg} \circ \theta_f$

(3) For each 2-morphism $\zeta: f \Rightarrow f'$ in $\mathcal{A}$, the equation

$$
1_{\theta_{A'}} \circ S\zeta + \theta_f = \theta_{f'} + T\zeta \circ 1_{\theta_A}
$$

holds as indicated below.

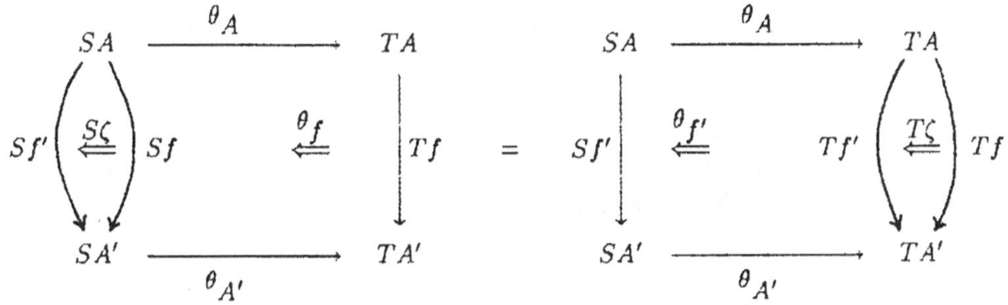

Note that axiom (3) is without effect if $\mathcal{A}$ is merely a category.

The 2-morphisms of $[\mathcal{A},\mathcal{C}]$ are modifications $\eta\colon \theta \Rightarrow \tau$.

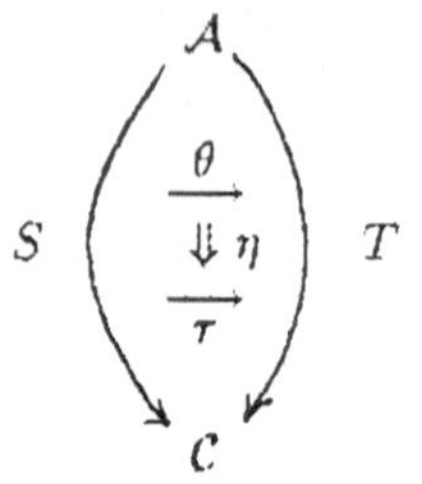

Such a modification assigns to each object A of $\mathcal{A}$ a 2-morphism $\eta_A : \theta_A \Rightarrow \tau_A$ in $\mathcal{C}$ satisfying: if $f\colon A \to B$ is a 1-morphism in $\mathcal{A}$ then $\eta_B \circ 1_{Sf} + \theta_f = \tau_f + 1_{Tf} \circ \eta_A$.

(2.1)

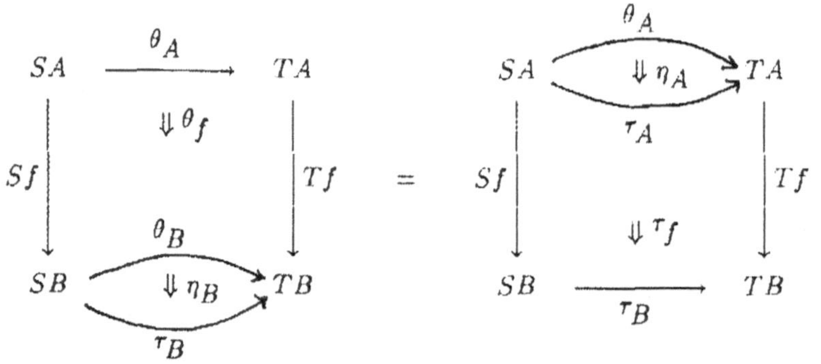

We refer to the $\{\eta_A\}$ as the components of $\eta\colon \theta \Rightarrow \tau$. Note that the identity 2-morphism $1_\theta : \theta \Rightarrow \theta$ has components $(1_\theta)_A = 1_{\theta_A} : \theta_A \Rightarrow \theta_A$.

(2.2) **Note.** One could require in the definition of $\theta\colon S \to T$ that the direction of θ_f be reversed. This defines an *op-lax natural transformation* $\theta\colon S \to T$. The resulting 2-category is denoted $[\mathcal{A},\mathcal{C}]''$.

(2.3) **Lemma.** Let $\eta\colon \theta \Rightarrow \tau$ be a 2-morphism in $[\mathcal{A},\mathcal{C}]$. Then η is invertible in $[\mathcal{A},\mathcal{C}]$ if and only if each of its components $\eta_A : \theta_A \Rightarrow \tau_A$ is an invertible 2-morphism in $\mathcal{C}$.

Proof. If $\eta\colon \theta \Rightarrow \tau$ is invertible then trivially so is each $\eta_A : \theta_A \Rightarrow \tau_A$. Conversely, for each A in $\mathcal{A}$, suppose that $-\eta_A : \tau_A \Rightarrow \theta_A$ denotes the inverse of $\eta_A : \theta_A \Rightarrow \tau_A$. From the coherence condition (2.1) for η we have

$$\eta_B \circ 1_{Sf} + \theta_f = \tau_f + 1_{Tf} \circ \eta_A$$

$$\theta_f - 1_{Tf} \circ \eta_A = -\eta_B \circ 1_{Sf} + \tau_f$$

or

$$\theta_f + 1_{Tf} \circ (-\eta_A) = (-\eta_B) \circ 1_{Sf} + \tau_f \ .$$

Now this last equation is just the coherence condition required for the $\{-\eta_A\}$ to be the components of a 2-morphism $-\eta\colon \tau \Rightarrow \theta$ in $[\mathcal{A}, \mathcal{C}]$. Clearly $-\eta$ is the inverse of η in $[\mathcal{A}, \mathcal{C}]$. ∎

(2.4) **Corollary.** $[\mathcal{A}, \mathcal{C}]$ is a groupoid enriched category if $\mathcal{C}$ is a groupoid enriched category.

(2.5) **Theorem.** Let $\theta\colon S \to T$ be a 1-morphism in $[\mathcal{A}, \mathcal{C}]$. Then θ is a homotopy equivalence in $[\mathcal{A}, \mathcal{C}]$ if and only if for each object A of $\mathcal{A}$ the 1-morphism $\theta_A : SA \to TA$ is a homotopy equivalence in $\mathcal{C}$ and for each 1-morphism $f\colon A \to B$ of $\mathcal{A}$ the 2-morphism $\theta_f : Tf \circ \theta_A \Rightarrow \theta_B \circ Sf$ is invertible in $\mathcal{C}$.

Proof. Suppose that $\theta\colon S \to T$ is a homotopy equivalence in $[\mathcal{A}, \mathcal{C}]$. Denote by $\overline{\theta}\colon T \to S$ a homotopy inverse of θ and let $\eta\colon \overline{\theta} \circ \theta \Rightarrow 1_S$ and $\rho\colon \theta \circ \overline{\theta} \Rightarrow 1_T$ be invertible 2-morphisms in $[\mathcal{A}, \mathcal{C}]$. By Vogt's Lemma we may assume that $1_\theta \circ \eta = \rho \circ 1_\theta$ and $\eta \circ 1_{\overline{\theta}} = 1_{\overline{\theta}} \circ \rho$. In particular for each A in $\mathcal{A}$ we have

$$(2.6) \qquad 1_{\theta_A} \circ \eta_A = \rho_A \circ 1_{\theta_A} \quad , \qquad \eta_A \circ 1_{\overline{\theta}_A} = 1_{\overline{\theta}_A} \circ \rho_A \ .$$

Now η and ρ are invertible in $[\mathcal{A}, \mathcal{C}]$ so by Lemma 2.3 each of $\eta_A : \overline{\theta}_A \circ \theta_A \Rightarrow 1_{SA}$ and $\rho_A : \theta_A \circ \overline{\theta}_A \Rightarrow 1_{TA}$ is invertible in $\mathcal{C}$. Thus θ_A and $\overline{\theta}_A$ are homotopy inverses in $\mathcal{C}$.

Next let $f\colon A \to B$ be a 1-morphism of $\mathcal{A}$. Then we have the following equality of composite 2-morphisms in $\mathcal{C}$.

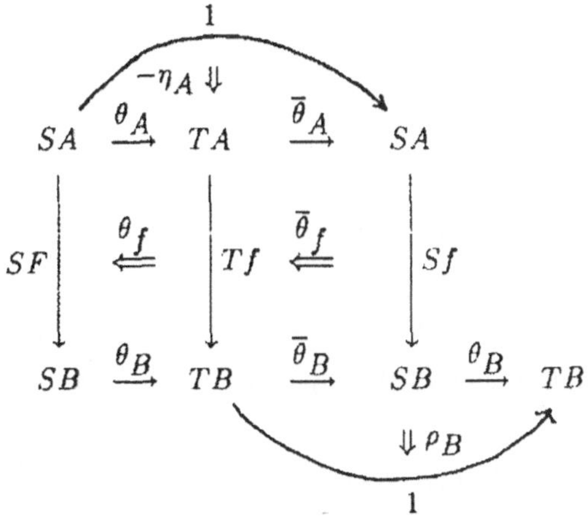

$=$ (1)

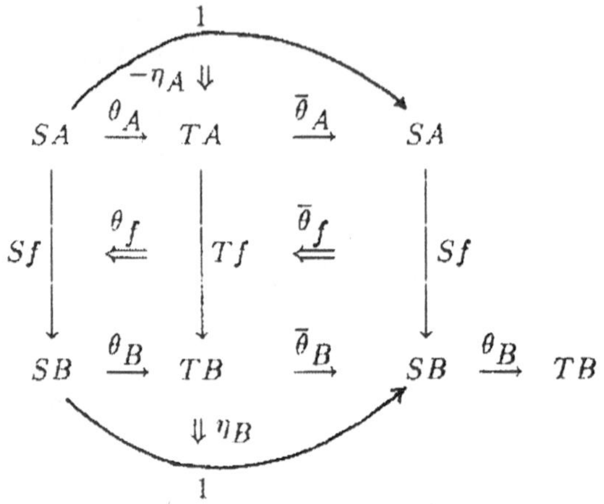

$=$ (2)

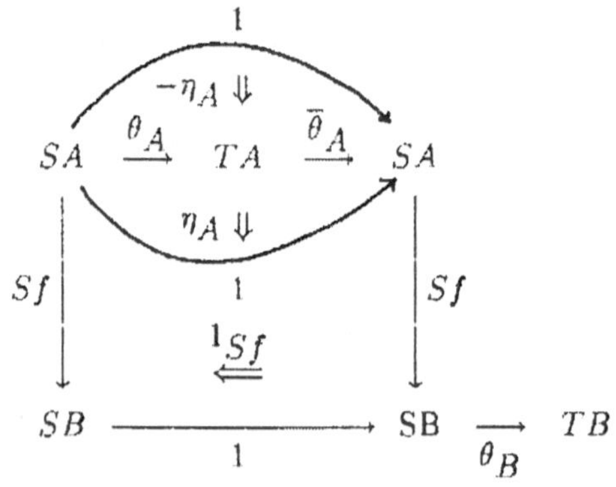

$=$

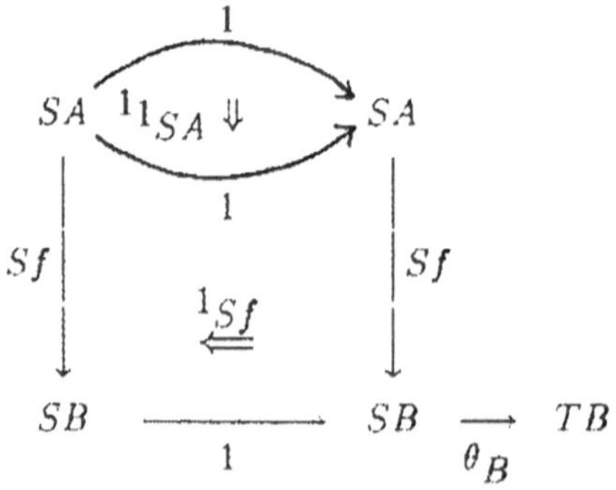

$= \quad 1_{(\theta_B \circ Sf)}$

Note that equality (1) is by (2.6) while equality (2) is by (2.1). Hence if we set

$$\nu = \rho_B \circ 1_{Tf} \circ 1_{\theta_A} + 1_{\theta_B} \circ \bar\theta_f \circ 1_{\theta_A} + 1_{\theta_B} \circ 1_{Sf} \circ (-\eta_A) \colon \theta_B \circ Sf \Rightarrow Tf \circ \theta_A$$

this shows that $\theta_f \circ \nu = 1_{(\theta_B \circ Sf)}$. By a similar computation we obtain $\nu \circ \theta_f = 1_{(Tf \circ \theta_A)}$. Therefore θ_f is invertible in $\mathcal{C}$.

Conversely, assume that for each object A of $\mathcal{A}$ the 1-morphism $\theta_A : SA \to TA$ is a homotopy equivalence in $\mathcal{C}$ and for each 1-morphism $f : A \to B$ of $\mathcal{A}$ the 2-morphism $\theta_f : Tf \circ \theta_A \Rightarrow \theta_B \circ Sf$ is invertible in $\mathcal{C}$. For each $A \in \mathcal{A}$, choose a 1-morphism $\overline{\theta}_A : TA \to SA$ in $\mathcal{C}$ and invertible 2-morphisms in $\mathcal{C}$,

$$\eta_A : \overline{\theta}_A \circ \theta_A \Rightarrow 1_{SA} \qquad , \qquad \rho_A : \theta_A \circ \overline{\theta}_A \Rightarrow 1_{TA} \quad .$$

By Vogt's Lemma we may assume

$$\text{(2.7)} \qquad \begin{cases} 1_{\theta_A} \circ \eta_A = \rho_A \circ 1_{\theta_A} \\ \eta_A \circ 1_{\overline{\theta}_A} = 1_{\overline{\theta}_A} \circ \rho_A \end{cases}$$

Now for each $f : A \to B$ in $\mathcal{A}$ we define $\overline{\theta}_f : Sf \circ \overline{\theta}_A \Rightarrow \overline{\theta}_B \circ Tf$ by

$$\overline{\theta}_f = 1_{\overline{\theta}_B} \circ 1_{Tf} \circ \rho_A + 1_{\overline{\theta}_B} \circ (-\theta_f) \circ 1_{\overline{\theta}_A} + (-\eta_B) \circ 1_{Sf} \circ 1_{\overline{\theta}_A} ,$$

as indicated below.

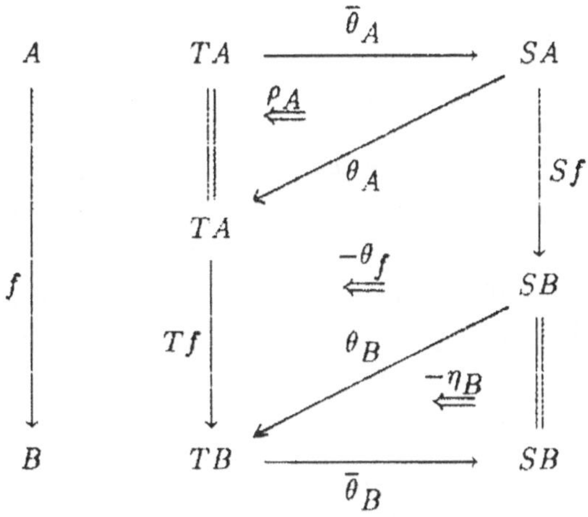

Note that this definition of $\overline{\theta}_f$ is possible because θ_f is invertible and that moreover $\overline{\theta}_f$ is itself an invertible 2-morphism of $\mathcal{C}$. Hence the assignments $\{\overline{\theta}_A, \overline{\theta}_f\}$ are seen to constitue a 1-morphism $\overline{\theta} : T \to S$ in $[\mathcal{A}, \mathcal{C}]$. Finally one verifies that the assignments $\{\eta_A\}$ and $\{\rho_A\}$ are the components for 2-morphisms $\eta : \overline{\theta} \circ \theta \Rightarrow 1_{1_S}$ and $\rho : \theta \circ \overline{\theta} \Rightarrow 1_{1_T}$ in $[\mathcal{A}, \mathcal{C}]$. For example, the coherence relation (2.1) for $\eta : \overline{\theta} \circ \theta \Rightarrow 1_{1_S}$

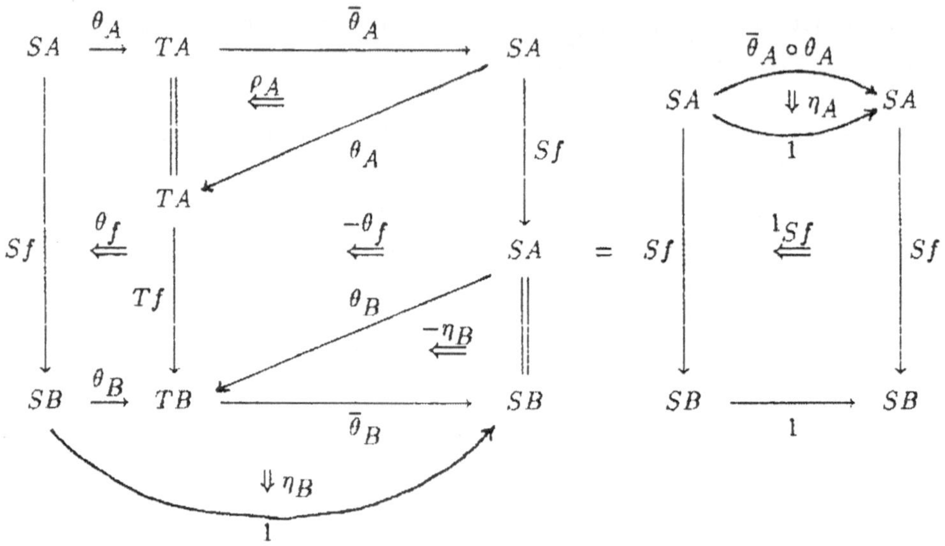

is valid as can be checked by using the relations (2.7). It is precisely in checking these coherence relations that Vogt's Lemma is needed but we omit further details. Therefore θ is a homotopy equivalence in $[\![\mathcal{A},\mathcal{C}]\!]$. ∎

(2.8) **Corollary.** Let $\mathcal{C}$ be a groupoid enriched 2-category. Then a 1-morphism $\theta \colon S \to T$ is a homotopy equivalence in $[\![\mathcal{A},\mathcal{C}]\!]$ if and only if $\theta_A \colon SA \to TA$ is a homotopy equivalence in $\mathcal{C}$ for each A in $\mathcal{A}$.

3. The Homotopy Pair Map Category.

(3.1) **Definition.** Let $\mathbf{2}$ be the category with two objects and one non-identity morphism. If $\mathcal{C}$ is a 2-category then the 2-category $[\![\mathbf{2},\mathcal{C}]\!]$ is called the ***homotopy pair map category*** associated to $\mathcal{C}$. The name is taken from Hardie's work. For $\mathcal{C} = \mathcal{T}\mathrm{op}_*$, the track 2-category of based topological spaces, Hardie [4] has proposed and studied the op-lax dual category $[\![\mathbf{2},\mathcal{T}\mathrm{op}_*]\!]''$ as the correct category in which to consider relative homotopy.

Denote $[\![\mathbf{2},\mathcal{C}]\!] = \mathrm{HPM}(\mathcal{C})$. The objects of $\mathrm{HPM}(\mathcal{C})$ are functors $\mathbf{2} \to \mathcal{C}$ but it is more convenient to regard them simply as 1-morphisms $f \colon X_0 \to X_1$ of $\mathcal{C}$. Then the 1-morphisms of $\mathrm{HPM}(\mathcal{C})$ correspond to squares

$$\begin{array}{ccc} X_0 & \xrightarrow{\theta_0} & Y_0 \\ f \downarrow & \overset{K}{\Leftarrow} & \downarrow g \\ X_1 & \xrightarrow[\theta_1]{} & Y_1 \end{array} \qquad (3.2)$$

Such a 1-morphism of HPM(C) is denoted $(\theta_0, K, \theta_1): f \to g$.

The category of fractions associated to HPM(C) as in Remark 1.4 is called the *homotopy pair class category* of C , denoted HPC(C) .

By Theorem 2.5 we have:

(3.3) **Theorem.** A 1-morphism $(\theta_0, K, \theta_1): f \to g$ in HPM(C) is a homotopy equivalence in HPM(C) if and only if θ_0 and θ_1 are homotopy equivalences in C and K is invertible in C .

Also Corollary 2.8 yields the following proposition (cf. [5, Theorem 1.3]).

(3.4) **Proposition (Hardie–Jansen).** Let C be a groupoid enriched category. Then a 1-morphism $(\theta_0, K, \theta_1): f \to g$ in HPM(C) is a homotopy equivalence in HPM(C) if and only if θ_0 and θ_1 are homotopy equivalences in C .

(3.5) **Notation.** We list three sub-2-categories of HPM(C) which we shall use.

(a) Let HPM$(C)_+$ denote the sub-2-category of HPM(C) each of whose 1-morphisms (3.2) has K invertible in C . Note that if C is groupoid enriched then HPM$(C)_+$ = HPM(C) .

(b) Let HPM$(C)_S$ be the sub-2-category of HPM(C) each of whose 1-morphisms (3.2) has K equal to the identity 2-morphism.

(c) We may regard C itself as a sub-2-category of HPM(C) . The identification is obtained by requiring in (3.2) that each of f and g be an identity 1-morphism and that K be the identity 2-morphism.

(3.6) **Proposition.** Let the sub-2-category $\mathcal{K}$ of HPM(C) be equal to one of C , HPM$(C)_S$ or HPM$(C)_+$. Then a 1-morphism $(\theta_0, K, \theta_1): f \to g$ in $\mathcal{K}$ is a homotopy equivalence in $\mathcal{K}$ if and only if θ_0 and θ_1 are homotopy equivalences in C .

Proof. A homotopy equivalence in *any* sub-2-category is always a homotopy equivalence in the full 2-category so by Theorem 2.5 the necessity of the condition holds. For the sufficiency, note that for each choice of $\mathcal{K}$ above, K is invertible so Theorem 2.5 is applicable. Moreover, by direct inspection, the homotopy inverse constructed in Theorem 2.5 is actually seen to belong to $\mathcal{K}$. This establishes the proposition. ∎

4. Lax Equalizers.

(4.1) **Definition.** Let $\mathcal{A}$ and $\mathcal{B}$ be 2-categories and $F, G: \mathcal{A} \to \mathcal{B}$ 2-functors. Let $\mathcal{K}$ be a fixed sub-2-category of $\mathrm{HPM}(\mathcal{B})$. We construct the *lax equalizer* $\mathcal{E}_{\mathcal{K}}(F,G)$ *of* F *and* G *relative to* $\mathcal{K}$. It is the sub-2-category of $\mathcal{A} \times \mathcal{K}$ described as follows. An object (A,u) of $\mathcal{A} \times \mathcal{K}$ belongs to $\mathcal{E}_{\mathcal{K}}(F,G)$ if $u\colon FA \to GA$. A 1-morphism $(f, K_f)\colon (A,u) \to (A',u')$ of $\mathcal{E}_{\mathcal{K}}(F,G)$ is determined by the data

$$(4.2) \qquad A \xrightarrow{f} A' \quad , \quad \begin{array}{ccc} FA & \xrightarrow{Ff} & FA' \\ u\big\downarrow & \overset{K_f}{\Longleftarrow} & \big\downarrow u' \\ GA & \underset{Gf}{\longrightarrow} & GA' \end{array}$$

consisting of a 1-morphism $f\colon A \to A'$ in $\mathcal{A}$ and a 1-morphism $(Ff, K_f, Gf)\colon u \to u'$ in $\mathcal{K}$.

The 2-morphisms of $\mathcal{E}_{\mathcal{K}}(F,G)$ are of the form

$$\lambda\colon (f, K_f) \Rightarrow (g, K_g)\colon (A,u) \to (A',u')$$

where $\lambda\colon f \Rightarrow g$ is a 2-morphism in $\mathcal{A}$ satisfying $G\lambda \circ 1_u + K_f = K_g + 1_{u'} \circ F\lambda$.

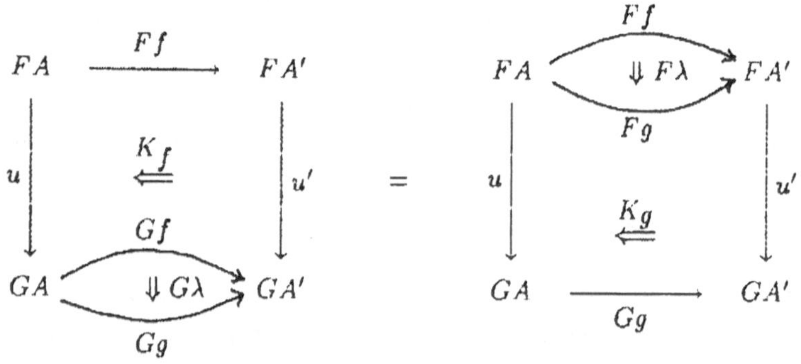

Let the 2-functor $i\colon \mathcal{E}_{\mathcal{K}}(F,G) \to \mathcal{A}$ be the projection functor. We may define a lax natural transformation $\theta\colon i \circ F \to i \circ G$ in $[\mathcal{E}_{\mathcal{K}}(F,G), \mathcal{B}]$ in the following way. On objects we let $\theta_{(A,u)} = u$ and if $(f, K_f)\colon (A,u) \to (A',u')$ is a 1-morphism of $\mathcal{E}_{\mathcal{K}}(F,G)$ we let $\theta_{(f,K_f)} = K_f$.

$$\begin{array}{ccc} \mathcal{E}_{\mathcal{K}}(F,G) & \xrightarrow{i} & \mathcal{A} \\ i\big\downarrow & \xrightarrow{\theta} & \big\downarrow G \\ \mathcal{A} & \underset{F}{\longrightarrow} & \mathcal{B} \end{array}$$

Observe that $\mathcal{E}_{\mathcal{K}}(F,G)$ will be groupoid enriched if $\mathcal{A}$ is groupoid enriched.

(4.3) **Notation.** For different choices of sub-2-category $\mathcal{K} \subset \mathrm{HPM}(\mathcal{B})$ we introduce notations for $\mathcal{E}_{\mathcal{K}}(F,G)$ as in the following table.

$\mathcal{K}$	$\mathcal{E}_{\mathcal{K}}(F,G)$
$\mathrm{HPM}(\mathcal{B})$	$\mathcal{E}(F,G)$
$\mathrm{HPM}(\mathcal{B})_+$	$\mathcal{E}_+(F,G)$
$\mathrm{HPM}(\mathcal{B})_\mathrm{S}$	$\mathcal{E}_\mathrm{S}(F,G)$
$\mathcal{B}$	$\mathcal{E}_\bullet(F,G)$

Of course if $\mathcal{B}$ is groupoid enriched then $\mathcal{E}_+(F,G) = \mathcal{E}(F,G)$.

(4.4) **Theorem.** Let $\mathcal{K}$ denote a given sub-2-category of $\mathrm{HPM}(\mathcal{B})$ and let $(f, K_f)\colon (A,u) \to (A',u')$ be a 1-morphism in $\mathcal{E}_{\mathcal{K}}(F,G)$.

(a) If (f,K_f) is a homotopy equivalence in $\mathcal{E}_{\mathcal{K}}(F,G)$ then $f\colon A \to A'$ is a homotopy equivalence in $\mathcal{A}$ and K_f is invertible in $\mathcal{B}$.

(b) Suppose that $\mathcal{K}$ is equal to one of $\mathcal{B}$, $\mathrm{HPM}(\mathcal{B})_\mathrm{S}$ or $\mathrm{HPM}(\mathcal{B})_+$. If f is a homotopy equivalence in $\mathcal{A}$ then (f,K_f) is a homotopy equivalence in $\mathcal{E}_{\mathcal{K}}(F,G)$.

Proof. (a) Note that $f\colon A \to A'$ in $\mathcal{A}$ and $(Ff, K_f, Gf)\colon u \to u'$ in $\mathcal{K}$ are the images of (f,K_f) under the projection 2-functors $i\colon \mathcal{E}_{\mathcal{K}}(F,G) \to \mathcal{A}$ and $\pi\colon \mathcal{E}_{\mathcal{K}}(F,G) \to \mathcal{K}$ respectively. Now these must be homotopy equivalences if (f,K_f) is a homotopy equivalence, because homotopy equivalences are preserved by 2-functors. Further by Theorem 2.5 K_f is invertible.

(b) Suppose that f is a homotopy equivalence with homotopy inverse $\bar{f}\colon A' \to A$. Then Ff and Gf are also homotopy equivalences. Consequently by Proposition 3.6, $(Ff, K_f, Gf)\colon u \to u'$ will be a homotopy equivalence in $\mathcal{K}$ for the stated choices of $\mathcal{K}$. Actually a more careful look at the proof of Theorem 2.5 shows that we may construct a 2-morphism $L\colon u \circ F\bar{f} \Rightarrow G\bar{f} \circ u'$ such that $(F\bar{f}, L, G\bar{f})\colon u' \to u$ is a homotopy inverse of $(Ff, K_f, Gf)\colon u \to u'$ in $\mathcal{K}$. But then $(\bar{f}, L)\colon (A',u') \to (A,u)$ will be in $\mathcal{E}_{\mathcal{K}}(F,G)$ and is clearly a homotopy inverse of (f,K_f) . ▮

(4.5) **Corollary.** Let $\mathcal{B}$ be a groupoid enriched category. Then a 1-morphism $(f, K_f)\colon (A,u) \to (A',u')$ is a homotopy equivalence in $\mathcal{E}(F,G)$ if and only if f is a homotopy equivalence in $\mathcal{A}$.

Proof. Because, if $\mathcal{B}$ is groupoid enriched then $\mathcal{E}_+(F,G) = \mathcal{E}(F,G)$ so Theorem 4.4.b also applies to $\mathcal{E}(F,G)$. ▮

(4.6) **Definition.** Let $\mathcal{A}$, $\mathcal{B}$, $\mathcal{C}$ be 2-categories and Γ, Δ 2-functors, as indicated:

$$A \xrightarrow{\;\Gamma\;} C \xleftarrow{\;\Delta\;} B$$

Also let $\pi_{\mathcal{A}} : \mathcal{A} \times \mathcal{B} \to \mathcal{A}$ and $\pi_{\mathcal{B}} : \mathcal{A} \times \mathcal{B} \to \mathcal{B}$ denote the projection 2-functors and let $\mathcal{K}$ be a sub-2-category of $\mathrm{HPM}(\mathcal{C})$. We define (cf. [3] and [9]) the *lax comma category of* Γ *and* Δ *relative to* $\mathcal{K}$ by $(\Gamma /\!/ \Delta)_{\mathcal{K}} = \mathcal{E}_{\mathcal{K}}(\Gamma \circ \pi_{\mathcal{A}}, \Delta \circ \pi_{\mathcal{B}})$. We may describe $(\Gamma /\!/ \Delta)_{\mathcal{K}}$ as follows.

The objects of $(\Gamma /\!/ \Delta)_{\mathcal{K}}$ are triples (A, u, B) such that $u : \Gamma A \to \Delta B$ is an object of $\mathcal{K} \subset \mathrm{HPM}(\mathcal{C})$. A 1-morphism $(f, K, g) : (A, u, B) \to (A', u', B')$ of $(\Gamma /\!/ \Delta)_{\mathcal{K}}$ consists of 1-morphisms $f : A \to A'$ in $\mathcal{A}$ and $g : B \to B'$ in $\mathcal{B}$ such that $(\Gamma f, K, \Delta g) : u \to u'$ is a 1-morphism in $\mathcal{K}$.

$$
\begin{array}{ccc}
\Gamma A & \xrightarrow{\;\Gamma f\;} & \Gamma A' \\[2pt]
u \downarrow & \overset{K}{\Leftarrow} & \downarrow u' \\[2pt]
\Delta B & \xrightarrow[\;\Delta g\;]{} & \Delta B'
\end{array}
$$

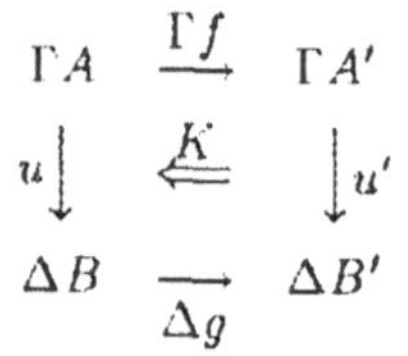

A 2-morphism in $(\Gamma /\!/ \Delta)_{\mathcal{K}}$, denoted

$$(A, u, B) \underset{(f', K', g')}{\overset{(f, K, g)}{\rightrightarrows}} {\Downarrow \lambda} (A', u', B') \qquad , \qquad \lambda = (\lambda_{f, f'}, \lambda_{g, g'}) \qquad ,$$

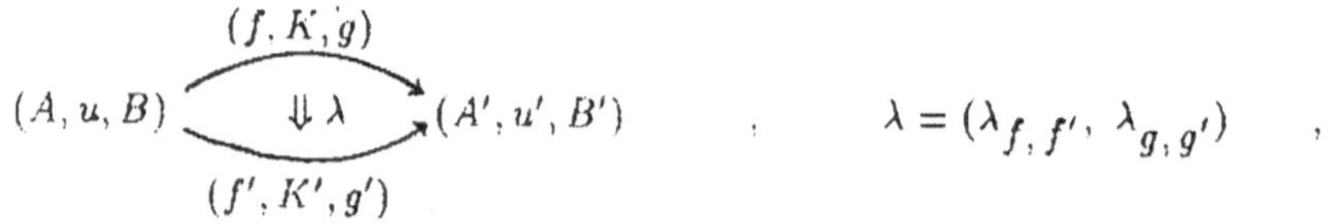

consists of a pair of 2-morphisms

$$A \underset{f'}{\overset{f}{\rightrightarrows}} {\Downarrow \lambda_{f, f'}} A' \qquad\qquad B \underset{g'}{\overset{g}{\rightrightarrows}} {\Downarrow \lambda_{g, g'}} B'$$

satisfying $\Delta(\lambda_{g, g'}) \circ 1_u + K = K' + 1_{u'} \circ \Gamma(\lambda_{f, f'})$.

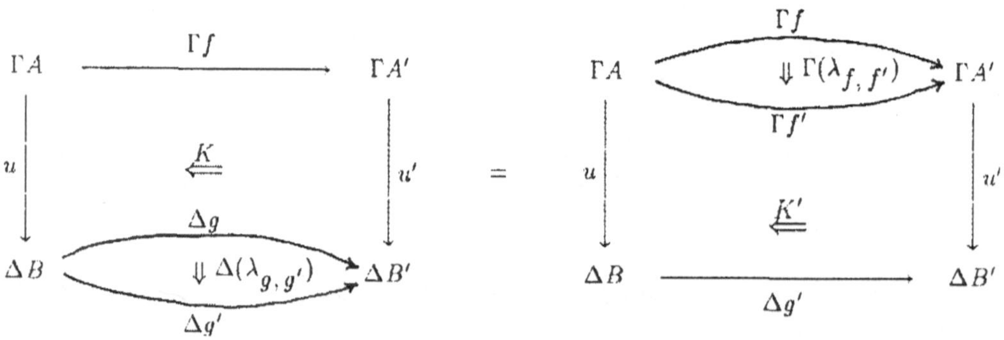

Observe that if $\mathcal{A}$ and $\mathcal{B}$ are groupoid enriched then $(\Gamma//\Delta)_{\mathcal{K}}$ is also groupoid enriched.

(4.7) **Notation.** We introduce notation as follows.

$$\Gamma//\Delta = \mathcal{E}(\Gamma \circ \pi_{\mathcal{A}}, \Delta \circ \pi_{\mathcal{B}})$$
$$(\Gamma//\Delta)_{+} = \mathcal{E}_{+}(\Gamma \circ \pi_{\mathcal{A}}, \Delta \circ \pi_{\mathcal{B}})$$
$$(\Gamma//\Delta)_{\mathrm{S}} = \mathcal{E}_{\mathrm{S}}(\Gamma \circ \pi_{\mathcal{A}}, \Delta \circ \pi_{\mathcal{B}})$$
$$(\Gamma//\Delta)_{\bullet} = \mathcal{E}_{\bullet}(\Gamma \circ \pi_{\mathcal{A}}, \Delta \circ \pi_{\mathcal{B}})$$

Observe that if $\mathcal{C}$ is groupoid enriched then $(\Gamma//\Delta)_{+} = \Gamma//\Delta$.

Applying Theorem 4.4 and Corollary 4.5 we obtain:

(4.8) **Theorem.** (a) Let (f, K, g) be a 1-morphism in $(\Gamma//\Delta)_{\mathcal{K}}$. If (f, K, g) is a homotopy equivalence in $(\Gamma//\Delta)_{\mathcal{K}}$ then f is a homotopy equivalence in $\mathcal{A}$, g is a homotopy equivalence in $\mathcal{B}$ and K is invertible in $\mathcal{C}$.

(b) Let (f, K, g) be a 1-morphism in any of the categories $(\Gamma//\Delta)_{\bullet}$, $(\Gamma//\Delta)_{\mathrm{S}}$ or $(\Gamma//\Delta)_{+}$. Then (f, K, g) is a homotopy equivalence if and only if f is a homotopy equivalence in $\mathcal{A}$ and g is a homotopy equivalence in $\mathcal{B}$.

(c) Let $\mathcal{C}$ be a groupoid enriched category. Then a 1-morphism (f, K, g) in $\Gamma//\Delta$ is a homotopy equivalence in $\Gamma//\Delta$ if and only if f is a homotopy equivalence in $\mathcal{A}$ and g is a homotopy equivalence in $\mathcal{B}$.

5. Examples.

In this last section we turn to a consideration of some examples which have been treated previously on a case-by-case basis by Hardie–Kamps ([6], [7], [8]). The interest is to fit them into the above framework. Actually (and we caution the reader to this point) Hardie–Kamps treat the op-lax duals of the categories given below; for example, the direction of K in Proposition 5.1 is reversed in [8], and similarly in Propositions 5.2 and 5.3.

Let $\mathbb{1}$ denote the category with one object and its identity morphism. Each object B of a 2-category $\mathcal{C}$ defines a functor $\ulcorner \mathrm{B} \urcorner : \mathbb{1} \to \mathcal{C}$ having constant value B. We consider the following functors and corresponding categories. Here P and Q are the respective projection 2-functors.

$\mathcal{C} \xrightarrow{\mathrm{Id}_{\mathcal{C}}} \mathcal{C} \xleftarrow{\ulcorner B \urcorner} \mathbb{1}$	$\mathrm{Id}_{\mathcal{C}} // \ulcorner B \urcorner = \mathcal{C}_B$
$\mathbb{1} \xrightarrow{\ulcorner B \urcorner} \mathcal{C} \xleftarrow{\mathrm{Id}_{\mathcal{C}}} \mathcal{C}$	$\ulcorner B \urcorner // \mathrm{Id}_{\mathcal{C}} = {}_B\mathcal{C}$
${}_X\mathcal{C} \xrightarrow{P} \mathcal{C} \xleftarrow{Q} \mathcal{C}_Y$	$(P//Q)_\bullet = {}_X\mathcal{C}_Y$

The homotopy categories (as in Remark 1.4) associated to $\mathcal{C}_B$, ${}_B\mathcal{C}$ and ${}_X\mathcal{C}_Y$ are called the *homotopy categories of spaces over B, spaces under B* and *spaces under X and over Y* respectively. Observe that each of these categories will be groupoid enriched whenever $\mathcal{C}$ is groupoid enriched.

The results of §4 together with Theorem 3.3 yield the following propositions. The precise details of the proofs are left to the reader.

(5.1) **Proposition.** A 1-morphism $(f, K)\colon u \to u'$ in $\mathcal{C}_B$ is specified by a diagram:

$$\begin{array}{ccc} C & \xrightarrow{f} & C' \\ {\scriptstyle u}\searrow & \overset{K}{\Leftarrow} & \swarrow{\scriptstyle u'} \\ & B & \end{array}$$

We have:

(a) (f, K) is a homotopy equivalence in $\mathcal{C}_B$ if and only if f is a homotopy equivalence in $\mathcal{C}$ and K is invertible.

(b) (Hardie–Kamps [8]) If $\mathcal{C}$ is groupoid enriched then (f, K) is a homotopy equivalence in $\mathcal{C}_B$ if and only if f is a homotopy equivalence in $\mathcal{C}$.

A similar proposition may be stated for ${}_B\mathcal{C}$.

(5.2) **Proposition.** A 1-morphism $(f, L, K)\colon (v, u) \to (v', u')$ of ${}_X\mathcal{C}_Y$ is specified by a diagram:

$$\begin{array}{ccc} & X & \\ {\scriptstyle v}\swarrow & \overset{L}{\Leftarrow} & \searrow{\scriptstyle v'} \\ C & \xrightarrow{f} & C' \\ {\scriptstyle u}\searrow & \overset{K}{\Leftarrow} & \swarrow{\scriptstyle u'} \\ & Y & \end{array}$$

We have:

(a) (f, L, K) is a homotopy equivalence in $_X C_Y$ if and only if f is a homotopy equivalence in C and L and K are invertible.

(b) (Hardie–Kamps [6]) If C is groupoid enriched then (f, L, K) is a homotopy equivalence in $_X C_Y$ if and only if f is a homotopy equivalence in C.

Let C be a 2-category with small hom-categories. Denote by **Cat** the 2-category of small categories, functors and natural transformations. We consider the functors

$$C \times C^{op} \xrightarrow{\ H\ } \mathbf{Cat}^{op} \xleftarrow{\ \ulcorner \mathbb{1} \urcorner\ } \mathbb{1}$$

in which $H(A, B) = hom_C(A, B)$ is the hom functor. The category $FC = H /\!/ \ulcorner \mathbb{1} \urcorner$ is called the *category of homotopy factorizations of C*. The associated homotopy category HFC is called the *homotopy category of homotopy factorizations of C* (see [7]).

We may consider an object of FC to be a 1-morphism $f\colon A \to B$ in C. A 1-morphism $(\phi, K, \psi)\colon f \to f'$ in FC corresponds to a diagram in C:

$$
\begin{array}{ccc}
A & \xrightarrow{\ \psi\ } & A' \\
f \downarrow & \overset{K}{\Leftarrow} & \downarrow f' \\
B & \xleftarrow{\ \phi\ } & B'
\end{array}
$$

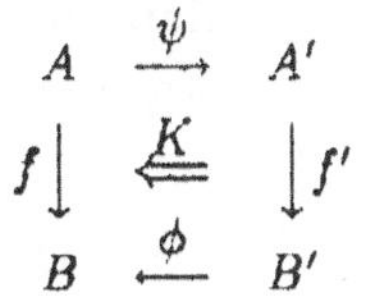

A 2-morphism $\lambda = (\lambda_{\phi, \phi'}, \lambda_{\psi, \psi'})\colon (\phi, K, \psi) \Rightarrow (\phi', K', \psi')\colon f \to f'$ in FC is a pair of 2-morphisms $\lambda_{\phi, \phi'}\colon \phi \to \phi'$ and $\lambda_{\psi, \psi'}\colon \psi \Rightarrow \psi'$ in C satisfying the equation $K' + \lambda_{\phi, \phi'} \circ 1_{f'} \circ \lambda_{\psi, \psi'} = K$.

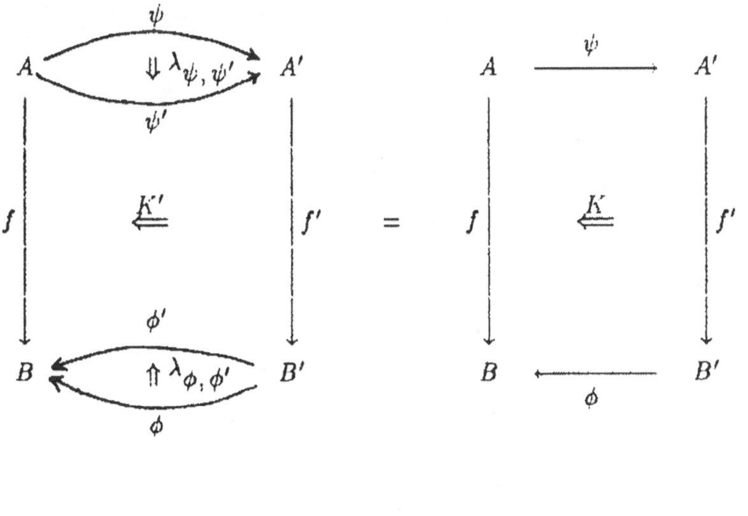

(5.3) **Proposition.** (a) A 1-morphism $(\phi, K, \psi): f \to f'$ is a homotopy equivalence in **F**C if and only if ϕ and ψ are homotopy equivalences in C and K is invertible.

(b) (Hardie–Kamps [7]) Let C be a groupoid enriched category. Then (ϕ, K, ψ) is a homotopy equivalence in **F**C if and only if ϕ and ψ are homotopy equivalences in C .

References

[1] P.H.H. Fantham and E.J. Moore, *Groupoid enriched categories and homotopy theory*, Can. J. Math. **35**(1983), 385–416.

[2] P. Gabriel and M. Zisman, Calculus of Fractions and Homotopy Theory, New York: Springer–Verlag, 1967.

[3] John W. Gray, *The categorical comprehension scheme*, Lecture Notes in Math. 99(1969), 242–312.

[4] K.A. Hardie, *On the category of homotopy pairs*, Topology and its Applications 14(1982), 59–69.

[5] K.A. Hardie and A.V. Jansen, *The Puppe and Nomura operators in the category of homotopy pairs*, Lecture Notes in Math. **915**(1982), 112–126.

[6] K.A. Hardie and K.H. Kamps, *Homotopy over B and under A*, Cahiers Top. et Géom. Diff. **28**(1987), 183–196.

[7] ______________________, *The homotopy category of homotopy factorizations*, Lecture Notes in Math. **1298**(1987), 162–170.

[8] ______________________, *Track homotopy over a fixed space*, Glasnik Mat. 24(1989), 161–179.

[9] G.M. Kelly, *On clubs and doctrines*, Lecture Notes in Math. **420**(1974), 181–256.

[10] T. Müller, *Zur Theorie der Würfelsätze*, Dissertation, Fernuniversität Hagen, 1982.

[11] R.M. Vogt, *A note on homotopy equivalences*, Proc. Amer. Math. Soc. **32**(1972), 627–629.

LOCALIZING $\varepsilon_{\#}(X)$

Ken-ichi Maruyama

Department of Mathematics, Faculty of Education,
Chiba University, Yayoicho Chiba Japan.

1. Introduction

Let $\varepsilon_{\#}^{m}(X)$ be the kernel of the obvious map from the group of self-homotopy equivalences of a space X to its automorphism group of homotopy:

$$\varepsilon(X) \rightarrow \prod_{j \leqq m} \mathrm{Aut}\,\pi_j(X).$$

We simply denote $\varepsilon_{\#}(X)$ when $m = \dim X$. It has been shown, by Dror and Zabrodsky that $\varepsilon_{\#}(X)$ is a nilpotent group in the case where X is a finite dimensional complex or a Postnikov piece (the number of non-trivial homotopy groups are finite)([2],Theorem A).

The theory of localization for nilpotent groups has been studied by several authors and proved to be a powerful tool in algebraic topology. The reader may consult the book [3].

The intention of our work is to study the localization of the nilpotent group $\varepsilon_{\#}(X)$. Actually $\varepsilon_{\#}$ commutes with localization (2.1 in § 2). This fact enables us to determine the group structures of $\varepsilon_{\#}(X)$ for some spaces, especially Lie groups. The article is organized as follows: in § 2 we recall the results in [4], in § 3 we give some examples.

2. Main Results in [4]

2.1. Main Theorem ([4]). Let X be a finite and simple C.W complex. Assume that $m \geqq \dim X$. Then the natural map:

$$\ell_{\#}: \varepsilon_{\#}^{m}(X) \rightarrow \varepsilon_{\#}^{m}(X_{(P)})$$

is the P-localization map for an arbitrary collection of primes P, where $X_{(P)}$ is the localization at P.

By making use of the main theorem we obtain the following.

Note: This paper is in final form and no version of it will be published elsewhere.

2.2 Theorem ([4]). Let X be a simple and finite rational H-space with $\beta_{2n_i-1}\text{-rank}_Q(\pi_{2n_i-1}(X)\otimes Q) \leq 1$ for $i \leq k$. Then $\varepsilon_\#(X)/\text{torsion} = Z+\cdots+Z$, the free abelian group of

$$\text{rank} = \Sigma_{i=1}^{k} \text{rank}_Q(\pi_{2n_i-1}(X)\otimes Q)\cdot(\beta_{2n_i-1}\text{-rank}_Q(\pi_{2n_i-1}(X)\otimes Q)),$$

where β_j is the j-th Betti number and $H^*(X;Q) = \wedge(x_1,\cdots,x_k)$ with $\deg x_i = 2n_i-1$.

Remark. (1). The finiteness condition of the main theorem is necessary. Let $M_i = S^n \cup_{2^i} e^{n+1}$ be the Moore space ($n > 1$). Then $[M_i, S^{n+1}] = Z/2^iZ$. We denote its generator by $f_i: M_i \to S^{n+1}$.
Let $X = (\vee_i M_i)\vee S^{n+1}$ ($\vee$: a wedge sum). A map $f = (1_{\vee_i M_i} + \vee_i f_i)\vee 1_{S^{n+1}}: X \to X$, + denotes the addition in $[\vee_i M_i, X]$. $f \in \varepsilon_\#(X)$ and it is not of finite order since $f^k = (1_{\vee_i M_i} + \vee_i kf_i) \vee 1_{S^{n+1}}$, and hence $\varepsilon_\#(X)_{(0)}$ is not trivial. On the other hand $\varepsilon_\#(X_{(0)}) = \varepsilon_\#(S^{n+1}_{(0)}) = 1$.

(2). In 2.2, the condition " $\beta_{2n_i-1}\text{-rank}_Q(\pi_{2n_i-1}(X)\otimes Q) \leq 1$ for $i \leq k$ " can be replaced by the following weaker condition (although it is less practical). Condition : $\varepsilon_\#(X)$ acts on decomposables of $H^{n_j}(X;Q)$ trivially for $j \leq \text{rank } X$.

3. Applications

In this section we give miscellaneous examples.

3.1. Example(of the main theorem). Let $\bar{2}$ be the odd primes.

(1). $\varepsilon_\#(SO(4))_{(\bar{2})} = \varepsilon_\#(S^3\times S^3)_{(\bar{2})} = Z_3 + Z_3$.

(2). $\varepsilon_\#(SO(5))_{(\bar{2})} = \varepsilon_\#(Sp(2))_{(\bar{2})} = Z_{15}$.

(3.[4]). $\varepsilon_\#(SO(6))_{(\bar{2})} = \varepsilon_\#(SU(4))_{(\bar{2})}$. Let us denote this group by G

Then $G = G_{(3)} \times Z_5 \times Z_5$ and $G_{(3)}$ has the following presentation.

$$G_{(3)} = \langle a, b, c \mid a^9, b^9, c^3, [a, b], [a, c], [b, c]a^{-3}\rangle$$

(4). $\varepsilon_{\#}(Spin(7))_{(5,7)} = \varepsilon_{\#}(SO(7))_{(5,7)} = Z_5 + Z_5 + Z_7 + Z_7$.

Proof. For an odd prime p, there are p-equivalences as follows. $SO(4) \underset{p}{\simeq} S^3 \times S^3$, $SO(5) \underset{p}{\simeq} Sp(2)$, $SO(6) \underset{p}{\simeq} SU(4)$, $SO(7) \underset{p}{\simeq} Spin(7)$. Thus the half parts of the assertions are trivial. Moreover there are p-equivalences $SO(6) \underset{p}{\simeq} Sp(2) \times S^5$, p : an odd prime and $Spin(7) \underset{p}{\simeq} G_2 \times S^7$ for $p \neq 3$. Finally, $\varepsilon(Sp(2))$ and $\varepsilon(G_2)$ are known ([5] [7]). Thus assertions on the group structures for (2),(3) and (4) are derived from above equivalences and the formula for $\varepsilon(X \times Y)$ by Sieradski [8]. For instance, we obtain the folowing.

$$\varepsilon_{\#}(Spin(7))_{(5)}$$

$$\parallel$$

$$0 \to [G_{2(5)} \wedge S^7_{(5)}, G_{2(5)}] \to \varepsilon_{\#}(G_{2(5)} \times S^7_{(5)}) \xrightarrow{i^{\#}} GL(2, \wedge_{ij}).$$

Im $i^{\#}$ is isomorhic to $[G_2, S^7]_{(5)} = Z_5$. The left term of the above exact sequence is also Z_5. Hence $\varepsilon_{\#}(Spin(7))_{(5)}$ is an abelian group of order 25. Let $p : G_{2(5)} \times S^7_{(5)} \to G_{2(5)}$ be a projection map, $j : S^7_{(5)} \to G_{2(5)} \times S^7_{(5)}$ an inclusion map. Then Im $i^{\#} = Z_5$ is naturally isomorphic to the subgroup $\langle j \alpha p + 1\rangle$ of $\varepsilon_{\#}(G_{2(5)} \times S^7_{(5)})$, where $\alpha \in [G_{2(5)}, S^7_{(5)}]$, and hence Im $i^{\#}$ splits. At a prime 7, we use the same argument.

3.2. Example(of 2.2). $\varepsilon_{\#}(E_6)/tor = Z$, $\varepsilon_{\#}(SU(8))/tor = Z$, $\varepsilon_{\#}(SU(9))/tor = Z + Z$. ($\varepsilon(SU(n))$ $n \leq 7$ are finite groups. See [1]).

Remark. (1). Further direct observation at (0) as in the proof of Proposition 2.1 in [4] shows that $\varepsilon_{\#}(SU(n))/tor$ is an abelian group if and only if $n \leq 17$. (2). There are other (rather obvious) isomorphisms $\varepsilon_{\#}(SU(n))_{(p)} = \varepsilon_{\#}(PSU(n))_{(p)}$ if $(p, n) = 1$, $\varepsilon_{\#}(Sp(n))_{(p)} =$

$\varepsilon_{\#}(PSp(n))_{(p)}$ if $p \neq 2$. Here PG denotes the projective group of G.

Concluding Remark. Recently, J. Moller has extended 2.1 to the case where a space is nilpotent [6].

References

[1] M. Arkowitz and C. R. Curjel, Groups of homotopy classes, Lecture Notes in Math., 4(1967).
[2] E. Dror and A. Zabrodsky, Unipotency and Nilpotency in Homotopy Equivalences, Topology, 18(1979), 187-197.
[3] P. Hilton, G. Mislin and J. Roitberg, Localization of Nilpotent Groups and Spaces, Mathematics Studies 15(1975), North Holland.
[4] K. Maruyama, Localization of a certain subgroup of self-homotopy equivalences, Pacific J. Math., 136(1989), 293-301.
[5] M. Mimura and N. Sawashita, On the group of self-homotopy equivalences of H-spaces of rank 2, J. Math. Kyoto Univ., 21(1981), 331-349.
[6] J. Moller, Self-homotopy Equivalences of $H_*(-;Z/p)$-Local Spaces, Kodai Math. J., 12(1989), 270-281.
[7] S. Oka, N. Sawashita and M. Sugawara, On the group of self-equivalences of a mapping cone, Hiroshima Math. J., 4(1974), 9-28.
[8] A. J. Sieradski, Twisted self-homotopy equivalences. Pacific J. Math 34(1970), 789-802.

Weak equivalences and quasifibrations *

J. P. May
Department of Mathematics
University of Chicago
Chicago, Il 60637

Quasifibrations are essentially fibrations up to weak homotopy. They play a fundamental role in homotopy theory since a variety of important constructions give rise to quasifibrations which fail to be fibrations. Quasifibrations were introduced in a basic 1958 paper by Dold and Thom [2], and some refinements of their work were added by Hardie in 1970 [4]. The importance of quasifibrations to the study of classifying spaces and fibrations was first established in a 1959 paper of Dold and Lashof [1], and a systematic account was given in [5]. Quasifibrations played an essential role in Quillen's 1973 paper [6] in which he introduced the higher algebraic K-groups of rings. They have been applied in quite a large number of more recent papers.

Despite their importance, quasifibrations have not been treated in any textbook, and I know of no better published reference than the original paper (in German) of Dold and Thom. Around 1972, I proved a new theorem about weak homotopy equivalences of pairs of spaces and observed that the basic facts about quasifibrations are very easy consequences of that result. I've never published this material, which was intended as part of a still projected volume on the homotopical foundations of algebraic topology. In view of its close connection to the theme of the Montreal conference, I thought that I would seize the occasion to give an exposition.

We give some preliminaries and state our theorem about weak equivalences in section 1. We explain the application to the theory of quasifibrations in section 2. We prove the theorem about weak equivalences in section 3.

---

* This paper is in final form and no version of it will be submitted for publication elsewhere.

§1. Weak equivalences of pairs

A map $f: X \to Y$ of spaces is said to be an n-equivalence if, for all $x \in X$, $f_*: \pi_q(X,x) \to \pi_q(Y,f(x))$ is a bijection for $0 \le q < n$ and a surjection for $q = n$. A map $f: (X,A) \to (Y,B)$ of pairs of spaces is said to be an n-equivalence if $(f_*)^{-1}\mathrm{Im}(\pi_0 B \to \pi_0 Y) = \mathrm{Im}(\pi_0 A \to \pi_0 X)$ and, for all $a \in A$, $f_*: \pi_q(X,A,a) \to \pi_q(Y,B,f(a))$ is a bijection for $1 \le q < n$ and a surjection for $q = n$. The condition on components means that if $f(x)$ can be connected to a point of B, then x can be connected to a point of A; it is automatically satisfied when X and Y are path connected. In both the absolute and the relative cases, f is said to be a weak equivalence if it is an n-equivalence for all n.

By the evident long exact sequences and the five lemma, plus some tedious extra details to handle fundamental groups, we have the following relationship between weak equivalences of pairs and of their constituent spaces.

Lemma 1.1. Let $f: (X,A) \to (Y,B)$ be a map of pairs such that both $f_*: \pi_0 A \to \pi_0 B$ and $f_*: \pi_0 X \to \pi_0 Y$ are bijections. If any two of the three maps $f: A \to B$, $f: X \to Y$, and $f: (X,A) \to (Y,B)$ are weak equivalences, then so is the third.

Our new theorem on weak equivalences of pairs is a kind of analog in the context of excisive triads. Recall that a triad $(X; A, B)$ is said to be excisive if X is the union of the interiors of A and B.

Theorem 1.2. Let $f: (X; X_1, X_2) \to (Y; Y_1, Y_2)$ be a map of excisive triads such that $f: (X_i, X_1 \cap X_2) \to (Y_i, Y_1 \cap Y_2)$ is an n-equivalence for $i = 1$ and $i = 2$. Then $f: (X, X_i) \to (Y, Y_i)$ is an n-equivalence for $i = 1$ and $i = 2$.

No useful conclusion could be derived with an assumption on only one of the pairs $(X_i, X_1 \cap X_2)$. While this result really does seem to be new, the following immediate consequence of the lemma and theorem is folklore; a proof appears in Gray [3, 16.24].

Corollary 1.3. Let $f: (X; X_1, X_2) \to (Y; Y_1, Y_2)$ be a map of excisive triads such that $f: X_1 \cap X_2 \to Y_1 \cap Y_2$, $f: X_1 \to Y_1$, and $f: X_2 \to Y_2$ are weak equivalences. Then $f: X \to Y$ is a weak equivalence.

In turn, this implies a local criterion for a map to be a weak equivalence.

Corollary 1.4. Let $f\colon X \to Y$ be a map and let $\mathcal{O}$ be an open cover of Y which is closed under finite intersections. If $f\colon f^{-1}U \to U$ is a weak equivalence for all $U \in \mathcal{O}$, then $f\colon X \to Y$ is a weak equivalence.

Proof. Let $\mathcal{C}$ be the collection of subspaces V of Y such that V is a union of spaces in $\mathcal{O}$, $f\colon f^{-1}V \to V$ is a weak equivalence, and $f\colon f^{-1}(U\cap V) \to U\cap V$ is a weak equivalence for all $U \in \mathcal{O}$. Order $\mathcal{C}$ by inclusion. The union of a chain in $\mathcal{C}$ is in $\mathcal{C}$ by an obvious colimit argument, and $\mathcal{C}$ is nonempty since it contains $\mathcal{O}$. Thus $\mathcal{C}$ has a maximal element V. Suppose $V \neq Y$. Then there is a $U \in \mathcal{O}$ which is not contained in V. The previous corollary implies that $U\cup V$ is in $\mathcal{C}$, contradicting the maximality of V.

§2. Quasifibrations

If $p\colon E \to B$ is a fibration, then $p\colon (E,p^{-1}A) \to (B,A)$ is a weak equivalence for all nonempty subspaces A of B; in particular, $p\colon (E,p^{-1}b) \to (B,b)$ is a weak equivalence for all $b \in B$ (e.g. [9, p.187]). The notion of a quasifibration turns this desirable property into a definition.

Definition 2.1. A surjective map $p\colon E \to B$ is a quasifibration if $p\colon (E,p^{-1}b) \to (B,b)$ is a weak equivalence for all $b \in B$.

It is to be emphasized that this notion does not properly belong to fibration theory since the pullback of a quasifibration need not be a quasifibration.

Assume given a fixed surjective map $p\colon E \to B$. We shall derive various criteria for p to be a quasifibration.

Clearly $p\colon E \to B$ is a quasifibration if and only if its restriction $p^{-1}C \to C$ is a quasifibration for each path component C of B. Thus we may as well restrict attention to path connected base spaces B.
Of course, if p is a quasifibration, then, for $b \in B$ and $x \in p^{-1}b$, the exact sequence of homotopy groups of the pair $(E,p^{-1}b)$ yields an exact sequence

$$\cdots \to \pi_{n+1}(B,b) \to \pi_n(p^{-1}b,x) \to \pi_n(E,x) \to \pi_n(B,b) \to \cdots \to \pi_0(B,b).$$

Let $Np = \{(x,\omega) \mid \omega: I \to B, \omega(1) = p(x)\} \subset E \times B^I$ and let $q: Np \to B$ be the fibration specified by $q(x,\omega) = \omega(0)$; thus $q^{-1}b$ is the usual homotopy theoretic fiber of p over b. If $\lambda: E \to Np$ is the natural equivalence, $\lambda(x) = (x, c_{p(x)})$, then $q \circ \lambda = p$ and λ restricts to a map $p^{-1}b \to q^{-1}b$ for each $b \in B$. Clearly p is a quasifibration if and only if $\lambda: (E, p^{-1}b) \to (Np, q^{-1}b)$ is a weak equivalence for all $b \in B$. By Lemma 1.1, this holds if and only $\lambda: p^{-1}b \to q^{-1}b$ is a weak equivalence for all $b \in B$. With B connected, the fibers $q^{-1}b$ all have the same homotopy type, hence the fibers $p^{-1}b$ all have the same weak homotopy type if p is a quasifibration.

Say that a subspace A of B is distinguished if the restriction $p: p^{-1}A \to A$ is a quasifibration. Since $p: (E, p^{-1}A, p^{-1}a) \to (B, A, a)$ induces a map of long exact sequences of homotopy groups of triples, the five lemma and some tedious verifications on the π_1 level give the following observation.

Lemma 2.2. Let A be a distinguished subspace of B. Then the maps $p: (E, p^{-1}a) \to (B, a)$ are weak equivalences for all $a \in A$ if and only if the map $p: (E, p^{-1}A) \to (B, A)$ is a weak equivalence.

The following analog of Corollary 1.3, which is the heart of the Dold-Thom theory of quasifibrations, is now a direct consequence of Theorem 1.2. This observation is perhaps the main point of our work.

Corollary 2.3. Let $(B; B_1, B_2)$ be an excisive triad. If $B_1 \cap B_2$, B_1, and B_2 are distinguished, then $p: E \to B$ is a quasifibration.

Proof. With (B, A) replaced by $(B_i, B_1 \cap B_2)$, Lemma 2.2 gives that

$$p: (p^{-1}B_i, p^{-1}B_1 \cap p^{-1}B_2) \to (B_i, B_1 \cap B_2)$$

is a weak equivalence for $i = 1$ and $i = 2$. By Theorem 1.2,

$$p: (E, p^{-1}B_i) \to (B, B_i)$$

is a weak equivalence for $i = 1$ and $i = 2$. By Lemma 2.2 applied with $A = B_i$, $p: (E, p^{-1}b) \to (B, b)$ is a weak equivalence for all $b \in B_i$, $i = 1$ and $i = 2$, and thus for all $b \in B$.

The proof of Corollary 1.4 applies to give the quasifibration analog of that result.

Corollary 2.4. Let $\mathcal{O}$ be an open cover of B which is closed under finite intersections. If each $U \in \mathcal{O}$ is distinguished, then $p: E \to B$ is a quasifibration.

These results are usually used in conjunction with the following observation. Recall that a homotopy $h_t: B \to B$ is a deformation of B onto A if $h_0 = \mathrm{Id}$, $h_t(a) = a$ for $a \in A$, and $h_1(B) \subset A$.

Lemma 2.5. Let A be a distinguished subspace of B. Suppose there exist deformations h of B onto A and H of E onto $p^{-1}A$ such that $p \circ H_1 = h_1 \circ p$ and $H_1: p^{-1}b \to p^{-1}h_1(b)$ is a weak equivalence for all $b \in B$. Then $p: E \to B$ is a quasifibration.

Proof. By Lemma 1.1, $H_1: (E, p^{-1}b) \to (p^{-1}A, p^{-1}h_1(b))$ is a weak equivalence for all $b \in B$. Passage to homotopy groups from the commutative diagram

$$\begin{array}{ccc} (E, p^{-1}b) & \xrightarrow{H_1} & (p^{-1}A, p^{-1}h_1(b)) \\ p\downarrow & & \downarrow p \\ (B, b) & \xrightarrow[h_1]{} & (A, h_1(b)) \end{array}$$

gives the conclusion.

Say that B is filtered if it is given as the union of an increasing sequence of subspaces F_nB such that each inclusion $F_nB \to F_{n+1}B$ is a cofibration. By an evident colimit argument, a map $p: E \to B$ is a quasifibration if each F_nB is distinguished. The following immediate inductive consequence of Corollary 2.3 and Lemma 2.5 is probably the most generally useful criterion for p to be a quasifibration.

Theorem 2.6. Let $p: E \to B$ be a map onto a filtered space B and suppose that the following conditions hold.

(i) F_0B and each open subset of each $F_nB - F_{n-1}B$ are distinguished.

(ii) For each $n \geq 1$, there is an open neighborhood U_n of $F_{n-1}B$ in F_nB and there are deformations h of U_n onto $F_{n-1}B$ and H of $p^{-1}U_n$ onto $p^{-1}F_{n-1}B$ such that $p \circ H_1 = h_1 \circ p$ and $H_1: p^{-1}b \to p^{-1}h_1(b)$ is a weak equivalence for each $b \in U_n$.

Then each F_nB is distinguished and $p: E \to B$ is a quasifibration.

There is an alternative criterion that often applies when E and B are built up from successive compatible pushout diagrams.

Theorem 2.7. Let $p: E \to B$ be a map of filtered spaces such that $F_nE = p^{-1}F_nB$ for $n \geq 0$ and, for $n \geq 1$, $p: F_nE \to F_nB$ is obtained by passage to pushouts from a commutative diagram of the form

$$\begin{array}{ccccc} & g_n & & j_n & \\ F_{n-1}E & \leftarrow & D_n & \rightarrow & E_n \\ p\downarrow & & \downarrow q_n & & \downarrow p_n \\ F_{n-1}B & \leftarrow & A_n & \rightarrow & B_n. \\ & f_n & & i_n & \end{array}$$

Suppose that the following conditions hold.

(i) F_0B is distinguished.

(ii) Each map $p_n: E_n \to B_n$ is a fibration.

(iii) Each map $i_n: A_n \to B_n$ is a cofibration.

(iv) Each right square is a pullback.

(v) $g_n: (q_n)^{-1}(a) \to p^{-1}f_n(a)$ is a weak equivalence for all $a \in A_n$.

Then each F_nB is distinguished and $p: E \to B$ is a quasifibration.

The inductive step here is a consequence of the second of the following two lemmas, which are due to Hardie [4]. Both refer to a commutative diagram

$$(*) \qquad \begin{array}{ccccc} & g & & j & \\ E & \leftarrow & D & \rightarrow & E' \\ p\downarrow & & \downarrow q & & \downarrow p' \\ B & \leftarrow & A & \rightarrow & B'. \\ & f & & i & \end{array}$$

Lemma 2.8. If, in $(*)$, p, q, and p' are quasifibrations and the maps $g: q^{-1}(a) \to p^{-1}f(a)$ and $j: q^{-1}(a) \to (p')^{-1}i(a)$ are weak equivalences for all $a \in A$, then the induced map $s: M(j,g) \to M(i,f)$ of double mapping cyclinders is a quasifibration.

Proof. $M(i,f) = B \cup_f (A \times I) \cup_i B'$ is the union of $B \cup_f (A \times [0,2/3])$ and $(A \times [1/3,1]) \cup_i B'$, and similarly for $M(j,g)$. The conclusion follows easily from Lemma 2.5 and Corollary 2.3.

Lemma 2.9. Suppose that $(*)$ satisfies the following conditions.

(i) p is a quasifibration.

(ii) p' is a fibration.

(iii) i is a cofibration.

(iv) The right square is a pullback.

(v) $g: q^{-1}(a) \to p^{-1}f(a)$ is a weak equivalence for all $a \in A$.

Then the map $r: E \cup_g E' \to B \cup_f B'$ induced by p and p' is a quasifibration.

Proof. We have the commutative diagram

$$\begin{array}{ccc} M(j,g) & \xrightarrow{\beta} & E \cup_g E' \\ s\downarrow & & \downarrow r \\ M(i,f) & \xrightarrow[\alpha]{} & B \cup_f B'. \end{array}$$

Since i and j are cofibrations (the latter by [9, I.7.14]), the quotient maps α and β are homotopy equivalences by a standard result on pushouts of equivalences. The map s is a quasifibration by the previous lemma. By Lemma 1.1 and a chase of the diagram, it suffices to show that $\beta: s^{-1}(x) \to r^{-1}(x)$ is a weak equivalence for each $x \in M(i,f)$. If $x \in B$ or $x \in B' - i(A)$, β is a homeomorphism. If $x = (a,s)$, where $a \in A$ and $0 < s \leq 1$, then it is easy to see that β can be identified with the weak equivalence $g: q^{-1}(a) \to p^{-1}f(a)$.

§3. The proof of Theorem 1.2

We begin with an analysis of the notion of an n-equivalence. In the absolute case, we have the following result. We omit the proof since a generalized version of the based analog is given in [6, Lemma 1] and we shall shortly be proving the more difficult relative analog.

Lemma 3.1. For each $n \geq 1$, the following statements about a map $f: X \to Y$ are equivalent.

(i) For each $x \in X$, $f_*: \pi_q(X,x) \to \pi_q(Y,fx)$ is an injection for $q = n-1$ and a surjection for $q = n$.

(ii) If $h: e \simeq fg$ on ∂I^n in the following diagram, then there exist $\tilde{g}$ and $\tilde{h}$ which make the diagram commute.

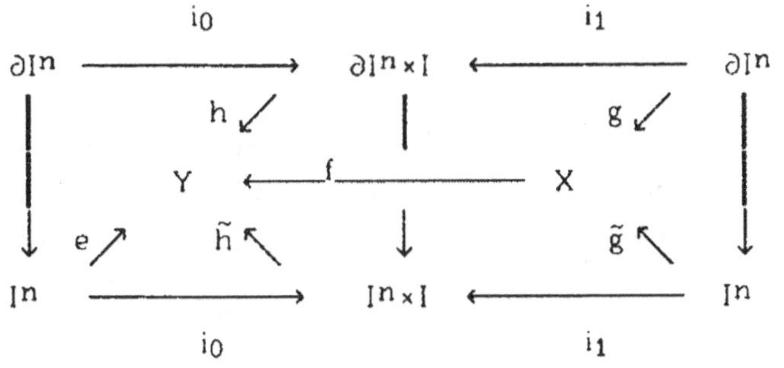

(iii) The conclusion of (ii) holds when $e = fg$ on ∂I^n and h is the constant homotopy.

In order to prove the relative analog, we will need the following relative homotopy extension property.

Limma 3.2 (relative HEP). Let $(L; J, K)$ be a triad such that the inclusions $J \cap K \to K$ and $J \cup K \to L$ are cofibrations. Then any homotopy $h: (J, J\cap K) \times I \to (X,A)$ of the restriction of a map $f: (L,K) \to (X,A)$ extends to a homotopy $H: (L,K) \times I \to (X,A)$ of f.

Proof. This holds by two applications of the usual HEP:

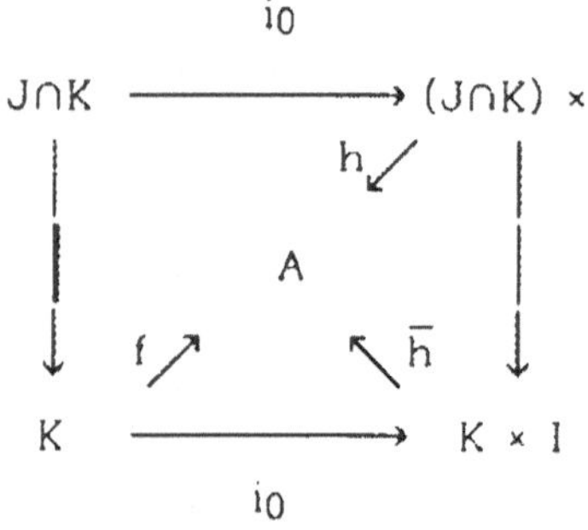

and

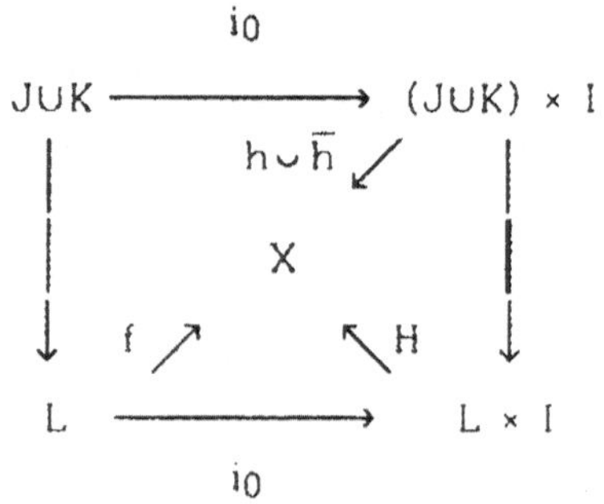

Before proceeding, we must fix some notations. Let

$$J^0 = \{0\} \text{ and } J^n = (\partial I^n \times I) \cup (I^n \times \{0\}) \subset I^{n+1} \text{ for } n \geq 1.$$

For a pair (X,A) with basepoint $a \in A$, we take

$$\pi_n(X, A, a) = [(I^n, \partial I^n, J^{n-1}), (X, A, a)] \text{ for } n \geq 1.$$

Let $\bar{I}^n = I^n \times \{1\}$ and $\partial \bar{I}^n = \partial I^n \times \{1\} = J^n \cap \bar{I}^n$. Define the negative of a homotopy h to be h traversed from 1 to 0 and define the sum $h_1 + \cdots + h_j$ of homotopies $h_i: f_{i-1} \simeq f_i$ to be the homotopy obtained traversing h_i on the interval $[(i-1)/j, i/j]$.

The following lemma and its proof are due to Sugawara [8].

Lemma 3.3. For each $n \geq 0$, the following statements about a map $f: (X,A) \to (Y,B)$ are equivalent.

(i) For each $a \in A$, $f_*: \pi_q(X, A, a) \to \pi_q(Y, B, fa)$ is an injection for $q = n$ and a surjection for $q = n+1$. (When $n = 0$, replace the injectivity statement by $(f_*)^{-1}\mathrm{Im}(\pi_0 B \to \pi_0 Y) = \mathrm{Im}(\pi_0 A \to \pi_0 X)$.)

(ii) If $h: e \simeq fg$ on J^n in the following diagram, then there exist $\tilde{g}$ and $\tilde{h}$ which make the diagram commute.

$$\begin{array}{ccccccc} (J^n,\partial\bar{I}^n) & \xrightarrow{\;i_0\;} & & (J^n,\partial\bar{I}^n)\times I & & \xleftarrow{\;i_1\;} & (J^n,\partial\bar{I}^n) \\ \Big\downarrow & & h\swarrow & \Big| & g\swarrow & & \Big\downarrow \\ & & (Y,B) & \xleftarrow{\quad f\quad} & (X,A) & & \\ & e\nearrow & \tilde{h}\nwarrow & \Big\downarrow & \tilde{g}\nwarrow & & \\ (I^{n+1},\bar{I}^n) & \xrightarrow{\;i_0\;} & & (I^{n+1},\bar{I}^n)\times I & & \xleftarrow{\;i_1\;} & (I^{n+1},\bar{I}^n) \end{array}$$

(iii) The conclusion of (ii) holds when $e = fg$ on J^n and h is the constant homotopy.

Proof. We shall leave to the reader the minor modifications of proofs needed when $n = 0$. Of course, (ii) implies (iii) trivially, and (iii) implies (i) by appropriate specializations. A direct proof that (i) implies (ii) is possible, but it is simpler to prove that (iii) implies (ii) and (i) implies (iii).

(iii) implies (ii). Assume given $h\colon e \simeq fg$ on J^n in the diagram of (ii). By application of relative HEP to the triad $(I^{n+1}; J^n, \bar{I}^n)$, there is a homotopy $j\colon (I^{n+1},\bar{I}^n)\times I \to (Y,B)$ of e which extends h. Since $j_1 = fg$ on J^n, (iii) gives a map $\tilde{g}\colon (I^{n+1},\bar{I}^n) \to (X,A)$ such that $\tilde{g} = g$ on J^n and a homotopy $k\colon j_1 \simeq f\tilde{g}$ such that k extends the constant homotopy h' at fg on J^n. Choose a homotopy $L\colon (J^n\times I, \partial\bar{I}^n\times I)\times I \to (Y,B)$ from $h+h'$ to h which is constant at fg on both $J^n\times\{0\}$ and $J^n\times\{1\}$. By application of relative HEP to the triad $(I^{n+2}; J^n\times I \cup I^{n+1}\times\partial I, I^n\times I)$, there is a homotopy $\tilde{L}\colon (I^{n+2}, \bar{I}^n\times I)\times I \to (Y,B)$ of $j+k$ which extends the union of L and the constant homotopies at e and f on $I^n\times\{0\}$ and $I^n\times\{1\}$. Let $\tilde{h} = \tilde{L}_1\colon e \simeq f\tilde{g}$. Then $\tilde{h}$ extends h, as required.

(i) implies (iii). Assume that $e = fg$ on J^n and that h is the constant homotopy in the diagram of (ii). Let $* = (0, \cdots, 0, 1) \in I^{n+1}$ and let $a = g(*)$ and $b = f(a)$. Since $(J^n, \partial\bar{I}^n, *)$ is equivalent to $(I^n, \partial I^n, J^{n-1})$, $g\colon (J^n, \partial\bar{I}^n, *) \to (X, A, a)$ may be regarded as representing an element of $\pi_n(X, A, a)$. Since e is defined on I^{n+1} with $e(\bar{I}^n) \subset B$, fg represents the trivial element of $\pi_n(Y, B, b)$. Since f is injective on π_n, there is a homotopy $j\colon (J^n, \partial\bar{I}^n, *)\times I \to (X, A, a)$ from g to the trivial map $\bar{a}$ at a. Relative HEP gives a homotopy $K\colon (I^{n+1},\bar{I}^n)\times I \to (Y,B)$ of e which extends fj. Since $fj_1 = \bar{b}$, $K_1\colon (I^{n+1}, \partial I^{n+1}, J^n) \to (Y, B, b)$ represents an element of $\pi_{n+1}(Y, B, b)$. Since f is surjective on π_{n+1}, there is a map $J_1\colon (I^{n+1}, \partial I^{n+1}, J^n) \to (X, A, a)$ and a homotopy $L\colon K_1 \simeq fJ_1$ of maps of triples. Another application of relative HEP (with unit interval reversed) gives a homotopy $J\colon (I^{n+1},\bar{I}^n)\times I \to (X,A)$ which ends at J_1 and extends j. Let $\tilde{g} = J_0$.

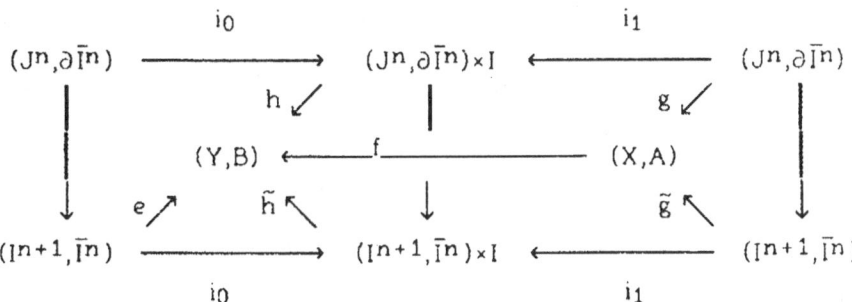

Certainly $\tilde{g}$ extends $j_0 = g$, and we have the homotopy $K+L-fJ$: $fg \simeq fg$ on $J^n \times I$. Choose any homotopy $M: (J^n \times I, \partial \bar{I}^n \times I) \times I \to (Y,B)$ from $fj+\bar{b}-fj$ to the constant homotopy at fg such that M is constant at fg on both $J^n \times \{0\}$ and $J^n \times \{1\}$. Relative HEP gives a homotopy $\tilde{M}: (I^{n+2}, \bar{I}^n \times I) \times I \to (Y,B)$ of $K+L-fJ$ which extends the union of M and the constant homotopies at e and $f\tilde{g}$ on $I^{n+1} \times \{0\}$ and $I^{n+1} \times \{1\}$. Let $\tilde{h} = \tilde{M}_1: e \simeq f\tilde{g}$; $\tilde{h}$ is constant at fg on J^n, as required.

Proof of Theorem 1.2. Replacing X by the mapping cylinder of f with its evident induced decomposition as an excisive triad, we may assume without loss of generality that f is an inclusion. Suppose given maps $g: (J^q, \partial \bar{I}^q) \to (X, X_i)$ and $e: (I^{q+1}, \bar{I}^q) \to (Y, Y_i)$ such that $fg = e$ on J^q, where $0 \le q \le n$ and $i = 1$ or $i = 2$. By the previous lemma, it suffices to construct an extension $\tilde{g}: (I^{q+1}, \bar{I}^q) \to (X, X_i)$ of g and a homotopy $\tilde{h}: e \simeq f\tilde{g}$ of maps $(I^{q+1}, \bar{I}^q) \to (Y, Y_i)$ such that $\tilde{h}$ restricts on J^q to the constant homotopy h at fg. Cubically subdivide I^{q+1} so finely that the image under e of each closed subcube lies entirely in the interior of Y_j, $j = 1$ or $j = 2$. Since f is an inclusion, the image under g of the intersection of each subcube with J^q lies entirely in the interior of X_j for the same j. Regard I^{q+1} as $I^q \times I$. The subdivision of I^{q+1} gives a cubical subdivision of I^q and a partition of I into subintervals $I_r = [v_{r-1}, v_r]$, where $0 = v_0 < v_1 < \cdots < v_s = 1$. We shall construct $\tilde{g}$ and $\tilde{h}_t$ on the spaces $K \times I_r$, where K runs through the cubical cells of I^q and $1 \le r \le s$, proceeding by induction on r and, for fixed r, by induction on the dimension of K. We shall so arrange things that

(a) $\tilde{g}(K \times I_r) \subset X_j$ and $\tilde{h}_t(K \times I_r) \subset Y_j$ if $e(K \times I_r) \subset \mathrm{Int}(Y_j)$;
(b) $\tilde{g}(K \times \{v_r\}) \subset X_1 \cap X_2$ and $\tilde{h}_t(K \times \{v_r\}) \subset Y_1 \cap Y_2$ if $e(K \times \{v_r\}) \subset Y_1 \cap Y_2$.

Since $e(\bar{I}^q) \subset X_i$, (a) and the case $r = s$ of (b) ensure that $\tilde{g}(\bar{I}^q) \subset X_i$ and $\tilde{h}_t(\bar{I}^q) \subset Y_i$. At each stage, the given maps g and $h_t = fg$ on J^q and the induction hypothesis specify maps $\tilde{g}$ and $\tilde{h}_t$ on $\partial K \times I_r \cup K \times \{v_{r-1}\}$, where ∂K is empty if K is a vertex. If either $e(K \times \{v_r\})$ is not contained in $Y_1 \cap Y_2$ or $e(K \times I_r)$ is contained in the intersection of the interiors of Y_1 and Y_2, we simply choose a representation (d,u) of $(K \times I_r, \partial K \times I_r \cup K \times \{v_{r-1}\})$ as a DR-pair and specify $\tilde{g}$ and $\tilde{h}_t$ on $K \times I_r$ by

$$\tilde{g}(x) = \tilde{g}d_1(x) \quad \text{and} \quad \tilde{h}_t(x) = \begin{cases} ed_{2t}(x) & \text{if } 0 \le t \le 1/2 \\ \tilde{h}_{2t-1}d_1(x) & \text{if } 1/2 \le t \le 1. \end{cases}$$

If $e(K \times \{v_r\})$ is contained in $Y_1 \cap Y_2$ and $e(K \times I_r)$ is contained in the interior of just one of the Y_j, the induction hypothesis gives

$$\tilde{g}: (\partial K \times I_r \cup I_r \times \{v_{r-1}\}, \partial K \times \{v_r\}) \to (X_j, X_1 \cap X_2)$$

and a homotopy $\tilde{h}: e \simeq f\tilde{g}$ of maps

$$(\partial K \times I_r \cup I_r \times \{v_{r-1}\}, \partial K \times \{v_r\}) \to (Y_j, Y_1 \cap Y_2).$$

Application of (ii) of Lemma 3.3 to the n-equivalence

$$f: (X_j, X_1 \cap X_2) \to (Y_j, Y_1 \cap Y_2)$$

gives the required extensions of $\tilde{g}$ and $\tilde{h}_t$ to $K \times I_r$.

Bibliography

1. A. Dold and R. K. Lashof. Principal quasifibrations and fibre homotopy equivalence of bundles. Ill. J. Math. 3(1959), 285-305.
2. A. Dold and R. Thom. Quasifaserungen und unendliche symmetrische Produkte. Annals of Math. 67(1958), 239-281.
3. Brayton Gray. Homotopy Theory. Academic Press. 1975.
4. K. A. Hardie. Quasifibrations and adjunction. Pacific J. Math. 35(1970), 389-397.
5. J. P. May. Classifying Spaces and Fibrations. Memoirs Amer. Math. Soc. 155. 1975.
6. J. P. May. The dual Whitehead theorems. London Math. Soc. Lecture Note Series Vol 86, 1983, 46-54.
7. D. Quillen. Higher algebraic K-theory I. Springer Lecture Notes in Mathematics Vol 341, 1973, 85-147.
8. M. Sugawara. On a condition that a space is an H-space. Math. J. Okayama Univ. 6(1957), 109-129.
9. G. W. Whitehead. Elements of Homotopy Theory. Springer. 1978.

TOPOLOGICAL AND ALGEBRAIC AUTOMORPHISMS OF 3-MANIFOLDS

Darryl McCullough*
University of Oklahoma

0. Introduction

The *mapping class group* of a manifold M is defined to be the group $\mathcal{H}(M)$ of isotopy classes of homeomorphisms of M. Many authors call this the *homeotopy group*, reserving the term mapping class group for the subgroup $\mathcal{H}_+(M)$ of $\mathcal{H}(M)$ consisting of the orientation-preserving classes, when M is orientable.

The close connection between the topology of 3-manifolds and the structural properties of their fundamental groups has been long known and heavily exploited. This connection extends to their automorphisms: Waldhausen [W] proved that if M is a closed orientable irreducible sufficiently large 3-manifold, then the natural homomorphism $\Phi: \mathcal{H}(M) \to Out(\pi_1(M))$, sending the isotopy class of a homeomorphism to the outer automorphism it induces on the fundamental group, is an isomorphism. For 3-manifolds with incompressible boundary (and satisfying the remaining hypotheses), the kernel of Φ is nontrivial only when M is an I-bundle, in which case it is of order 2 generated by the reflection in the I-fibers, and its image is as large as the fundamental group allows— it is the outer automorphisms that take the image of the fundamental group of each boundary component to a conjugate of the image of the fundamental group of a boundary component. These automorphisms are called the automorphisms that preserve the peripheral structure, and they form a subgroup $Out_\partial(\pi_1(M))$ of $Out(\pi_1(M))$. These results have been extended to nonorientable irreducible sufficiently large 3-manifolds which do not contain 2-sided projective planes [H1], [L].

When the boundary is compressible, it is necessary to replace the fundamental group by its peripheral group system; then, similar results hold [T1].

Our goal is to extend these results to larger classes of compact 3-manifolds. Some indication of the kernel of Φ for 3-manifolds which are nontrivial connected sums is given in the following (paraphrased) result [K-M, Theorem 4.3.2] (which relies on work of Hendriks [H3]):

THEOREM: *Let M be a compact 3-manifold containing no two-sided projective planes, and let $f:(M,\partial M,x_0) \to (M,\partial M,x_0)$ be a map, inducing the identity automorphism on $\pi_1(M,x_0)$, and having local degree $+1$ at x_0. Then f is properly homotopic to a composite of rotations about 2-spheres, Dehn twists about 2-discs, homeomorphisms which permute D^3 summands, and homeomorphisms which slide D^3 summands around loops in M.*

---

* Research supported in part by the National Science Foundation
This paper is in final form and no version of it will be submitted for publication elsewhere.

Note that a D^3 summand is just a neighborhood of a 2-sphere boundary component. The "rotations about 2-spheres" and "slide" homeomorphisms mentioned in the theorem are defined in §1. We remark that the concerns expressed in Remark 4.3.3 of [K-M] have been resolved by Gilbert [G], whose results verify the presentation for $Aut(G)$ given in [F-R2].

When dealing, as in the present paper, with homeomorphisms and isotopies rather than proper homotopy equivalences and proper homotopy, one must allow the additional possibilities of interchanging homotopy 3-sphere summands, sliding them around loops, or performing homeomorphisms on summands ("factor homeomorphisms") which are properly homotopic but not isotopic to the identity. Thus the description of the orientation-preserving elements of the kernel of Φ is the following:

THEOREM 1.5: *Suppose M is compact and orientable. Then the kernel of the homomorphism $\Phi: \mathcal{H}_+(M) \to Out(\pi_1(M))$ is generated by the rotations about the sum 2-spheres, factor homeomorphisms using homeomorphisms which induce the identity automorphism on the fundamental groups of the summands, interchanges of homotopy 3-sphere summands, interchanges of D^3 summands, and slides of homotopy 3-sphere and D^3 summands.*

The proof, given in §1, relies on [K-M] and results of Laudenbach [L] which state that in the absence of fake 3-cells, homotopic imbedded 2-spheres are isotopic. As an application of Theorem 1.5, we give some sufficient conditions for the kernel to be finitely generated, finitely presented, or finite.

To understand the image of Φ, we make use of the known presentation for the automorphism group of a free product of finitely many indecomposable groups, due to Fuks-Rabinovitch [F-R1], [F-R2], see also [M-M], [G]. We construct homeomorphisms realizing the generators (that preserve the peripheral structure). In order to realize all generators, it is necessary to adapt an idea used in [M2] and later exploited in [M-M]. Roughly speaking, a finite family of 3-manifolds is constructed by attaching the irreducible summands M_i to a punctured 3-cell in all possible ways (there are two ways for each summand, corresponding to the two isotopy classes of homeomorphism of S^2 used to make the identification in the connected sum). An automorphism of the fundamental group which geometrically reverses orientation of some of the summands but not others can be realized by a "uniform" homeomorphism of this family which permutes the components. This leads to our main result concerning the image of Φ:

COROLLARY 2.2: *Suppose M is compact. If each $\mathcal{H}(M_i) \to Out_\partial(\pi_1(M_i))$ has image of finite index, then $\mathcal{H}(M) \to Out_\partial(\pi_1(M))$ has image of finite index.*

1. The kernel of Φ in the orientable case.

To fix notation, let M be a compact 3-manifold with incompressible boundary. As is well-known (see for example [H2]), M can be factored as a connected sum $M = M_1 \# M_2 \# \ldots \# M_r \# (\#_{j=1}^{s} N_j)$ where each M_i is irreducible and each N_j is one of the two 2-sphere bundles over S^1. The factors are more-or-less unique up to homeomorphism (but not up to isotopy in M), although there are some complications in the nonorientable case (see [T]). We introduce an explicit notation describing a construction of M, as follows. Let Σ be the result of removing from a 3-sphere the interiors of $r+2s$ disjoint (smoothly imbedded) 3-balls $B_1, B_2, \ldots, B_r, D_1, E_1, D_2, E_2, \ldots, D_s, E_s$. For $1 \le i \le r$, let M_i' result from removing the interior of a small 3-ball B_i' centered at a chosen basepoint $b_i \in M_i$. Then M can be constructed from the disjoint union of Σ and the M_i' and s copies $S_j \times I$ of $S^2 \times I$ by identifying each ∂B_i with $\partial B_i'$, each $S_j \times \{0\}$ with ∂D_j, and each $S_j \times \{1\}$ with ∂E_j. Because of orientation considerations, it may be necessary to choose (isotopy classes of) the identifying homeomorphisms with some care in order to obtain M; since there are only two isotopy classes of homeomorphisms from S^2 to S^2, and isotopic attaching homeomorphisms yield homeomorphic results, there are only finitely many manifolds which can result from a given set of pieces.

We can now describe five kinds of homeomorphisms of M.

(1) Rotations about the sum 2-spheres

For an imbedded 2-sphere $S = S^2 \times \{0\} \subseteq S^2 \times I \subseteq M$, one can define a homeomorphism called a *rotation about* S as follows. Let $\tau\colon I \to SO(3, \mathbb{R})$ be a loop based at the identity rotation which generates $\pi_1(SO(3, \mathbb{R})) \cong \mathbb{Z}/2$. Define $r\colon M \to M$ by $r(x,t) = (\tau(t)(x), t)$ for $(x,t) \in S^2 \times I$ and $r(m) = m$ for $m \notin S^2 \times I$. Because product neighborhoods are unique up to isotopy, the mapping class of this rotation is well-defined, and since τ has order 2 in $\pi_1(SO(3, \mathbb{R}))$, the square of r is isotopic to the identity. Define $\mathcal{R}(M)$ to be the subgroup of $\mathcal{H}(M)$ generated by rotations about imbedded 2-spheres in M. Notice that if g is any homeomorphism of M, and r is a rotation about the 2-sphere S, the product grg^{-1} is a rotation about the 2-sphere $g(S)$; therefore $\mathcal{R}(M)$ is a normal subgroup of $\mathcal{H}(M)$. From §3 of [M1], we have

LEMMA 1.1: *Let r_S be a rotation about the 2-sphere $S \subseteq M$ and let F be an incompressible surface in the interior of M. Then there is a product of rotations about 2-spheres disjoint from F so that $\langle r_S \rangle = \langle r \rangle$ in $\mathcal{H}(M)$.*

Take F equal to the union of the spheres ∂B_i, ∂D_j, and ∂E_j; then Lemma 1.1 shows that $\mathcal{R}(M)$ is generated by rotations about spheres disjoint from F. Since each complementary component of F in M is a punctured irreducible 3-manifold, a rotation about any 2-sphere disjoint from F is isotopic to a product of rotations about some subset of the 2-spheres in F. Therefore the rotations about the 2-spheres in F generate $\mathcal{R}(M)$. Since these rotations commute, and each has order at most 2, we have proven the following.

PROPOSITION 1.2: *$\mathcal{R}(M)$ is generated by the rotations about the 2-spheres ∂B_i, ∂D_j, and ∂E_j. It is a normal subgroup of $\mathcal{H}(M)$ isomorphic to $(\mathbb{Z}/2)^k$ for some nonnegative integer k.*

(2) Factor homeomorphisms

There is a subgroup of index at most 2 in the basepoint-preserving full mapping class group $\mathcal{H}(M_i, b_i)$ consisting of those mapping classes containing a homeomorphism which is the identity on B_i' (see for example [M2] for details). For such a homeomorphism h_i, define a homeomorphism h of M by taking the identity on $M - M_i'$ and taking the restriction of h_i on M_i'. Such an h is called a *factor homeomorphism*. (We do not assert that the isotopy class of h is well-defined for elements of $\mathcal{H}(M_i, b_i)$; in fact, the isotopy class of h can vary by a rotation about ∂B_i.)

(3) Slide homeomorphisms

Let α be an arc properly imbedded in $M - int(M_i')$, both of whose endpoints lie in ∂B_i. Let M' be the manifold resulting from M by replacing M_i' by the 3-ball B_i. Choose an isotopy j_t of M', satisfying

(a) $j_0 = 1_{M'}$,
(b) j_1 is the identity on B_i,
(c) there is a regular neighborhood of $B_i \cup \alpha$ such that each j_t is the identity outside this neighborhood
(d) the isotopy j_t slides B_i around α.

Define a homeomorphism h of M by taking j_1 on $M - M_i'$ and the identity on M_i'. We call h a *slide homeomorphism* which *slides* M_i *around* α. A change of the choice of α in its homotopy class in $\pi_1(M', B_i)$ changes h by an isotopy (see for example §2.1 of [M-M]) and possibly by a rotation about ∂B_i, consequently a choice of homotopy class of α determines at most two isotopy classes of slide homeomorphism. A similar construction can be performed using the manifold obtained from M by replacing $S_j \times I$ by the balls D_j and E_j for some fixed j (the isotopy slides one of the 3-balls while keeping the other fixed). If the isotopy moves D_j (respectively E_j) around α, we say h *slides the left end* (respectively, *the right end*) *of the* j^{th} *handle around* α.

(4) Permutations

If M_i is homeomorphic to M_j, preserving orientation if they are oriented, then homeomophisms can be defined which are the identity on all M_k', $k \neq i, j$, and on all $S_k \times I$, and which interchange M_i' and M_j'. Similarly, for each $1 \leq i, j \leq s$ there are homeomorphisms which interchange $S_i \times I$ and $S_j \times I$. These are called *transpositions*, and products of transpositions are called *permutation homeomorphisms*.

(5) Spins

Using a homeomorphism which interchanges D_j and E_j, one constructs a homeomorphism which reverses the direction of an arc in M crossing $S_j \times I$ from ∂D_j to ∂E_j. This is called a *spin* of the jth 1-handle.

The next result is well-known.

PROPOSITION 1.3: *If M is compact and orientable, then the group of orientation-preserving mapping classes $\mathcal{H}_+(M)$ is generated by the isotopy classes of rotations about the sum 2-spheres, factor homeomorphisms, slide homeomorphisms, permutation homeomorphisms, and spins.*

PROOF: Since the inverse of each of the five kinds of homeomorphism is a homeomorphism of the same kind, it suffices to take an arbitrary orientation-preserving homeomorphism and compose it with a sequence of the specified kinds of homeomorphisms, and isotopies, until the identity homeomorphism is obtained. The proof is based on following lemma, which is essentially due to M. Scharlemann (see Appendix A of [B])

LEMMA 1.4: *Let T be a collection of pairwise disjoint essential 2-spheres in M. Then there is a composite g of slide homeomorphisms such that $g(T) \subseteq \Sigma$.*

PROOF SKETCH: An intersection curve of $\partial\Sigma \cap T$, innermost on T, bounds a disc D in T whose interior is disjoint from $\partial\Sigma$. If D lies in the closure of $M - \Sigma$, then there is an isotopy that pulls D into Σ, eliminating the intersection curve ∂D and possibly others. Eliminate all such intersections possible. Then, when D lies in Σ, there is a punctured 3-cell Σ_0 in B bounded by D and a disc in $\partial\Sigma$, and a loop α which starts in Σ_0, moves into $M - \Sigma$ and reenters Σ in the complement of Σ_0, without touching T (follow along the sphere that contains D) and then moves through Σ, intersecting T, to form a closed loop. Using this loop as a guide, apply slide homeomophisms to slide the summands attached in Σ_0 around loops that pass through $M - \Sigma$ without intersecting T; the image of D after these slides can be isotoped to decrease the number of intersection curves. Repeat to eliminate all intersections. Then, each sphere in the image of T either lies in Σ, or is parallel into Σ, and another isotopy moves all of the image of T into Σ.

To complete the proof of Proposition 1.3, we first use Lemma 1.4 on the image of the boundary of Σ. This yields a product of slide homeomorphisms whose composite with h preserves Σ. Apply permutation and spin homeomorphisms to make the homeomorphism preserve each component of $\partial\Sigma$, then apply factor homeomorphisms and rotations to make the homeomorphism the identity. This completes the proof.

THEOREM 1.5: *Suppose M is compact and orientable. Then the kernel of the homomorphism $\Phi: \mathcal{H}_+(M) \to Out(\pi_1(M))$ is generated by the rotations about the sum 2-spheres, factor homeomorphisms using homeomorphisms which induce the identity automorphism on the fundamental groups of the summands, interchanges of homotopy 3-sphere summands, interchanges of D^3 summands, and slides of homotopy 3-sphere and D^3 summands.*

PROOF: If $r = 1$ and $s = 0$, then M is irreducible and the theorem holds. If $r = 0$ and $s = 1$, then $M = S^2 \times S^1$ and the theorem follows from [G1]. So assume that $r + s \geq 2$.

Let (h) be an element of the kernel. Changing h by isotopy, we may assume that h preserves a basepoint of M and induces the identity automorphism on the fundamental group of M. Applying Theorem 4.3.2 of [K-M] (which was stated above in the introduction) and using Proposition 1.2 shows that after composition with a product of homeomorphisms of the kinds listed in the Theorem, we may assume that h is properly homotopic to the identity automorphism. It follows that for any 2-sphere S imbedded in M, the restriction of h to S is homotopic to the inclusion.

Assume for now that M has no homotopy 3-sphere summands (other than S^3). For this case we can prove the following lemma.

LEMMA 1.6: *Suppose that $T_1, T_2, \ldots, T_n$ is a collection of pairwise disjoint pairwise nonisotopic essential imbedded 2-spheres in M, and assume that M contains no fake 3-cells. Let h be a homeomorphism of M such that $h(T_i) = T_i$ for $1 \leq i \leq m$ and $h(T_i)$ is homotopic to T_i for $m + 1 \leq i \leq n$. Then h is isotopic preserving $T_1, \ldots, T_m$ to a homeomorphism h' such that $h'(T_j) = T_j$ for $1 \leq j \leq n$.*

PROOF: An easy extension of the Lemma on p. 124 of [L] to collections of disjoint 2-spheres shows that $h(T_{m+1})$ is homotopic to T_{m+1} by a homotopy that avoids $\cup_{i=1}^{m} T_i$. Since in the absence of fake 3-cells homotopic 2-spheres are isotopic (Theorem III.1.3 of [L]), $h(T_{m+1})$ is isotopic to T_{m+1} in the complement of $\cup_{i=1}^{m} T_i$. Induction completes the proof.

If $s > 0$ or $r > 2$, apply Lemma 1.6 to the spheres $\partial B_1, \ldots, \partial B_r, S_1 \times \{\frac{1}{2}\}, \ldots, S_s \times \{\frac{1}{2}\}$. If $r = 2$ and $s = 0$, then ∂B_1 and ∂B_2 are parallel; in this case apply Lemma 1.6 to ∂B_1. Since h induces the identity automorphism, it cannot interchange the sides of any of these 2-spheres (when $M = D^3 \# D^3 = S^2 \times I$, this does not follow from considering the fundamental group, but it is implied because h must preserve the boundary components of M) so h is isotopic to a map which is the identity on Σ. The listed generators can now be applied to make the homeomorphism equal to the identity.

Suppose now that M has homotopy 3-sphere summands. Each of the generators of $\mathcal{H}_+(M)$, except for slide homeomorphisms, is defined so as to preserve the union of the homotopy 3-sphere summands. Every arc α used to define a slide homeomorphism is homotopic to an arc whose interior is disjoint from all homotopy 3-sphere summands, so every slide homeomorphism is also isotopic to a homeomorphism preserving their union. Therefore we may assume at the outset that this union is preserved by the homeomorphism h. Applying interchanges of homotopy 3-sphere summands and factor homeomorphisms supported on them, we may assume that h restricts to the identity homeomorphism on each homotopy 3-sphere summand. Split M along the boundaries of these summands, and apply the previous case to the component corresponding to their complement. The 2-sphere boundary components corresponding to the deleted summands are fixed by h, so need not be moved during the process. All other homeomorphisms needed will extend to the allowable kinds of homeomorphisms; in particular, a slide of one of these D^3 summands extends using the identity on the homotopy 3-cell. The theorem follows.

REMARK 1.7: Theorem 4.3.4 of [K-M] gives some information about the manifolds which may admit orientation-reversing homeomorphisms inducing the identity automorphism.

Let $K(X)$ denote the kernel of $\Phi: \mathcal{H}_+(X) \to Out(\pi_1(X))$. Recall that a 3-manifold is said to *satisfy the Poincaré Conjecture* when every homotopy 3-cell in the manifold is diffeomorphic to the standard 3-cell. As an application of Theorem 1.5, we will prove the following.

THEOREM 1.8: *Suppose that the universal cover of M satisfies the Poincaré conjecture. If each $K(M_i)$ is finitely generated (respectively, finitely presented) then $K(M)$ is finitely generated (respectively, finitely presented). If moreover M has no D^3 summands, then if each $K(M_i)$ is finite, $K(M)$ is also finite.*

The hypothesis that the universal cover of M satisfies the Poincaré Conjecture may be stronger than the assumption that M satisfies the Poincaré Conjecture, since it excludes the possibility that M has irreducible summands covered by fake 3-spheres.

We remark that when M_i has nonempty incompressible boundary, $K(M_i)$ is trivial [W], but when M_i has compressible boundary, $K(M_i)$ is typically not finitely generated [M1]. When M_i is closed, $K(M_i)$ is trivial in all known cases.

The proof of Theorem 1.8 will use the following two lemmas.

LEMMA 1.9: *Let $\widehat{M}$ be the manifold that results from filling in a 2-sphere boundary component S of a compact 3-manifold M with a 3-cell. Then there is an exact sequence*

$$1 \longrightarrow L \longrightarrow K(M,S) \longrightarrow K(\widehat{M}) \longrightarrow 1$$

in which L is finitely presented. Consquently, $K(M)$ is finitely generated (respectively, finitely presented) if and only if $K(\widehat{M})$ is finitely generated (finitely presented).

PROOF: Let $KHomeo(X)$ denote the subspace of $Homeo(X)$ consisting of orientation-preserving homeomorphisms that induce the identity outer automorphism, so that $K(X)$ is the group of path components of $KHomeo(X)$. Let E be the 3-ball filled in to give $\widehat{M}$, and let e_0 be its center. From the restriction fibration $KHomeo(\widehat{M}) \to int(\widehat{M})$, there is an exact sequence

$$\pi_1(\widehat{M}) \longrightarrow \mathcal{K}(\widehat{M}, x_0) \longrightarrow \mathcal{K}(\widehat{M}) \longrightarrow 1.$$

There are homotopy equivalences

$$KHomeo(\widehat{M}, x_0) \simeq KHomeo(\widehat{M}, E) \simeq KHomeo(M, \partial E),$$

and hence isomorphisms $\mathcal{K}(\widehat{M}, e_0) \cong \mathcal{K}(\widehat{M}, E) \cong \mathcal{K}(M, \partial E)$. The kernel of $\pi_1(\widehat{M}) \longrightarrow \mathcal{K}(\widehat{M}, x_0)$ is the subgroup of traces of circular isotopies, which are central in $\pi_1(\widehat{M})$. By a recent result of G. Mess [M], the center must be finitely generated. Therefore the kernel is a finitely generated abelian group, and the quotient of $\pi_1(\widehat{M})$ by this subgroup is the group L satisfying the statement of the lemma. Since M is compact, it has finitely many 2-sphere boundary components, so the last sentence of the lemma follows.

In the next lemma, let $\mathcal{K}_1(M_i, b_i)$ denote the subgroup of finite index in $\mathcal{K}(M_i, b_i)$ consisting of the mapping classes that preserve B_i' and whose restriction to B_i' is orientation-preserving. Let $\overline{\mathcal{K}}(X)$ denote the quotient of $\mathcal{K}(X)$ by $\mathcal{R}(X)$. There is a homomorphism $j : \prod_{i=1}^{n} \overline{\mathcal{K}}_1(M_i, b_i) \to \overline{\mathcal{K}}(M)$ induced by deforming a homeomorphism to the identity on B_i', restricting to M_i', and extending to M using the identity map. This is well defined since two choices of map representing a class in $\overline{\mathcal{K}}_1(M_i, b_i)$ and equal to the identity on B_i' must differ by a rotation about $\partial B_i'$.

LEMMA 1.10: *Suppose that $r + s \geq 2$, that M has no D^3 summands, and that the universal cover of M satisfies the Poincaré Conjecture. Then the homomorphism j is an isomorphism.*

PROOF: The hypothesis that the universal cover of M satisfies the Poincaré Conjecture implies the same for M, therefore Theorem 1.5 shows that j is surjective.

An element of the kernel of j can be represented by a pieced-together homeomorphism $h = h_1 \cup h_2 \cup \cdots \cup h_m$ which is isotopic to a rotation about a 2-sphere in M. By Proposition 1.2, this rotation is isotopic to a product of rotations about spheres disjoint from the ∂B_i, so after changing the h_j by rotations we may assume that h is isotopic to the identity. To prove the lemma, we must show that h is isotopic to the identity by an isotopy that preserves Σ.

Choose a 3-ball $\Sigma_0 \subset \Sigma$. Since $h(\Sigma) = \Sigma$, the trace of the isotopy from h to the identity is trivial at a basepoint in Σ_0 (under the hypotheses of lemma 1.10. the fundamental group $\pi_1(M)$ is a nontrivial free product, and any inner automorphism which preserves the free factors must be the identity; since $h(\Sigma) = \Sigma$, its induced automorphism preserves the free factors, and since h is isotopic to the identity, it induces conjugation by the trace of the isotopy) Therefore h is isotopic preserving Σ_0 to the identity, and isotopic (rel Σ_0) to the identity or to a rotation about $\partial\Sigma_0$. In the latter case, we may change h by rotations about all boundary components of Σ (the product of these is isotopic preserving Σ and preserving Σ_0 to the rotation about $\partial\Sigma_0$) to obtain h isotopic to the identity (rel Σ_0). Changing h by further isotopy near the boundary of M, we may obtain that it is isotopic to the identity (rel $\Sigma_0 \cup \partial M$).

We are now in a position to apply the following result, Theorem 2 of [H-M]:

THEOREM: *Suppose none of the summands M_i has universal cover a homotopy 3-sphere nondiffeomorphic to S^3. Then the inclusion map*

$$Diff(M - int(\Sigma_0), \Sigma - int(\Sigma_0) \text{ rel } \partial(M - int(\Sigma_0))) \longrightarrow$$
$$Diff(M - int(\Sigma_0) \text{ rel } \partial(M - int(\Sigma_0)))$$

is a homotopy equivalence.

Applying this to the restriction of h to $M - int(\Sigma_0)$ shows that h is isotopic to the identity preserving Σ, and hence preserving the M_i'. This completes the proof of Lemma 1.10.

The assertions in Theorem 1.8 for finite generation and finite presentation follow easily from Proposition 1.2 and Lemmas 1.9 and 1.10. In case there are no D^3 summands, Lemma 1.9 is not needed and the last assertion in Theorem 1.8 follows as well.

2. The image of Φ.

Let $Z = (\mathbb{Z}/2)^r \times (\mathbb{Z}/2)^s$. Elements of Z will be written as $(\sigma, \tau) = (\sigma_1, \sigma_2, \ldots, \sigma_r, \tau_1, \tau_2, \ldots, \tau_s)$, where each σ_i and τ_j is 1 or -1. If $\tau = (\tau_1, \tau_2, \ldots, \tau_r)$, then the notation $(1, \tau)$ means the element $((1, 1, \ldots, 1), \tau)$ and similarly for $(\sigma, 1)$; the identity element can be written as $(1, 1)$.

We will construct a family of 3-manifolds indexed by the elements of Z. These are all the manifolds that can be constructed from the summands (M_i, b_i) and s 2-sphere bundles over S^1. For each i, fix an orientation for the ball $B_i' \subseteq M_i$. Fix once and for all an orientation-reversing reflection $\rho: S^2 \to S^2$, and fix identifications of ∂B_i, ∂D_j, ∂E_j, and $\partial B_i'$ with S^2, compatible with the orientations of Σ and B_i'. Let $M(\sigma, \tau)$ be constructed from Σ and the M_i as in §1, in the following way. Attach each M_i to Σ so that the identification of $\partial B_i'$ with ∂B_i is ρ if $\sigma_i = 1$ and is the identity if $\sigma_i = -1$. Attach each $S_j \times \{0\}$ to ∂D_j by the identity, and attach each $S_j \times \{1\}$ to ∂E_j by ρ if $\tau_j = 1$ and the identity if $\tau_j = -1$. The resulting 3-manifold $M(\sigma, \tau)$ will be orientable if and only if all M_i are orientable and all $\tau_j = 1$. Notice that each $M(\sigma, \tau)$ contains (an identical copy of) Σ.

Let W be the cell-complex that results from $M(1,1)$ by collapsing Σ to a point. The quotient map $M(1,1) \to W$ induces an isomorphism on fundamental groups (with respect to a basepoint in the interior of Σ). There is a canonical identification of W with the corresponding quotients of the other $M(\sigma, \tau)$, hence there are canonical isomorphisms of the fundamental groups of the $M(\sigma, \tau)$ with the fundamental group of $M(1,1)$. For a homeomorphism h from one of these manifolds to another, we will speak of the *automorphism* that h induces on the fundamental group; this means the automorphism on $\pi_1(M(1,1))$ induced by h after making the canonical identifications of the fundmental groups of the domain and range with $\pi_1(M(1,1))$.

The disjoint union of the 2^{r+s} manifolds $M(\sigma, \tau)$ will be denoted by $\mathcal{M}$, and $\pi_1(M(1,1))$, identified with the fundamental group of each component as described in the previous paragraph, will be denoted by π.

A little more notation will be convenient. In either $(\mathbb{Z}/2)^r$ or $(\mathbb{Z}/2)^s$, denote by ϵ_i the element which has 1's in all places except the ith, and has ϵ in the ith place (where ϵ is 1 or -1).

Let $\overline{\mathcal{H}}(\mathcal{M})$ denote the quotient of the ordinary mapping class group of the manifold $\mathcal{M}$ by the normal subgroup generated by rotations about 2-spheres. We are going to define a subgroup of $\overline{\mathcal{H}}(\mathcal{M})$ called the *uniform mapping class group* and denoted by $\mathcal{U}(\mathcal{M})$. The idea of a uniform homeomorphism is that insofar as possible it "does the same thing" to each component of $\mathcal{M}$. We will now make this precise; the main complication is that the components of $\mathcal{M}$ must be permuted in a complicated way to achieve these homeomorphisms.

Consider a homeomorphism f_i of M_i that preserves B_i'. Define a homeomorphism F_i of $\mathcal{M}$ as follows. Deform f_i by isotopy preserving B_i' so that its restriction to $\partial B_i'$ is the identity, if the restriction of f_i to B_i' is orientation-preserving, or is ρ, if the restriction to B_i' is orientation-reversing. The homeomorphism F_i carries each $M(\sigma,\tau)$ to $M((\deg(f_i))_i\cdot\sigma,\tau)$ using the restriction of f_i to each M_i', while carrying the closure of the complement of M_i' in $M(\sigma,\tau)$ to the closure of the complement of M_i' in $M((\deg(f_i))_i\cdot(\sigma,\tau)$ using the "identity." A different choice of isotopy making f_i equal to the identity or ρ on $\partial B_i'$ may change the resulting F_i by rotations about the ∂B_i's, but the class (F_i) is well-defined in the quotient $\overline{\mathcal{H}}(\mathcal{M})$. The homeomorphism F_i is called a *uniform factor homeomorphism* and its class (F_i) lies in $\mathcal{U}(\mathcal{M})$.

If $w_{ij}:(M_i,b_i)\to(M_j,b_j)$ is a homeomorphism, let $\deg(w_{ij})$ denote the degree of w_{ij} with respect to the local orientations at b_i and b_j. A similar construction leads to a *uniform interchange* $\langle W_{ij}\rangle\in\mathcal{U}(\mathcal{M})$ such that

$$W_{ij}(M(\sigma_1,\sigma_2,\ldots,\sigma_i,\ldots,\sigma_j,\ldots,\sigma_r,\tau)) = M(\sigma_1,\sigma_2,\ldots,\deg(w_{ij})\sigma_j,\ldots,\deg(w_{ij})\sigma_i,\ldots,\sigma_r,\tau)$$

A similar construction using the "identity" homeomorphism from $S_i\times I$ to $S_j\times I$ yields a uniform interchange of 1-handles denoted by $W_{r+i,r+j}$.

A *uniform spin* T_j preserves each $M(\sigma,\tau)$; on $S_j\times I$, T_j sends (x,t) to $(\rho(x),-t)$ if $\tau_j=1$ and to $(x,-t)$ if $\tau_j=-1$.

Uniform slides are defined using an arc α in $M(1,1)$ such that $\alpha\subseteq\Sigma\cup M_k$ or $\Sigma\cup S_k\times I$ (and the endpoints of α lie in ∂B_i, in ∂D_i, or in ∂E_i). In each $M(\sigma,\tau)$, there is a corresponding arc whose intersections with Σ and M_k and $S_k\times I$ are properly homotopic to those of α. The *degree* of α *in* $M(\sigma,\tau)$ is defined to be 1 if traveling around α in $(M(\sigma,\tau)-M_i')\cup B_i$ preserves the local orientation on B_i and to be -1 if not. The uniform slide defined using α carries $M(\sigma,\tau)$ to $M((\deg_{M(\sigma,\tau)}(\alpha))_i\cdot(\sigma,\tau))$. We denote the uniform slide by $M_{ik}(\alpha)$ if $1\le i\le r$, by $L_{ik}(\alpha)$ if it slides the left end of $S_i\times I$, and by $R_{ik}(\alpha)$ if it slides the right end of $S_i\times I$.

The automorphisms of π induced by the uniform homeomorphisms are the following. Write $\pi=G_1*G_2*\cdots*G_r*(*_{j=1}^{s}\mathbb{Z})$, where $G_i=\pi_1(M_i,b_i)\subseteq\pi_1(M(1,1))$ and the jth infinite cyclic factor is generated by the homotopy class a_j of a loop in $\Sigma\cup S_j\times I$ that travels once across $S_j\times I$. Let g_i stand for an arbitrary element of G_i. By considering the constructions of the uniform homeomorphisms, we can read of their induced automorphisms as follows (we will list only their effects on the generators that they may change; all other generators are fixed by the automorphism).

(1) $(F_i)_\# = \phi_i$ where $\phi_i(g_i) = (f_i)_\#(g_i)$.
(2) $(W_{ij})_\# = \omega_{ij}$, where if $1 \le i, j \le r$, $\omega_{ij}(g_i) = (w_{ij})_\#(g_i) \in G_j$ and $\omega_{ij}(g_j) = (w_{ij})_\#^{-1}(g_j) \in G_i$; similarly for interchanges of handles.
(3) $(T_j)_\#(a_j) = t_j(a_j) = a_j^{-1}$.
(4) If a denotes the element of G_k or the kth infinite cyclic factor represented by α, then $(M_{ik}(\alpha))_\# = \mu_{ik}(a)$, where $\mu_{ik}(a)(g_i) = a^{-1}g_i a$. Similarly, $(L_{ik}(\alpha))_\# = \lambda_{ik}(a)$, where $\lambda_{ik}(a)(a_i) = a^{-1}a_i$, and $(R_{ik}(\alpha))_\# = \rho_{ik}(a)$, where $\rho_{ik}(a)(a_i) = a_i a$.

By [F-R2] (see §5 of [M-M], or [G]), the automorphisms listed here, as i, j, k, and a range over all possibilities, generate the automorphism group $Aut(\pi)$ and hence also $Out(\pi)$. In fact, it is enough to take the a from a list of generators for G_k (and the a_k), and the ϕ_i from a list of generators of $Aut(G_i)$, so one can obtain a finite generating set when the $Aut(G_i)$ are finitely generated. Define $Out_\partial(\pi)$ to be the subgroup generated by ϕ_i where ϕ_i preserves the peripheral structure of $\pi_1(M_i)$, the ω_{ij} where $w_{ij} : \pi_1(M_i) \to \pi_1(M_j)$ preserves peripheral structure, all $\omega_{k\ell}$ with $r+1 \le k, \ell \le r+s$, and all t_k, $\mu_{ik}(a)$, $\lambda_{ik}(a)$, and $\rho_{ik}(a)$.

There is one last uniform homeomorphism to be defined. The *uniform reflection* R is defined by using a reflection of Σ that restricts to the standard reflection ρ on each boundary component of Σ, and taking R equal to the identity on each M_i', and $\rho \times 1_I$ on each $S_j \times I$. It sends each $M(\sigma, \tau)$ to $M(-\sigma, \tau)$, and induces the identity automorphism on π.

After all these preliminaries, we are ready for the main result of this section. The uniform mapping class group acts on the finite set of components of $\mathcal{M}$, so for each $(\sigma, \tau) \in Z$, the stabilizer of $M(\sigma, \tau)$, denoted by $\text{stab}_{\mathcal{U}(\mathcal{M})}(M(\sigma, \tau))$, has finite index in $\mathcal{U}(\mathcal{M})$.

THEOREM 2.1: *For each $(\sigma, \tau) \in Z$, the homomorphism $\text{stab}_{\mathcal{U}(\mathcal{M})}(M(\sigma, \tau)) \to \mathcal{H}(M(\sigma, \tau))$ induced by restriction is surjective.*

PROOF: Let h be a homeomorphism of $M(\sigma, \tau)$. We will find a product k of uniform homeomorphisms so that kh is the identity on $M(\sigma, \tau)$; since the inverse of a uniform homeomorphism is a uniform homeomorphism, this will prove Theorem 2.1.

Let F denote the union of the ∂B_i, the ∂D_j, and the ∂E_j in $M(\sigma, \tau)$. Using uniform slide homeomorphisms to adapt the proof of Lemma 1.4 to the current context, we find a product k_1 of uniform slide homeomorphisms so that $k_1 h(F)$ is the copy of F in some $M(\sigma_1, \tau_1)$. By applying some product k_2 of uniform interchanges and uniform spins, we can obtain that $k_2 k_1 h$ takes each component of F to the corresponding component in some $M(\sigma_2, \tau_2)$, and hence carries the copy of M_i' in $M(\sigma, \tau)$ homeomorphically to the copy of M_i' in $M(\sigma_2, \tau_2)$. Applying the corresponding product k_3 of factor homeomorphisms, we make $k_3 k_2 k_1 h$ equal to the identity on each M_i and $S_j \times I$. It restriction to Σ is isotopic either to the identity or to the standard reflection; in the latter case, apply the uniform reflection R. The composite is now isotopic to the identity on each M_i, each $S_j \times I$, and on Σ. It follows that the image of $M(\sigma, \tau)$ under this homeomorphism is $M(\sigma, \tau)$, and that the homeomorphism is isotopic to a product of rotations about the 2-spheres in F. Therefore the desired uniform homeomorphism has been produced, proving Theorem 2.1.

COROLLARY 2.2: *Suppose M is compact. If each $\mathcal{H}(M_i) \to Out_\partial(\pi_1(M_i))$ has image of finite index, then $\mathcal{H}(M) \to Out_\partial(\pi_1(M))$ has image of finite index.*

PROOF: Regard M as one of the components $M(\sigma, \tau)$ of $\mathcal{M}$. We will need more information about the presentation of $Aut(\pi)$ given by Fuks-Rabinovitch. A convenient list of the relations, consistent with the notation we are using here, is given in [M-M]. Let N be the subgroup of $Out_\partial(\pi)$ generated by all $\mu_{ij}(a)$, $\lambda_{ij}(a)$, $\rho_{ij}(a)$, all ω_{ij} with $r+1 \leq i, j \leq r+s$, all t_j, and all ϕ_i for which ϕ_i is an inner automorphism of $\pi_1(M_i)$. It is readily verified using the relations that N is a normal subgroup of $Out_\partial(\pi)$ whose quotient $Out_\partial(\pi)/N$ is isomorphic to a semidirect product $(\prod_{i=1}^{r} Out_\partial(\pi_1(M_i))) \circ \Omega$, where Ω is a finite permutation group generated by one interchange for each pair $\pi_1(M_i) \cong \pi_1(M_j)$ which are isomorphic by an isomorphism preserving peripheral structure.

Observe that each generator of N is induced by a uniform homeomorphism. Therefore N is in the image of $\mathcal{U}(\mathcal{M}) \to Out_\partial(\pi)$. Moreover, under the hypothesis of Corollary 2.2, the image of the composite of this homomorphism with the quotient homomorphism has finite index in the quotient. Using Theorem 2.1, the corollary follows (since the restriction of $\mathcal{U}(\mathcal{M}) \to Out_\partial(\pi)$ to the stabilizer factors through $\mathcal{H}(M(\sigma, \tau))$).

By examining the size of the orbits of $M(\sigma, \tau)$ under the action of $\mathcal{U}(\mathcal{M})$, one may obtain information about the index of the image of Φ in $Out_\partial(\pi_1(M))$. The general statement is somewhat complicated (see [M3] for a similar result).

3. The kernel of Φ in the nonorientable case.

Using the family $\mathcal{M}$ constructed in section 2, the main theorem of section 1 may be extended to a larger class of 3-manifolds.

THEOREM 3.1: *Suppose M is compact, nonorientable, and contains no two-sided projective planes. Let h be a basepoint-preserving homeomorphism of M which induces the identity automorphism on $\pi_1(M)$, and has local degree $+1$ at the basepoint. Then h is isotopic to a product of rotations about the sum 2-spheres, factor homeomorphisms using homeomorphisms which induce the identity automorphism on the fundamental groups of the summands, interchanges of homotopy 3-sphere summands, interchanges of D^3 summands, and slides of homotopy 3-sphere and D^3 summands.*

PROOF SKETCH: One adapts the proof of Theorem 1.5. Since M contains no two-sided projective planes, Theorem 4.3.2 of [K-M] still applies, as do the results needed from [L]. Proposition 1.3 no longer holds; it was used only to ensure that the union of the homotopy 3-sphere summands is preserved up to isotopy by any homeomorphism. Since any *uniform* homeomorphism of $\mathcal{M}$ is isotopic to one that preserves all M_i' that are homotopy 3-spheres, Theorem 2.1 implies this in the nonorientable case. Consequently, the rest of the proof of Theorem 1.5 goes through without difficulty.

The theorem from [H-M] needed in the proof of Lemma 1.10 relies on the main result of [H-L], whose proof makes heavy use of orientability. It seems likely that the results in both of these papers could be extended at least to the case of nonorientable 3-manifolds containing no two-sided projective planes; if so then our approach would yield a version of Theorem 1.8 for the manifolds as in Theorem 3.1.

References

[B] F. Bonahon, Cobordism of automorphisms of surfaces, *Ann. École Norm. Sup.* (4) 16 (1983), 237-270.

[F-R1] D. I. Fuks-Rabinovitch, On the automorphism groups of free products I, *Math. Sb.* 8 (50) (1940), 265-276.

[F-R2] D. I. Fuks-Rabinovitch, On the automorphism groups of free products II, *Math. Sb.* 9 (51) (1941), 183-220.

[G] N. Gilbert, Presentations of the automorphism group of a free product, *Proc. London Math. Soc.* (3) 54 (1987), 115-140.

[Gl] H. Gluck, The imbedding of two-spheres in four-spheres, *Bull. Amer. Math. Soc.* 67(1961), 586-589.

[H1] W. Heil, On $\mathbb{P}^2$-irreducible 3-manifolds, *Bull. Amer. Math. Soc.* 75 (1969), 772-775.

[H2] J. Hempel, *3-manifolds*, Annals of Math. Study No. 86, Princeton University Press, Princeton (1976).

[H3] H. Hendriks, Applications de la théorie d'obstruction en dimension 3, *Bull. Soc. Math. France Mémoire* 53 (1977), 81-196.

[H-L] H. Hendriks and F. Laudenbach, Difféomorphismes des sommes connexes en dimension trois, *Topology* 23 (1984), 423-443.

[H-M] H. Hendriks and D. McCullough, On the diffeomorphism group of a reducible 3-manifold, *Topology Appl.* 26 (1987), 25-31.

[K-M] J. Kalliongis and D. McCullough, π_1-injective mappings of compact 3-manifolds, *Proc. London Math. Soc.* (3), 52 (1986), 173-192.

[L] F. Laudenbach, Topologie de la dimension trois. Homotopie et isotopie, *Astérisque* 12 (1974), 1-152.

[M1] D. McCullough, Twist groups of compact 3-manifolds, *Topology* 24 (1985), 461-474.

[M2] D. McCullough, The group of homotopy equivalences for a connected sum of closed aspherical manifolds, *Indiana University Math. J.* 30 (1981), 249-260.

[M3] D. McCullough, Errata: The group of homotopy equivalences for a connected sum of closed aspherical manifolds, *Indiana University Math. J.* 34 (1985), 201-203.

[M-M] D. McCullough and A. Miller, Homeomorphisms of 3-manifolds with compressible boundary, *Mem. Amer. Math. Soc.* 344 (1986), 1-100.

[M] G. Mess, Centers of 3-manifold groups and groups which are coarse quasiisometric to planes, preprint.

[T] B. Trace, Two comments concerning the uniqueness of prime factorizations for 3-manifolds, *Bull. London Math. Soc.* 19 (1987), 75-77.

[T1] T. Tucker, Boundary-reducible 3-manifolds and Waldhausen's theorem, *Mich. Math. J.* 20 (1973), 321-327.

[W] F. Waldhausen, On irreducible 3-manifolds which are sufficiently large, *Annals of Math.* 87 (1968), 56-88.

PROJECTING HOMEOMORPHISMS FROM COVERING SPACES

Andy Miller
Department of Mathematics, University of Oklahoma
Norman, Oklahoma 73019

If X and Y are locally compact Hausdorff spaces then their homeomorphism groups $Homeo(X)$ and $Homeo(Y)$ are Hausdorff topological groups when endowed with the compact–open topology. Let $p : X \to Y$ be a continuous surjection. The group of *p-projectable homeomorphisms* of X is

$$Homeo_p(X) = \left\{ \tilde{f} \in Homeo(X) \mid \exists f \in Homeo(Y) \text{ with } fp = p\tilde{f} \right\} .$$

Thus $\tilde{f}$ is an element of $Homeo_p(X)$ if and only if $p\tilde{f}p^{-1}$ defines a homeomorphism on Y. Clearly the correspondence $\tilde{f} \mapsto p\tilde{f}p^{-1}$ defines a group homomorphism $\rho[p] : Homeo_p(X) \to Homeo(Y)$, which we call the *projecting homomorphism.* The image of $\rho[p]$ will be denoted by $Homeo^p(Y)$, and we refer to this as the group of *p-liftable homeomorphisms* of Y. In summary, one has

$$Homeo^p(Y) = \left\{ f \in Homeo(Y) \mid \exists \tilde{f} \in Homeo(X) \text{ with } p\tilde{f} = fp \right\} ,$$

and an epimorphism

$$\rho[p] : Homeo_p(X) \longrightarrow Homeo^p(Y) .$$

Of course $Homeo_p(X)$ and $Homeo^p(Y)$ are themselves topological groups when given the subspace topology. In this paper we pose the question of whether the projecting homomorphism $\rho[p]$ is a topological group quotient map. The motivation behind this question stems from the fact that topological group quotient maps preserve much structural information, so that strong bonds between the topological groups $Homeo(X)$ and $Homeo(Y)$ can be deduced. As a more immediate goal we are interested in understanding the local structure of $Homeo_p(X)$.

In §1 we show that $\rho[p]$ is a continuous map under very mild hypotheses on p. However the question of whether it is an open map (which is equivalent to being a quotient map) is much more delicate. The remainder of the paper is devoted to a study of the special case where p is a covering map. Even in this setting our

---

I am greatly endebted to my Ph.D. advisor Jeff Tollefson of the University of Connecticut for his suggestions and encouragement of this work.

This paper is in final form and no version of it will be submitted for publication elsewhere.

question can have a negative resolution: in §5 we give an example of a covering map p between noncompact 2–manifolds where $\rho[p]$ fails to be open. On the other hand, the main theorem (theorem 3.9) asserts that $\rho[p]$ is an open map for a wide range of coverings including regular coverings and coverings over a compact base. Under different hypotheses the restriction of $\rho[p]$ to the path component of id_X in $Homeo_p(X)$ is an open map (theorem 3.10). At any rate the projecting homomorphism is always a fibration with unique path lifting (corollary 2.4), and if it is an open map then it must be a covering map. In §4 we discuss properties of $\rho[p]$ when it is a covering map and we describe some specific examples in detail. We note that some of our results overlap with theorems about lifting Lie group actions to covering spaces given in [B: §I.9]. In fact we may recover these as special corollaries of the present work.

Closely related to the above is an analagous question concerning the projecting homomorphism between spaces of self–homotopy equivalences, we consider this question in §6. While a number of authors have studied spaces of p–projectable self–equivalences, most of these studies have focused on homotopy or categorical properties of the spaces rather than on the structure of the projecting homomorphism. For more information see [J], [BHMP], [R] and the references cited therein. We consider locally compact Hausdorff spaces X and Y and their self–homotopy equivalence monoids $E(X)$ and $E(Y)$, which are Hausdorff topological semigroups when endowed with the compact–open topology. Given a continuous surjection $p: X \to Y$ we define the submonoid of *p-projectable homotopy equivalences* of X to be

$$E_p(X) = \left\{ \tilde{f} \in E(X) \mid \exists f \in E(Y) \text{ with } fp = p\tilde{f} \right\} .$$

As before there is also a monoid of *p–liftable homotopy equivalences* of Y defined by

$$E^p(Y) = \left\{ f \in E(Y) \mid \exists \tilde{f} \in E(X) \text{ with } p\tilde{f} = fp \right\} ,$$

and a monoid epimorphism

$$\rho^E[p] : E_p(X) \longrightarrow E^p(Y) .$$

In §6 we show that $\rho^E[p]$ is a topological semigroup quotient map and a covering map when p is a regular covering map.

Throughout the paper we shall adapt the following conventions. We denote a basepoint of X by x_0 and a basepoint of Y by y_0, and we will always implicitly assume that the continuous surjection p is a map of pairs $(X, x_0) \to (Y, y_0)$. When fundamental groups of X and Y are considered they will usually be based at x_0 and y_0 respectively, and so we abbreviate $\pi_1(X) = \pi_1(X, x_0)$ and $\pi_1(Y) = \pi_1(Y, y_0)$. As a basic reference for the compact–open topology on function spaces we refer to [D]. For instance one result we will use frequently without comment is that if X is locally compact then a path in $E(X)$ is the same as a homotopy. Similarly, a path in $Homeo(X)$ is an isotopy between homeomorphisms. For basic definitions and facts concerning covering maps the reader might consult [Sp] or [M].

1. Continuity of the Projecting Homomorphism

In this section we address the continuity of the projecting homomorphism for a continuous surjection

(1.1) $p : X \to Y$ where X and Y are locally compact Hausdorff spaces.

We say that p is *quasiproper* if and only if for each compact subset C of Y there is a compact set $\tilde{C}$ in X so that $p(\tilde{C}) = C$. If $C \subset Y$ is compact and $O \subset Y$ is open then

$$U(C, O) = \{f \in Homeo(Y) \mid f(C) \subset O\} \ .$$

The collection of all such sets $U(C,O)$ then forms a subbase for the compact-open topology on $Homeo(Y)$. Therefore the sets

$$U^p(C,O) = U(C,O) \bigcap Homeo^p(Y)$$

form a subbase for the topology on $Homeo^p(Y)$. Likewise the sets

$$U_p(\tilde{C}, \tilde{O}) = U(\tilde{C}, \tilde{O}) \bigcap Homeo_p(X)$$

where $\tilde{C} \subset X$ is compact and $\tilde{O} \subset X$ is open forms a subbase for the topology on $Homeo_p(X)$.

Lemma 1.2 *If* p *is quasiproper then* $\rho[p]$ *is continuous.*

<u>proof</u>: Let $\rho = \rho[p]$. Consider a subbasic open set $U^p(C,O)$ in $Homeo^p(Y)$ where $C \subset Y$ is compact and $O \subset Y$ is open. Choose a compact set $\tilde{C} \subset X$ with $p(\tilde{C}) = C$. Then

$$\begin{aligned} \rho^{-1}(U^p(C,O)) &= \{\tilde{f} \in Homeo_p(X) \mid \rho(\tilde{f})(C) \subset O\} \\ &= \{\tilde{f} \in Homeo_p(X) \mid \tilde{f} p^{-1}(C) \subset p^{-1}(O)\} \\ &= \{\tilde{f} \in Homeo_p(X) \mid \tilde{f}(\tilde{C}) \subset p^{-1}(O)\} \\ &= U_p(\tilde{C}, p^{-1}(O)) . \end{aligned}$$

(For the verification of the second to last equality observe that if $\tilde{f}(\tilde{C}) \subset p^{-1}(O)$ then $\tilde{f}(p^{-1}(C)) = p^{-1}\rho(\tilde{f})(C) = p^{-1}\rho(\tilde{f})p(\tilde{C}) = p^{-1}p\tilde{f}(\tilde{C}) \subset p^{-1}(O)$.) Thus we have shown that the inverse image of each subbasic open set in $Homeo^p(Y)$ is a subbasic open set in $Homeo_p(X)$ and the lemma follows. □

Being quasiproper is a relatively weak hypothesis on p. For example if p is either a proper map, or a simplicial map between simplicial complexes, or a topological group homomorphism then it is quasiproper. We also note that locally trivial maps between locally compact Hausdorff spaces are quasiproper. To show this suppose that p is such a map, and let $C \subset Y$ be compact. Then C can be covered by finitely many open sets O_i so that $cl(O_i)$ is compact and p is trivial over $cl(O_i)$. For each i there can then be found a compact $\tilde{C}_i \subset X$ for which $p(\tilde{C}_i) = cl(O_i) \cap C$, and the union of these sets is a compact subset of X whose image is C. As a special case of this we note that $\rho[p]$ is continuous if p is a covering map.

Given a quasiproper map p satisfying (1.1) then $\rho[p]$ is a continuous homomorphism between topological groups, its kernel is

$$A_p(X) = \{f \in Homeo(X) \mid pf = p\} \subset Homeo_p(X)$$

which is a closed subgroup. The fundamental question we have posed is: under what conditions can $\rho[p]$ be identified with the topological group quotient map $Homeo_p(X) \to Homeo_p(X)/A_p(X)$? Since the latter is an open map, this is equivalent to asking when $\rho[p]$ is an open map. Starting in the next section we will exclusively study this problem in the case where p is a covering map. One important comment which pertains to this special case is that if p is a (Hurewicz) fibration then so is $\rho^E[p]$. This follows as a standard application of the exponential law for function spaces with the compact–open topology.

Before ending this section we mention one variation in which the question we have posed has an affirmative answer. Suppose that $\tilde{G}$ is a compact subgroup of $Homeo_p(X)$ and that $G = \rho[p](\tilde{G})$. Then the restriction of $\rho[p]$ gives a closed map $\tilde{G} \to G$ and so it is a topological group quotient map. This result also holds when $\tilde{G}$ and G are locally compact and σ–compact (using a well-known topological group consequence of the Baire category theorem). To put this in perspective, we note that even the nicest spaces do not have locally compact homeomorphism groups, so the hypotheses we have mentioned here are far too restrictive. What we really desire is to find hypotheses on the surjection p which guarantee the openness of $\rho[p]$.

2. Preliminary Results

In this section we develope some key lemmas concerning the projecting homomorphism for covering maps. In particular we show that such projecting homomorphisms are always fibrations with unique path lifting. We consider covering maps

(2.1) $p : X \to Y$ between connected, locally path connected spaces which satisfy (1.1).

A subset of X is said to be *elementary* provided that the restriction of p to that subset is a homeomorphism onto its image. We also refer to the image set as being *elementary in* Y.

LEMMA 2.2 *If p is a covering map satisfying (2.1) then $A_p(X)$ is a closed discrete subgroup of $Homeo_p(X)$. Furthermore the following three statements are equivalent:*

(i) *$\rho[p]$ is a topological group quotient map,*

(ii) $\rho[p]$ *is a (regular) covering map,*
(iii) $\rho[p]$ *is an open map.*

proof: The group $A_p(X)$, which is the group of covering transformations for p, acts freely on X. Let $\tau \in A_p(X)$ be a covering transformation and choose an elementary open neighborhood $\tilde{O}$ of $\tau(x_0)$. Then $A_p(X) \cap U_p(x_0, \tilde{O}) = \{\tau\}$, which shows that $A_p(X)$ is discrete. The rest follows readily since a topological group quotient map with discrete kernel is a covering map. □

The next lemma is very central in our study. Its proof shows that each isotopy starting at id_Y lifts uniquely to an isotopy starting at id_X. In other words this says that $\rho[p]$ has unique path lifting.

LEMMA 2.3 *Given* $p : X \to Y$ *as in (2.1).*
(a) *If* $\tilde{f} \in E_p(X)$ *and* $\rho^E[p](\tilde{f}) = f \in Homeo(Y)$ *then* $\tilde{f} \in Homeo_p(X)$.
(b) $Homeo^p(Y)$ *is a union of path components of* $Homeo(Y)$.

proof: (a) By a standard covering space argument (Lemma 5.6.7 in [M]) $\tilde{f}$ must be a covering map. Now both $p_\#$ and $p_\# \tilde{f}_\# = f_\# p_\#$ are isomorphisms from $\pi_1(X)$ to $p_\#(\pi_1(X)) \subset \pi_1(Y)$. Thus

$$\tilde{f}_\# = \left(p_\#^{-1}\right)\left(p_\# \tilde{f}_\#\right)$$

is an isomorphism. In particular $\tilde{f}_\#$ is surjective, and it follows that $\tilde{f}$ is a homeomorphism. (b) Consider a path in $Homeo(Y)$ starting at $f \in Homeo^p(Y)$ and passing through $g \in Homeo(Y)$. This is an isotopy from f to g. If $\tilde{f} \in Homeo_p(X)$ is a lift of f then the isotopy lifts to a unique homotopy starting at $\tilde{f}$ and ending at $\tilde{g}$ where $\tilde{g}$ is a homotopy equivalence satisfying $\rho^E[p](\tilde{g}) = g$. By (a), $\tilde{g}$ is a p-projectable homeomorphism and $g \in Homeo^p(Y)$. Thus the entire path is contained in $Homeo^p(Y)$. □

COROLLARY 2.4 *If* p *is a covering map as in (2.1) then* $\rho[p] : Homeo_p(X) \to Homeo^p(Y)$ *is a (Hurewicz) fibration with unique path lifting.*

proof: Since p is a fibration we know that $\rho^E[p] : E_p(X) \to E^p(Y)$ is a fibration. By lemma 2.3,

$$\rho^E[p]^{-1}(Homeo^p(Y)) = Homeo_p(X)$$

and thus $\rho[p]$ is the pullback of $\rho^E[p]$ via the inclusion $Homeo^p(Y) \hookrightarrow E^p(Y)$. This implies that $\rho[p]$ is a fibration. Additionally it has unique path lifting since the fibers, being translates of $A_p(X)$, are discrete. □

COROLLARY 2.5 *Let* $p : X \to Y$ *be a covering map satisfying (2.1). If* G *is a locally path connected, semi-locally simply connected subgroup of* $Homeo^p(Y)$ *then* $\rho = \rho[p]$ *restricts to a covering map* $\rho^{-1}(G) \to G$.

proof: By corollary 2.4 the restriction of ρ to $\rho^{-1}(G)$ is a fibration with unique path lifting over G. Using the given hypotheses on the base space G the result follows by theorem 2.4.10 of [Sp]. □

Thus, for example, each Lie group action on Y is covered by a Lie group action on X (cf. theorem 9.1 of [B]).

3. Main Results

In this section we consider covering maps

(3.1) $p : X \to Y$ where X and Y satisfy (2.1) and are paracompact.

The goal is to prove the main theorem (3.9) which states that $\rho[p]$ is a covering map if either the image of $p_\#$ in $\pi_1(Y)$ is "compactly supported" or the covering map p is at "most finitely irregular". We say that a subset H of $\pi_1(Y)$ is *compactly supported* if there is a compact subset C of Y so that the inclusion induced image of $\pi_1(C)$ in $\pi_1(Y)$ contains H. In this case we also say that the compact set C *supports* H. If H is contained in a finitely generated subgroup then it is compactly supported. Conversely, if Y is semi-locally simply connected and metrizable and H is compactly supported then it is contained in a finitely generated subgroup (cf. §5.12 of [M]). We say that p is *infinitely irregular* if the normalizer of $p_\#(\pi_1(X))$ in $\pi_1(Y)$ has infinite index in $\pi_1(Y)$. Otherwise, p is *at most finitely irregular*. Note that a regular covering map is at most finitely irregular (as the normalizer has index 1). We now begin developing the necessary ideas to prove the main theorem.

Lemma 3.2 *If $\mathcal{B}$ is a basis for the topology on a locally compact locally path connected paracompact Hausdorff space X then there is a basis $\check{\mathcal{B}}$ which is a barycentric refinement of $\mathcal{B}$ so that the sets*

$$\{U(C,\mathcal{O}) \mid C \text{ is compact and path connected, and, } \mathcal{O} \in \check{\mathcal{B}}\ \}$$

form a subbasis for the topology of $Homeo(X)$.

proof: Since X is paracompact the basis $\mathcal{B}$ has an open barycentric refinement. Take $\check{\mathcal{B}}$ to consist of all elements of $\mathcal{B}$ which are contained in an element of this barycentric refinement. This is a basis for X, and it is itself a barycentric refinement of $\mathcal{B}$. That the given collection of sets forms a subbasis is an easy consequence of standard results concerning the compact open topology, see section XII.5 of [D]. □

The next lemma gives the key reduction in the proof of the main theorem. We say that $\rho[p]$ is *weakly pointwise open at id_X* provided that there is an elementary open neighborhood $\tilde{\mathcal{O}}_0$ of x_0 in X so that $\rho[p](U_p(x_0,\tilde{\mathcal{O}}_0))$ is a neighborhood of id_Y in $Homeo^p(Y)$.

LEMMA 3.3 *If $p : X \to Y$ is a covering map as in (3.1) then $\rho[p]$ is an open map if and only if it is weakly pointwise open at id_X.*

proof: If $\rho = \rho[p]$ is open then it is certainly weakly pointwise open at id_X. Thus we focus on the converse and assume that ρ is weakly pointwise open at id_X. Choose an open elementary set $\tilde{O}_0$ with $x_0 \in \tilde{O}_0$ and an open set $W \subset Homeo^p(Y)$ with $id_Y \in W \subset \rho(U_p(x_0, \tilde{O}_0))$. We shall show that for each basic open neighborhood U of id_X in $Homeo_p(X)$ there is an open set V with $id_Y \in V \subset \rho(U)$. By standard topological group arguments (translating open sets to include the identity element and etc.) this statement is sufficient to guarantee the openness of ρ.

The subbasis for $Homeo_p(X)$ we will use here is that which is obtained from lemma 3.2 by choosing $\mathcal{B}$ to be the basis consisting of all elementary open sets. The basic open neighborhood U of id_X in $Homeo_p(X)$ may then be assumed to have the form

$$U = \bigcap_{i=1}^{k} U_p(C_i, \tilde{O}_i)$$

where C_i is compact and path connected, $\tilde{O}_i \in \check{\mathcal{B}}$, and $C_i \subset \tilde{O}_i$. Next we describe how to construct the open set $V \subset Homeo^p(Y)$.

Let each compact set C_i be joined to the basepoint x_0 by a path. Since the union of these paths is compact, it can be expressed as a union of compact subsets $C_{k+1}, \ldots, C_n$ where for each $i = k+1, \ldots, n$ there is an $\tilde{O}_i \in \check{\mathcal{B}}$ with $C_i \subset \tilde{O}_i$. We may additionally arrange that $x_0 \in C_n$ and that $\tilde{O}_n \subset \tilde{O}_0$. Now define

$$V = W \cap \left(\bigcap_{i=1}^{n} U^p \left(p(C_i), p(\tilde{O}_i) \right) \right) .$$

Clearly V is an open set in $Homeo^p(Y)$. Furthermore, V contains id_Y since $p(C_i) \subset p(\tilde{O}_i)$ for $i = 1, \ldots n$.

To complete the proof we shall show that $V \subset \rho(U)$. Let $f \in V$. Since $f \in W$ there is $\tilde{f} \in U_p(x_0, \tilde{O}_0)$ with $\rho(\tilde{f}) = f$ (and it is unique). Note that, for each $i \leq n$, $\tilde{f}(C_i)$ is contained in $p^{-1}(p(\tilde{O}_i))$ as $f \in U(p(C_i), p(\tilde{O}_i))$. We claim that in fact $\tilde{f}(C_i) \subset \tilde{O}_i$. First consider the case where $i = n$. Then

$$\tilde{f}(x_0) \in \tilde{O}_0 \cap \tilde{f}(C_n) \subset \tilde{O}_0 \cap p^{-1}(p(\tilde{O}_n)) = \tilde{O}_n .$$

Since C_n is path connected and $\tilde{O}_n$ is elementary we conclude that $\tilde{f}(C_n) \subset \tilde{O}_n$. For the case where $i \neq n$ we choose a sequence $C_n = C_{j_0}, C_{j_1}, \ldots, C_{j_t} = C_i$ so that $C_{j_{m-1}} \cap C_{j_m} \neq \emptyset$ for $1 \leq m \leq t$. Suppose that $\tilde{f}(C_{j_{m-1}}) \subset \tilde{O}_{j_{m-1}}$ (which we have seen holds at least for $m = 1$). Let x_m be an element of $C_{j_{m-1}} \cap C_{j_m}$. Then $\tilde{f}(x_m) \in \tilde{O}_{j_{m-1}} \cup \tilde{O}_{j_m}$. Since $\tilde{O}_{j_{m-1}} \cup \tilde{O}_{j_m}$ is elementary ($\check{\mathcal{B}}$ is a barycentric refinement of $\mathcal{B}$) and C_{j_m} is path connected, it must hold that

$$\tilde{f}(C_{j_m}) \subset \left(\tilde{O}_{j_{m-1}} \cup \tilde{O}_{j_m} \right) \cap p^{-1} p(\tilde{O}_{j_m}) = \tilde{O}_{j_m} .$$

From this we conclude inductively that $\tilde{f}(C_i) \subset \tilde{\mathcal{O}}_i$ as claimed. Thus

$$\tilde{f} \in \bigcap_{i=1}^{k} U_p(C_i, \tilde{\mathcal{O}}_i) = U$$

and $f = \rho(\tilde{f}) \in \rho(U)$. Since f is an arbitrary element of V, this shows that $V \subset \rho(U)$ and completes the proof. □

In the next lemma and corollary we digress from our proof of the main theorem to give an application which illustrates the use of the previous lemma. Suppose that $\tilde{G}$ is a subgroup of $Homeo_p(X)$ and that $G = \rho[p](\tilde{G})$. Then $\rho[p]$ induces a continuous topological group epimorphism $\rho_{\tilde{G}} : \tilde{G} \to G$ which has discrete kernel. We say that $\rho_{\tilde{G}}$ is *weakly pointwise open at* id_X if there is an elementary open set $\tilde{\mathcal{O}}_0$ containing x_0 so that $\rho_{\tilde{G}}(U_p(x_0, \tilde{\mathcal{O}}_0) \cap \tilde{G})$ is a neighborhood of id_Y in G.

LEMMA 3.4 *If* $p : X \to Y$ *satisfies (3.1) and* $\rho_{\tilde{G}} : \tilde{G} \to G$ *is the restriction of* $\rho[p]$ *to* $\tilde{G}$ *then* $\rho_{\tilde{G}}$ *is a covering map if and only if it is weakly pointwise open at* id_X.

proof: The proof of lemma 3.3 directly adapts to this situation. To be specific, the neighborhood V constructed there should be replaced with $V \cap G$, and then the rest of the argument is as before. □

Let $Homeo(X, x_0)$ be the group of homeomorphisms of X which fix x_0, and let

$$Homeo_p(X, x_0) = Homeo_p(X) \cap Homeo(X, x_0), \text{ and}$$
$$Homeo^p(Y, y_0) = \rho[p](Homeo_p(X, x_0)).$$

We caution that $Homeo^p(Y, y_0) \neq Homeo^p(Y) \cap Homeo(Y, y_0)$, but instead it consists only of the elements of $Homeo^p(Y) \cap Homeo(Y, y_0)$ whose induced homomorphism on $\pi_1(Y)$ leaves the image of $p_\#$ invariant (rather than just invariant up to conjugacy).

COROLLARY 3.5 *Given* $p : X \to Y$ *satisfying (3.1), then* $\rho[p] : Homeo_p(X, x_0) \to Homeo^p(Y, y_0)$ *is a topological group isomorphism.*

proof: If $\tilde{\mathcal{O}}_0$ is an elementary open neighborhood of x_0 then

$$Homeo_p(X, x_0) \cap U_p(x_0, \tilde{\mathcal{O}}_0) = Homeo_p(X, x_0).$$

Thus $\rho[p]\left(Homeo_p(X, x_0) \cap U_p(x_0, \tilde{\mathcal{O}}_0)\right) = Homeo^p(Y, y_0)$ is a neighborhood of id_Y, and $\rho[p]$ is weakly pointwise open at id_X. The result follows from lemma 3.4 by taking $\tilde{G} = Homeo_p(X, x_0)$. □

We now continue towards a proof of the main theorem. The advantage of the reduction to weak pointwise openness at id_X is that this property may be entirely understood in terms of fundamental groups as the next lemma shows. We recall that if ω is a path from y_0 to $\overline{y}_0$ in a space Y then there is a isomorphism

$$h_{[\omega]} : \pi_1(Y, y_0) \longrightarrow \pi_1(Y, \overline{y}_0)$$

defined by $h_{[\omega]}([\gamma]) = [\omega^{-1}\gamma\omega]$.

LEMMA 3.6 *Let $p : X \to Y$ be a covering map as in (3.1) and let $\rho = \rho[p]$. Suppose that $f \in Homeo(Y)$. If $\widetilde{O}_0$ is a path connected open neighborhood of x_0 in X then $f \in \rho(U_p(x_0, \widetilde{O}_0))$ if and only if $p(\widetilde{O}_0)$ contains a path ω from y_0 to $f(y_0)$ with $f_\#(p_\#(\pi_1(X))) = h_{[\omega]}(p_\#(\pi_1(X)))$.*

proof: First suppose that $f = \rho(\widetilde{f})$ where $\widetilde{f} \in U_p(x_0, \widetilde{O}_0)$. Let $\tilde{\omega}$ be a path in $\widetilde{O}_0$ from x_0 to $\overline{x}_0 = \widetilde{f}(x_0)$. Then

$$f_\#(p_\#(\pi_1(X))) = p_\#\widetilde{f}_\#(\pi_1(X)) = p_\#(\pi_1(X, \overline{x}_0)) = p_\# h_{[\tilde{\omega}]}(\pi_1(X)) = h_{[\omega]}(p_\#(\pi_1(X)))$$

where $\omega = p(\tilde{\omega})$.

For the converse, suppose that $f_\#(p_\#(\pi_1(X))) = h_{[\omega]}(p_\#\pi_1(X)))$. Let $\tilde{\omega}$ be the lift of ω starting at x_0 and ending at $\overline{x}_0 \in \widetilde{O}_0$. Then

$$f_\#(p_\#(\pi_1(X))) = p_\# h_{[\tilde{\omega}]}(\pi_1(X)) = p_\#(\pi_1(X, \overline{x}_0)) \subset \pi_1(Y, f(y_0)) .$$

Therefore, by the lifting theorem for covering maps, $fp : (X, x_0) \to (Y, f(y_0))$ can be lifted to $\widetilde{f} : (X, x_0) \to (X, \overline{x}_0)$. Since $\widetilde{f}$ covers the homeomorphism f it is a covering map. Moreover $\widetilde{f}_\# = p_\#^{-1} f_\# p_\#$ is an isomorphism, which implies that $\widetilde{f} \in Homeo_p(X)$. Thus $f = \rho(\widetilde{f}) \in \rho(U_p(x_0, \widetilde{O}_0))$. □

For the next lemma we assume that $p : X \to Y$ is a covering map where

(3.7) X and Y satisfy (3.1), and are semi-locally simply connected.

LEMMA 3.8 *Let Y be a space as in (3.7) which has a fixed basis. If H is a compactly supported subset of $\pi_1(Y)$ then there is a basic open neighborhood O_0 of y_0 and an open neighborhood V_H of id_Y in $Homeo(Y)$ such that for each $f \in V_H$ there is a path ω from $f(y_0)$ to y_0 in O_0 with $f_\#|_H = h_{[\omega]}|_H$.*

proof: The collection $\mathcal{B}$ of all basic open sets whose induced image in $\pi_1(Y)$ is trivial forms a basis for Y. As in lemma 3.2, $\mathcal{B}$ has a barycentric refinement $\dot{\mathcal{B}}$ which is also a basis. We may further refine $\dot{\mathcal{B}}$ to $\ddot{\mathcal{B}}$ which has the property that the closure of each element is compact and contained in a $\dot{\mathcal{B}}$–basic open set.

Let C be a compact set in Y which supports H. Choose a finite covering $O'_0, O'_1, \ldots, O'_m$ of C by elements of $\ddot{\mathcal{B}}$ where $y_0 \in O'_0$. For each i, $0 \le i \le m$, let C_i be the compact set $cl(O'_i)$ and let O_i be a $\check{\mathcal{B}}$–basic open set with $C_i \subset O_i$. We now define

$$V_H = \bigcap_{i=1}^{m} U(C_i, O_i),$$

which is an open neighborhood of id_Y in $Homeo(Y)$.

Suppose that $f \in V_H$, and that γ is a loop in C which is based at y_0 and represents an element $[\gamma]$ of H. Since $C \subset \cup_{i=1}^{m} C_i$, we may express γ as a product of paths $\gamma = \gamma_0\gamma_1\cdots\gamma_k$ where each γ_j is contained in one of the C_i's; say $\gamma_j \subset C_{s_j}$ where C_{s_0} is chosen to equal C_0. For $j = 0,\ldots,k$ let y_j be the initial point of γ_j (which is also the terminal point of γ_{j-1}). Let ω_j be a path in O_{s_j} from $f(y_j)$ to y_j for each $j = 0,\ldots,k$. Put $\omega = \omega_0$. For convenience we also define $y_{k+1} = y_0$, $C_{s_{k+1}} = C_{s_0} = C_0$, and $\omega_{k+1} = \omega$. Now observe that

$$\omega_j\gamma_j\omega_{j+1}^{-1}f(\gamma_j)^{-1}$$

is a loop based at $f(y_j)$ which is contained in $O_{s_j} \cup O_{s_{j+1}}$ for each $j = 0,\ldots,k$. (See figure 1.) Since $O_{s_j} \cup O_{s_{j+1}}$ is contained in an element of $\mathcal{B}$ ($\check{\mathcal{B}}$ is a barycentric refinement of $\mathcal{B}$) this loop is homotopically trivial in Y. Writing

$$\omega\gamma\omega^{-1}f(\gamma)^{-1} = \prod_{j=0}^{k} A_j\ (\omega_j\gamma_j\omega_{j+1}^{-1}f(\gamma_j)^{-1})\, A_j^{-1}$$

where $A_j = \prod_{i=0}^{j-1} f(\gamma_j)$, it follows that $[\omega\gamma\omega^{-1}f(\gamma)^{-1}] = 1$ in $\pi_1(Y, f(y_0))$. Thus for each $[\gamma] \in H$ we have $[\omega\gamma\omega^{-1}]f_\#[\gamma]^{-1} = 1$ and so $f_\#[\gamma] = [\omega\gamma\omega^{-1}] = h_{[\omega]}[\gamma]$. □

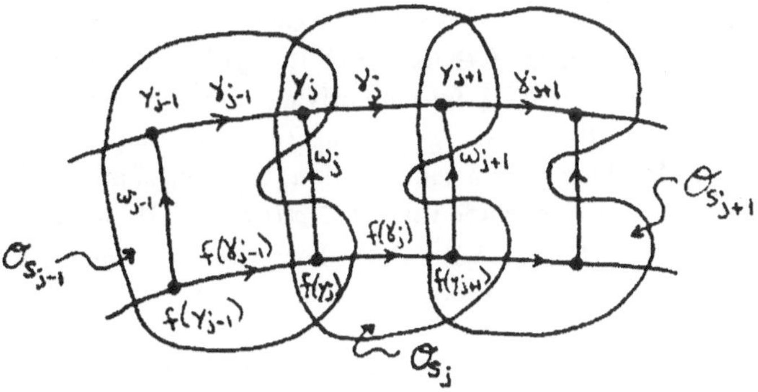

FIGURE 1

THEOREM 3.9 *Let $p: X \to Y$ be a covering map satisfying (3.7). Suppose that either*
(a) *$p_\#(\pi_1(X))$ is compactly supported in $\pi_1(Y)$, or*
(b) *p is at most finitely irregular.*
Then $\rho[p]: Homeo_p(X) \to Homeo^P(Y)$ is a regular covering map. If p is regular then the conclusion holds even if p only satisfies (3.1).

proof: For convenience we shall write $\rho = \rho[p]$ and $P = p_\#(\pi_1(X)) \subset \pi_1(Y)$.

Assume first that p is regular. Let $\widetilde{\mathcal{O}}_0$ be an elementary open neighborhood of x_0 in X. Suppose that $f \in U^P(y_0, p(\widetilde{\mathcal{O}}_0))$. Then f lifts to $\widetilde{f} \in U_p(x_0, \widetilde{\mathcal{O}}_0)$ (take any lift of f and then correct by composition with a covering transformation so that $\widetilde{f}(x_0) \in \widetilde{\mathcal{O}}_0$). This shows that $U^P(y_0, p(\widetilde{\mathcal{O}}_0)) \subset \rho(U_p(x_0, \widetilde{\mathcal{O}}_0))$ and therefore ρ is weakly pointwise open at id_X. By lemmas 3.3 and 2.2, ρ is a covering map only assuming that p satisfies (3.1). In the remainder of the proof we assume that p satisfies (3.7).

Let us next assume that p satisfies condition (a). By applying lemma 3.8 using a basis for Y which consists of open elementary sets we may find an open neighborhood V_P of id_Y in $Homeo^P(Y)$ and an elementary open set $\mathcal{O}_0$ with $y_0 \in \mathcal{O}_0$ so that for each $f \in V_P$, $f_\#(P) = h_{[\omega(P)]}$ for some path ω in $\mathcal{O}_0$. Let $\widetilde{\mathcal{O}}_0$ be an elementary open set in X with $x_0 \in \widetilde{\mathcal{O}}_0$ and $p(\widetilde{\mathcal{O}}_0) = \mathcal{O}_0$. It follows from lemma 3.6 that $V_P \subset \rho(U_p(x_0, \widetilde{\mathcal{O}}_0))$. This shows that ρ is weakly pointwise open at id_X. By lemmas 3.3 and 2.2, ρ is a regular covering map.

Now we assume that p satisfies condition (b). Choose a set $\{1, a_1, a_2, \ldots, a_n\}$ of left coset representatives for $N(P)$ in $\pi_1(Y)$. Suppose that, for one of the nontrivial coset representatives a_i, P is properly contained in $a_i P a_i^{-1}$. Then for any $k > \ell$, P is properly contained in $a_i^{k-\ell} P a_i^{\ell-k}$ which implies that $a_i^k a_i^{-\ell} \notin N(P)$. Consequently, $\{a_i^k\}_{k=1}^\infty$ forms an infinite set of distinct coset representatives for $N(P)$ in $\pi_1(Y)$ in contradiction to our assumption of finite irregularity. Therefore P is not contained in $a_i P a_i^{-1}$ for $i = 1, \ldots, n$. For each i choose an element $b_i \in P \setminus a_i P a_i^{-1}$, and now define $H = \{b_i\}_{i=1}^n$. Since H is compactly supported in $\pi_1(Y)$ lemma 3.8 implies that there is an open neighborhood V_H of id_Y in $Homeo^P(Y)$ and an elementary open neighborhood $\mathcal{O}_0$ of y_0 in Y so that for each $f \in V_H$ there is a path ω in $\mathcal{O}_0$ with $f_\#|_H = h_{[\omega]}|_H$. Since $f \in Homeo^P(Y)$, $f_\#(P)$ is conjugate to $h_{[\omega]}(P)$. Suppose that $f_\#(P) = h_{[\omega]}(a_i P a_i^{-1})$ for a nontrivial coset representative a_i of $N(P)$ in $\pi_1(Y)$. Then $b_i \in H \subset P$ so that $f_\#(b_i) = h_{[\omega]}(b_i)$ is an element of $h_{[\omega]}(a_i P a_i^{-1})$. By choice $b_i \notin a_i P a_i^{-1}$ which gives a contradiction, implying that $f_\#(P)$ must equal $h_{[\omega]}(P)$. From lemma 3.6 we conclude that $f \in \rho(U_p(x_0, \widetilde{\mathcal{O}}_0))$ where $x_0 \in \widetilde{\mathcal{O}}_0$ and $p(\widetilde{\mathcal{O}}_0) = \mathcal{O}_0$. Thus $V_H \subset \rho(U_p(x_0, \widetilde{\mathcal{O}}_0))$, which shows that ρ is weakly pointwise open at id_X. Lemma 2.2 completes the proof. □

We denote the path component of the identity in $Homeo_p(X)$ by $D_p(X)$. It consists of all projectable homeomorphisms which can be deformed to id_X through an isotopy consisting of projectable homeomorphisms. By lemma 2.3 the image of $D_p(X)$ under $\rho[p]$ is $D(Y)$, the path component of id_Y in $Homeo(Y)$. With these definitions, $\rho[p]$ restricts to give

$$\rho_0[p] : D_p(X) \longrightarrow D(Y)$$

which is a continuous topological group epimorphism. If $\rho[p]$ is a covering map and if $D(Y)$ is locally path connected then this restriction to path components is also a covering map, but in fact $\rho_0[p]$ will still be a covering map even if $\rho[p]$ is not.

THEOREM 3.10 *Let $p : X \to Y$ be a covering map satisfying (3.7). If $D(Y)$ is locally path connected then $\rho_0[p]$ is a regular covering map.*

proof: Let $q : U \to X$ be a universal covering map and let $r = pq$, which is a universal covering of Y. By theorem 3.9, $\rho[q] : Homeo_q(U) \to Homeo^q(X)$ and $\rho[r] : Homeo_r(U) \to Homeo^r(Y)$ are coverings. Since $D(Y)$ is locally path connected these coverings restrict to coverings $\rho_0[q] : D_q(U) \to D(X)$ and $\rho_0[r] : D_r(U) \to D(Y)$ (theorem 2.1.14 of [Sp]). As will be shown below, $D_r(U)$ is the identity path component in $\rho_0[q]^{-1}(D_p(X))$. The group $D_r(U)$ is locally homeomorphic to $D(Y)$ (by $\rho_0[r]$) and to $D_p(X)$ (by $\rho_0[q]$), and so all of these groups are locally path connected. Therefore the restriction of $\rho_0[q] : \rho_0[q]^{-1}(D_r(U)) \to D_p(X)$ gives a covering map $\rho_0'[q] : D_r(U) \to D_p(X)$ which satisfies $\rho_0'[q]\rho_0[p] = \rho_0[r]$. Since $\rho_0'[q]$ and $\rho_0[r]$ are coverings then so is $\rho_0[p]$.

It remains to show that $D_r(U)$ is the path component of id_U in $\rho_0[q]^{-1}(D_p(X))$. First suppose that f' is an element of $D_r(U)$. Then there is an isotopy $id_U \simeq f'$ which projects to an isotopy $id_Y \simeq f$ where $f = \rho_0[r](f')$. The latter isotopy lifts (using lemma 2.3) first (via p) to an isotopy $id_X \simeq \bar{f} \in D_p(X)$, and then (via q) to an isotopy $id_U \simeq f'' \in \rho_0[q]^{-1}(D_p(X))$. This covers the same isoptopy on Y. By uniqueness of isotopy lifting we have $f' = f''$ and this shows that $D_r(U) \subset \rho_0[q]^{-1}(D_p(X))$. For the reverse inclusion suppose that $id_U \simeq f'$ is a path in $\rho_0[q]^{-1}(D_p(X))$ (so it is an isotopy on U). Applying $\rho_0[p]\rho_0[q]$ we get an isotopy $id_Y \simeq f$ on Y, and this lifts via r back to an isotopy $id_U \simeq f''$ on U. By uniqueness $f' = f'' \in D_r(U)$, which completes the proof. □

4. APPLICATIONS

In this section we work out some consequences of the previous results. In particular we give a constructive description of $\rho_0[p]$ when it is a covering map (theorem 4.1) and an explicit description of it in the case where Y is a closed manifold of dimension at most 2 (theorems 4.5 and 4.6). Similar results are possible for $\rho[p]$ but we mostly avoid them here because of the difficulties inherent in classifying covering spaces between non-connected spaces. In theorem 4.4 we do give a description of the local structure of $Homeo_p(X)$ when Y is a compact polyhedron or manifold.

The *evaluation map* $v = v^Y : Homeo(Y) \to Y$ is defined by $v(f) = f(y_0)$. It is continuous and it restricts to a continuous map $v : D(Y) \to Y$. The image $v_\#(\pi_1(D(Y)))$ of $v_\#$ in $\pi_1(Y)$ forms the *isotopy trace subgroup* $T(Y)$ of $\pi_1(Y)$. This subgroup $T(Y)$ consists of elements of $\pi_1(Y)$ which can be represented by loops which are traces of circular isotopies on Y. (A *circular isotopy* on Y is an isotopy

$\{f_t\}$ with $f_0 = id_Y = f_1$, and its *trace* is the loop $f_t(y_0)$.) The isotopy trace subgroup is contained in the *homotopy trace subgroup* $G(Y)$ of $\pi_1(Y)$ (which is also known as the *Gottlieb subgroup*). Since $G(Y)$ is contained in the center of $\pi_1(Y)$ we infer that $T(Y)$ is a central subgroup.

THEOREM 4.1 *Let $p : X \to Y$ be a covering map which satisfies (3.1) and suppose that $\rho = \rho_0[p] : D_p(X) \to D(Y)$ is a covering map. Then*

$$\rho_\#(\pi_1(D_p(X))) = \upsilon_\#^{-1}(p_\#(\pi_1(X)))$$

and its group of covering transformations is isomorphic to

$$\frac{T(Y)}{T(Y) \cap p_\#(\pi_1(X))} .$$

In short, ρ is the pullback of p along $\upsilon : D(Y) \to Y$.

proof: The key observation is that $p\upsilon^X = \upsilon^Y \rho$, so that $p_\#\upsilon_\# = \upsilon_\#\rho_\#$. An immediate consequence is that

$$\rho_\#(\pi_1(D_p(X)) \subset \upsilon_\#^{-1} p_\# \upsilon_\#(\pi_1(D_p(X))) = \upsilon_\#^{-1}(p_\#(\pi_1(X))) .$$

For the reverse inclusion, let $[\gamma]$ be an element of $\upsilon_\#^{-1}(p_\#(\pi_1(X))) \subset \pi_1(D(Y))$. Then γ is a circular isotopy on Y with trace $\upsilon(\gamma)$. This isotopy lifts uniquely to an isotopy $\tilde{\gamma}$ on X where $\tilde{\gamma}_0 = id_X$ and $\tilde{\gamma}_1 \in A_p(X)$. The trace of $\tilde{\gamma}$ is the lifting of the trace of γ starting at x_0. Since $[\upsilon(\gamma)] = \upsilon_\#[\gamma] \in p_\#(\pi_1(X))$, the trace of $\tilde{\gamma}$ must be a loop. Therefore $\tilde{\gamma}_1(x_0) = x_0$, which implies that $\tilde{\gamma}_1 = id_X$, and $\tilde{\gamma}$ is a circular isotopy. This shows that $[\gamma] = \rho_\#[\tilde{\gamma}] \in \rho_\#\pi_1(D_p(X))$ and the first statement of the theorem follows. The covering transformation group is

$$\begin{aligned}\pi_1(D(Y))/\rho_\#(\pi_1(D_p(X))) &= \pi_1(D(Y))/\upsilon_\#^{-1}(T(Y) \cap p_\#(\pi_1(X))) \\ &\cong T(Y)/T(Y) \cap p_\#(\pi_1(X))\end{aligned}$$

which proves the second statement. □

COROLLARY 4.2 *Let $p : X \to Y$ be a universal covering map and suppose that $\rho = \rho_0[p] : D_p(X) \to D(Y)$ is a covering map. Then $\rho_\#(\pi_1(D_p(X))) = ker(\upsilon_\#^Y)$ and the covering transformation group of $\rho_0[p]$ is $T(Y)$.*

proof: All that is needed is that $T(Y) \cap \pi_1(X) = 1$, which certainly holds here. □

COROLLARY 4.3 *Suppose that Y satisfies (3.1) and that $D(Y)$ is locally path connected. Then a covering space of $D(Y)$ arises as $\rho_0[p]$ for a covering map $p : X \to Y$ if and only if the associated subgroup of $\pi_1(D(Y))$ contains $ker(\upsilon_\#)$.*

proof: If p is a covering map then the image of $\rho_0[p]_\#$ contains $ker(\upsilon_\#)$ by theorem 4.1. Conversely, if H is a subgroup of $\pi_1(D(Y))$ which contains $ker(\upsilon_\#)$ then let

$p : X \to Y$ be a covering map with $p_\#(\pi_1(X)) = v_\#(H)$. Then $\rho_0[p]_\#(\pi_1(D_p(X))) = v_\#^{-1}(p_\#(\pi_1(X))) = v_\#^{-1}(v_\#(H)) = H$ and the corollary follows. □

In the following theorem $PL(X)$ denotes the group of PL homeomorphisms of X (a polyhedron).

THEOREM 4.4 *Let $p : X \to Y$ be a covering map.*
(a) If Y is a compact connected polyhedron then $Homeo_p(X)$ is a locally contractible subgroup of $Homeo(X)$, and $\rho_0[p] : D_p(X) \to D(Y)$ is a covering map.
(b) If Y is a closed connected n-manifold then $\ell_2 \times Homeo_p(X)$ is homeomorphic to $Homeo_p(X)$. If $Homeo(Y)$ is also an ANR then $Homeo_p(X)$ is an ℓ_2-manifold.
(c) If Y is a closed connected PL-manifold and p is a PL-map then $PL_p(X)$ is an ℓ_2^f-manifold.

proof: In each case $p : X \to Y$ satisfies (3.1) and Y is semi-locally simply connected. Moreover $p_\#(\pi_1(X))$ is compactly supported since Y is compact. Therefore $\rho = \rho[p] : Homeo(X) \to Homeo^p(Y)$ is a covering map by theorem 3.9. If Y is a compact polyhedron then $Homeo(Y)$ (and hence $Homeo_p(X)$) is locally contractible [Si]. In particular $Homeo(Y)$ is locally path connected so that the restriction $\rho_0[p]$ of ρ to path components is also a covering map. (This also follows from 3.10.) This implies statement (a). If Y is a closed n-manifold it is known that $Homeo(Y) \times \ell_2$ is homeomorphic to $Homeo(Y)$ [G]. The same property then holds for $Homeo^p(Y)$, and also for its covering space $Homeo_p(X)$. If $Homeo(Y)$ is known to be an ANR then it is an ℓ_2-manifold [T]; in this case $Homeo^p(Y)$ and $Homeo_p(X)$ must also be ℓ_2-manifolds. Thus statement (b) follows. Statement (c) follows similarly using the result of [KW] that the space $PL(Y)$ of PL-homeomorphisms of a closed PL-manifold is an ℓ_2^f-manifold. (Of course $\rho[p]^{-1}(PL^p(Y)) = PL_p(X)$.) □

In the remainder of this section we will give concrete descriptions of the covering map $\rho_0[p]$ when $p : X \to Y$ is a covering map over a closed manifold of dimension 1 or 2.

THEOREM 4.5 *Let $p : X \to S^1$ be a covering map over the circle. Then $D_p(X) \approx X \times \ell_2$ and the covering map $\rho_0[p]$ may be identified with the covering map $p \times id : X \times \ell_2 \to S^1 \times \ell_2$.*

proof: By theorem 4.6(a) $\rho_0[p]$ is a covering map. It has been shown by R.D. Anderson (cf. [KW]) that $D(S^1) \approx S^1 \times \ell_2$, and the evaluation map $v : D(S^1) \to S^1$ can be identified with projection of $S^1 \times \ell_2$ onto the first factor. The result follows directly from theorem 4.1. □

THEOREM 4.6 *Let $p : X \to Y$ be a covering map over a closed 2-manifold Y.*
(a) If $\chi(Y) < 0$ then $D_p(X) \approx \ell_2$ and $\rho_0[p]$ is a topological group isomorphism.
(b) If $Y = T^2$ then $D_p(X) \approx X \times \ell_2$ and $\rho_0[p]$ may be identified with the covering map $p \times id : X \times \ell_2 \to T^2 \times \ell_2$.

(c) If $Y = K^2$ *then* $D_p(X) \approx X' \times \ell_2$ *where* $p' : X' \to S^1$ *is a covering map with covering transformation group* $Cent(\pi_1(K^2))/Cent(\pi_1(K^2)) \cap p_\#(\pi_1(X))$. *Also* $\rho_0[p]$ *may be identified with the covering map* $p' \times id : X' \times \ell_2 \to S^1 \times \ell_2$.
(d) If $Y = P^2$ *then* $D_p(X) \approx P^3 \times \ell_2$ *and* $\rho_0[p]$ *is a topological group isomorphism.*

proof: Again $\rho_0[p]$ is a covering map by theorem 4.4. If $\chi(Y) < 0$ then $D(Y) \approx \ell_2$ (cf. [H]); since ℓ_2 is simply connected, $\rho_0[p]$ must be a homeomorphism and (a) follows. If $Y = T^2$ then $D(Y) \approx T^2 \times \ell_2$ and the evaluation map $D(Y) \to Y$ can be identified with projection onto the first factor. Thus (b) follows from theorem 4.1. If $Y = K^2$ then $D(Y) \approx S^1 \times \ell_2$ (cf. [H]) and the evaluation map induces an isomorphism from $\pi_1(D(K^2))$ onto $T(K^2) = \text{Cent}(\pi_1(K^2))$. By theorem 4.1 $\rho_0[p]$ may be identified with a covering of $S^1 \times \ell_2$ having covering transformation group isomorphic to $\text{Cent}(\pi_1(K^2))/\text{Cent}(\pi_1(K^2)) \cap p_\#(\pi_1(X))$, and (c) follows. If $Y = P^2$ then we may assume that p is the universal cover $S^2 \to P^2$. It is known that $D(S^2) \approx P^3 \times \ell_2$. Since $T(S^2) = 1$ part (d) follows from corollary 4.2. □

5. A Counterexample

In this section we give an example of a covering map $p : X \to Y$ whose projecting homomorphism is not a covering map. Of course the example does not satisfy the hypotheses of theorem 3.9.

Let Y be the open planar surface obtained by removing from the complex plane the closed disks of radius one–half centered at the positive integers. We represent Y as in figure 2, choosing points $y_0, y_1, y_2, \ldots$ and simple closed curves $\alpha_1, \alpha_2, \ldots$ as indicated there.

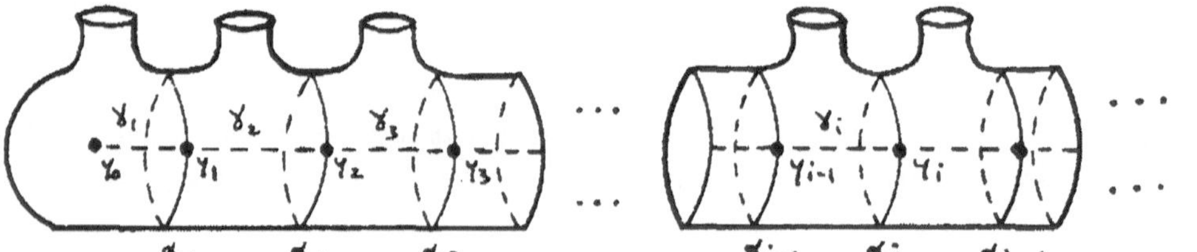

FIGURE 2

The fundamental group $\pi_1(Y)$ is a free group $\langle a_1, a_2, \ldots \rangle$ on countably many generators. The element a_i may be represented by the loop $\gamma_i \alpha_i \gamma_i^{-1}$ where γ_i is a path from y_0 to y_i. Let P be the subgroup of $\pi_1(Y)$ generated by the set

$$\{ba_1b^{-1} \mid b = a_i^{k_i} a_{i-1}^{k_{i-1}} \cdots a_3^{k_3} a_2^{k_2} \text{ for some integers } i > 1,\ k_2, \ldots, k_i\}.$$

Define $p : (X, x_0) \to (Y, y_0)$ to be a covering map with $p_\#(\pi_1(X)) = P$. We shall show that $\rho = \rho[p] : \mathit{Homeo}_p(X) \to \mathit{Homeo}^P(Y)$ is not an open map. (So that ρ fails to be a covering map.)

Let f_n be the homeomorphism of Y which is a Dehn twist about the simple closed curve α_n. This homeomorphism fixes the basepoint y_0 and has support in an annular neighborhood of α_n. The induced automorphism on $\pi_1(Y)$ is given by

$$f_{n\#}(a_i) = \begin{cases} a_i & \text{for } 1 \le i \le n \\ a_n a_i a_n^{-1} & \text{for } i > n \end{cases}.$$

The sequence $\{f_n\}_{n=1}^{\infty}$ converges to id_Y in the compact open topology of $\mathit{Homeo}(Y)$ because if $U = \bigcap_{i=1}^{K} U(C_i, \mathcal{O}_i)$ is any basic open neighborhood of id_Y then $f_n \in U$ whenever n is large enough so that α_n "lies to the right" of $\bigcup_{i=1}^{K} C_i$. We also note the following.

LEMMA 5.1 *With the above terminology,* $f_{n\#}(P) = a_n P a_n^{-1} \neq P$.

<u>proof</u>: Let ba_1b^{-1} be a generator for P where

$$b = a_i^{k_i} \cdots a_{n+1}^{k_{n+1}} a_n^{k_n} \cdots a_2^{k_2}$$

(without loss i may be assumed to be larger than n). Observe that

$$\begin{aligned} f_{n\#}(b) &= (a_n a_i a_n^{-1})^{k_i} \cdots (a_n a_{n+1} a_n^{-1})^{k_{n+1}} a_n^{k_n} \cdots a_2^{k_2} \\ &= a_n (a_i^{k_i} \cdots a_{n+1}^{k_{n+1}}) a_n^{-1} a_n^{k_n} \cdots a_2^{k_2} \\ &= a_n b' \end{aligned}$$

where $b' a_1 b'^{-1}$ is another generator for P. We now see that

$$f_{n\#}(ba_1b^{-1}) = a_n(b' a_1 b'^{-1}) a_n^{-1} \in a_n P a_n^{-1}.$$

Similarly, each generator $a_n b' a_1 b'^{-1} a_n^{-1}$ for $a_n P a_n^{-1}$ equals $f_{n\#}(ba_1b^{-1})$ for some generator ba_1b^{-1} of P and this shows that $f_{n\#}(P) = a_n P a_n^{-1}$.

We now argue that $a_n P a_n^{-1}$ is not equal to P. Any element of P can be expressed as a product

$$(b_1 a_1^{\epsilon_1} b_1^{-1})(b_2 a_1^{\epsilon_2} b_2^{-1}) \cdots (b_N a_1^{\epsilon_N} b_N^{-1})$$

of generators and their inverses where successive terms are not inverses of each other. Writing this product as a reduced word in $\pi_1(Y)$ gives something of the form

$$b_1 a_1^{\epsilon_1} b_2' a_1^{\epsilon_2} b_3' \cdots b_{N-1}' a_1^{\epsilon_N} b_N^{-1}.$$

For example, the element

$$(a_n a_{n+1}) a_1 (a_n a_{n+1})^{-1} = a_n (a_{n+1} a_1 a_{n+1}^{-1}) a_n^{-1} \in a_n P a_n^{-1}$$

does not have this form and thus it is not in P. This completes the proof. □

The lemma implies that $pf_n : (X, x_0) \to (Y, y_0)$ lifts to a map $\widetilde{f}_n : (X, x_0) \to (X, a_n \cdot x_0)$. The lifting $\widetilde{f}_n$ is a covering map and induces a surjection on $\pi_1(X)$. Thus $f_n = \rho(\widetilde{f}_n) \in Homeo_p(X)$. Let $\widetilde{O}_0$ be an elementary open 2-disk in X which contains x_0. If ω is any path from y_0 to $f_n(y_0) = y_0$ in $p(\widetilde{O}_0)$ then $h_{[\omega]} = id_Y$. Therefore, if f_n is in $\rho(U_p(x_0, \widetilde{O}_0))$ then $f_{n\#}(P)$ must equal P by lemma 3.6; however this is not possible since $a_n P a_n^{-1} \neq P$. We conclude that f_n is not an element of $\rho(U_p(x_0, \widetilde{O}_0))$. Since $\lim f_n = id_Y \in \rho(U_p(x_0, \widetilde{O}_0))$ it follows that $\rho(U_p(x_0, \widetilde{O}_0))$ is not open in $Homeo^p(Y)$. Therefore ρ is not an open map. (In fact we have shown that ρ is not weakly pointwise open at id_X.) Note that the restriction $\rho_0[p]$ of ρ to $D_p(X)$ is a covering map by theorem 3.10.

6. Projectable Homotopy Equivalences

In this section we discuss properties of the topological semigroup epimorphism $\rho^E[p] : E_p(X) \to E^p(Y)$ where $p : X \to Y$ is a covering map. The basic approach is parallel to that used in studying $\rho[p]$ except that the lack of a group structure on $E_p(X)$ causes some additional complications.

LEMMA 6.1 *Let $p : X \to Y$ be a regular covering map satisfying (3.1). The operation on $E_p(X)$ induces a monoid operation on the right coset space $A_p(X)\backslash E_p(X)$ which is algebraically isomorphic to $E^p(Y)$.*

proof: Suppose that $\widetilde{f}$ and $\widetilde{g}$ are in $E_p(X)$ and that $\rho^E[p](\widetilde{f}) = \rho^E[p](\widetilde{g})$. Since $\widetilde{f}(x_0)$ and $\widetilde{g}(x_0)$ are in the same fiber there is $\tau \in A_p(X)$ (a covering transformation) so that $\widetilde{g}(x_0) = \tau\widetilde{f}(x_0)$. Since $\widetilde{g}$ and $\tau\widetilde{f}$ cover the same map on X it follows that $\widetilde{g} = \tau\widetilde{f}$. The lemma follows directly. □

The projecting homomorphism $\rho^E[p]$ is said to be *weakly pointwise open* if and only if for each $\widetilde{f} \in E_p(X)$ there is an elementary open neighborhood $\widetilde{O}_0$ of $\widetilde{f}(x_0)$ so that $\rho^E[p](U_p(x_0, \widetilde{O}_0))$ is a neighborhood of $\rho^E[p](\widetilde{f})$. Here we have adapted the notation $U_p(C, \widetilde{O}) = U(C, \widetilde{O}) \cap E_p(X)$, and we will also write $U^p(C, O) = U(C, O) \cap E^p(Y)$.

LEMMA 6.2 *Let $p : X \to Y$ be a regular covering map satisfying (3.1). Then $\rho^E[p]$ is an open map if and only if it is weakly pointwise open.*

proof: The proof is based on the proof of lemma 3.3, however the translation trick used in the first paragraph of that proof cannot be employed in the present setting. Supposing that $\rho = \rho^E[p]$ is weakly pointwise open and that $\widetilde{f} \in E_p(X)$, let $\widetilde{O}_0$ be an elementary open neighborhood of $\widetilde{f}(x_0)$ and let $W \subset E^p(Y)$ be an open neighborhood of $\rho(\tilde{f})$ with $W \subset \rho(U_p(x_0, \tilde{\mathcal{O}}_0))$. Let $U = \cap_{i=1}^{k} U_p(C_i, \tilde{\mathcal{O}}_i)$ be a basic open neighborhood of $\tilde{f}$ in $E_p(X)$ where C_i is compact and path connected and $\tilde{f}(C_i) \subset \tilde{\mathcal{O}}_i$. Choose compact sets $C_{k+1}, \ldots, C_n$ and basic open sets $\tilde{\mathcal{O}}_{k+1}, \ldots, \tilde{\mathcal{O}}_n$ with $\tilde{f}(C_i) \subset \tilde{\mathcal{O}}_i$ so that $\cup_{i=1}^{n} C_i$ is connected, as was done in the prof of lemma 3.3. Then

$$V = W \cap \left(\bigcap_{i=1}^{n} U^p(p(C_i), p(\tilde{\mathcal{O}}_i)) \right)$$

is an open neighborhood of $\rho(\tilde{f})$ and $V \subset \rho(U)$. (For the latter, the argument in the last paragraph of the proof of lemma 3.3 applies again.) This shows that ρ is open at $\tilde{f}$, and thus it is an open map since $\tilde{f}$ was chosen arbitrarily. □

COROLLARY 6.3 *Let $p : X \to Y$ be a regular covering map satisfying (3.1). Then $\rho^E[p]$ is a covering map.*

proof: If $\tilde{f} \in E_p(X)$ and $\tilde{\mathcal{O}}_0$ is an elementary open neighborhood of $\tilde{f}(x_0)$ then $\rho^E[p](\tilde{f}) \in U^p(y_0, p(\tilde{\mathcal{O}}_0)) \subset \rho^E[p](U_p(x_0, \tilde{\mathcal{O}}_0))$. Therefore $\rho^E[p]$ is weakly pointwise open, and so it is open by lemma 6.2. Using the argument in §1, it is also continuous. This means that $\rho^E[p]$ can be identified with the topological semigroup homomorphism $E_p(X) \to A_p(X)\backslash E_p(X)$. The latter can be thought of as the orbit space projection under the action of $A_p(X)$ on $E_p(X)$ by left translation. Since this action is free and properly discontinuous the orbit space projection is a covering map. □

Let $DE(Y)$ be the path component of id_Y in $E(Y)$ and let $DE_p(X)$ be the path component of id_X in $E_p(X)$. The projecting homomorphism $\rho^E[p]$ restricts to a topological semigroup epimorphism

$$\rho_0^E[p] : DE_p(X) \longrightarrow DE(Y) .$$

Also the evaluation map restricts to $v : DE(Y) \to Y$ given by $v(f) = f(y_0)$, and the image of $v_\#$ is the homotopy trace subgroup $G(Y) \subset \mathrm{Cent}(\pi_1(Y))$.

COROLLARY 6.4 *Let $p : X \to Y$ be a covering map satisfying (3.7). If $DE(Y)$ is locally path connected then $\rho_0^E[p]$ is a covering map. It is the pullback of p along $v : DE(Y) \to Y$ and its group of covering transformations is isomorphic to*

$$\frac{G(Y)}{G(Y) \cap p_\#(\pi_1(X))} .$$

proof: See the proofs of corollary 3.10 and theorem 4.1. □

COROLLARY 6.5 *Let Y be a connected aspherical polyhedron for which $DE(Y)$ is locally path connected. If $p : X \to Y$ is a covering map with $G(Y) \cap p_\#(\pi_1(X)) = 1$ then $\rho_0^E[p]$ is a universal covering map.*

proof: It is known that $v_{\#}$ is an isomorphism onto $G(Y) = \mathrm{Cent}(\pi_1(Y))$ under these hypotheses. By corollary 6.4 the image of $\rho_0^E[p]_{\#}$ is $v_{\#}^{-1}p_{\#}(\pi_1(X))$, which is trivial by hypothesis. □

7. References

[BHMP] P. Booth, P. Heath, C. Morgan and R. Piccinini, H–spaces of self–equivalences of fibrations and bundles, *Proc. London Math. Soc.* (3) 49 (1984), 111-127.

[B] G.Bredon, *Introduction to Compact Transformation Groups*, Academic Press, New York, 1972.

[D] J. Dugundji, *Topology*, Allyn and Bacon, Boston, 1966.

[G] R. Geoghegan, On spaces of homeomorphisms, embeddings and functions. I, *Topology* 11 (1972), 159-177.

[H] W. Haver, Topological description of the space of homeomorphisms on closed 2–manifolds, *Ill. J. Math.* 19 (1976), 632-635.

[J] I.M. James, The space of bundle maps, *Topology* 2 (1963), 45-59.

[KW] J. Keesling and D. Wilson, The group of PL–homeomorphisms of a compact PL–manifold is an ℓ_2^f-manifold, *Trans. Amer. Math. Soc.* 193 (1974), 249-256.

[M] W. S. Massey, *Algebraic Topology: An Introduction*, Harcourt, Brace & World Inc., New York, 1967; Springer–Verlag, 1977.

[R] J. Rutter, Self–equivalences and principal morphisms, *Proc. London Math. Soc.*(3) 20 (1970), 644-658.

[Si] L. Siebenmann, Deformation of homeomorphisms on stratified sets, *Comment. Math. Helv.* 47 (1972), 123-163.

[Sp] E. H. Spanier, *Algebraic Topology*, Springer–Verlag, New York, 1966.

[T] H. Torunczck, Absolute retracts as factors of normed linear spaces, *Fund. Math.* 86 (1974), 53-67.

Equivariant Self-homotopy Equivalences of 2-stage G-spaces

JESPER MICHAEL MØLLER
Matematisk Institut, Københavns Universitet
Universitetsparken 5, DK-2100 København Ø, Denmark

1. Introduction.

For any finite group, G, and any based G-space, $(X, *)$, the G-homotopy classes of G-self-homotopy equivalences of $(X, *)$ form a group, denoted $\varepsilon(X, *)^G$. The purpose of this paper is to algebraicize the computation of $\varepsilon(X, *)^G$ in case X is a 2-stage G-space, i.e. the mapping fibre of a G-map between two Eilenberg-MacLane G-spaces.

Among all 2-stage G-spaces, the non G-simply-connected ones are the most intractable because of the action of the fundamental group on the higher homotopy groups. This paper deals in detail only with this worst case. The other cases can be handled similarly.

In outline the contents of this paper are as follows: Let $\underline{\pi}$ and $\underline{A}$ be group valued functors on the orbit category for G. Suppose further that $\underline{A}$ is abelian and equipped with a $\underline{\pi}$-module structure. Chapter 2 deals with an analysis of the very simple homotopy properties of spaces of equivariant maps into the aspherical G-space $K(\underline{\pi}, 1)$. (The corresponding non-equivariant results can be found in Gottlieb [3]; see also [5] and [10].) Chapter 3 contains the definition of the graded group $H_G^*(\underline{\pi}, \underline{A})$, the Bredon cohomology of $\underline{\pi}$ with coefficients in $\underline{A}$. For each homogeneous cohomology class $\theta \in H_G^{n+1}(\underline{\pi}, \underline{A})$, $n > 1$, there is a unique G-homotopy type, K_θ, with θ as its only non-trivial Postnikov invariant. The main result, Theorem 4.1, concerns the space of G-self-maps of K_θ, in particular the group $\varepsilon(K_\theta, *)^G$ of G-self-homotopy equivalences which is shown to fit into a short exact sequence

$$0 \to H_G^n(\underline{\pi}, \underline{A}) \to \varepsilon(K_\theta, *)^G \to E_\theta^G \to 1$$

where E_θ^G is a certain subgroup of the product of the automorphism groups of $\underline{\pi}$ and $\underline{A}$. The congruence class of this group extension is an element of the second cohomology group of E_θ^G with coefficients in $H_G^n(\underline{\pi}, \underline{A})$. Lemma 4.4 shows how to compute this class from an E_2-level differential in a spectral sequence defined in purely algebraic terms.

In the non-equivariant and simply-connected case, a short exact sequence similar to the one above was first constructed by Shih [13]; see also the later papers [15] and [9].

The following notational conventions are in force throughout the paper: G is a finite group with canonical orbit category [1] $\mathcal{O}$, $(\mathcal{G}^{ab})$ $\mathcal{G}$, is the category of (abelian) groups, an (abelian) $\mathcal{O}$-group is a contravariant functor from $\mathcal{O}$ into $(\mathcal{G}^{ab})$ $\mathcal{G}$, $\mathrm{Hom}_{\mathcal{O}}(\underline{M}, \underline{N})$, where $\underline{M}$ and $\underline{N}$ are $\mathcal{O}$-groups, is the set of natural transformations $\underline{M} \to \underline{N}$, and, finally, all function spaces are equipped with the compactly generated topology associated to the compact-open topology and denoted as in Switzer [14]: If $u : X \to Y$ is a map, $A \subset X$, and $p : Y \to B$ a fibration, $F_u(X, A; Y, B)$ is the space (with base point u) of all maps $f : X \to Y$ with $f|A = u|A$ and $pf = pu$.

This paper is in final form and no version of it will be submitted for publication elsewhere.

2. Spaces of equivariant maps into $K(\underline{\pi},1)$.

Fix a contravariant functor $\underline{\pi}: \mathcal{O} \to \mathcal{G}$ and let $K := K(\underline{\pi},1)$ be the associated based Eilenberg-MacLane G-space [2] for which $\underline{\pi}_1 K(\underline{\pi},1) = \underline{\pi}$ and $\underline{\pi}_i K(\underline{\pi},1) = 0$ for $i \geq 2$. Let also (X,A) be a $G-CW$-pair and $u: X \to K(\underline{\pi},1)$ a G-map. We shall in this chapter describe the homotopy groups $\pi_*(F_u(X,A;K)^H, u)$, the most interesting case being $A = \emptyset$, for any subgroup H of G. See [3], [5], [10] for the non-equivariant case.

The orbit category $\mathcal{O}_H$ of the subgroup $H \leq G$ is a subcategory of $\mathcal{O} = \mathcal{O}_G$: The inclusion $\mathcal{O}_H \to \mathcal{O}$ takes H/I, $I \leq H$, to G/I and is the inclusion

$$\mathcal{O}_H(H/I\,,\, H/J) = (H/J)^I \subset (G/J)^I = \mathcal{O}(G/I\,,\, G/J)$$

on the sets of morphisms. If F is any functor on $\mathcal{O}$, $F \mid H$ will denote F pre-composed with this inclusion $\mathcal{O}_H \subset \mathcal{O}$.

We shall assume that X is a pointed G-space with base point $x_0 \in X^G$ such that the fundamental $\mathcal{O}$-group $\underline{\pi}_1(X) = \underline{\pi}_1(X,x_0)$ is defined. Note that if $f: X \to K$ is a based H-homotopy class, then the family $(\pi_1(f^I))_{I \leq H}$ is a natural transformation of the $\mathcal{O}_H$-group $\underline{\pi}_1(X) \mid H$ into the $\mathcal{O}_H$-group $\underline{\pi} \mid H$. Hence we have a map from the set of path-components of $F_u(X,x_0;K)^H$ into the set of natural transformation of $\underline{\pi}_1(X) \mid H$ into $\underline{\pi} \mid H$. Actually,

PROPOSITION 2.1. *The homotopy groups of $F_u(X,x_0;K)^H$ are*

$$\pi_i F_u(X,x_0;K)^H = \begin{cases} \mathrm{Hom}_{\mathcal{O}_H}(\underline{\pi}_1(X)|H, \underline{\pi}|H) & i = 0 \\ 0 & i > 0 \end{cases}$$

PROOF: Apply ([2], Theorem 3) with the i-fold suspension of X as domain and K as range. □

A special case occurs when also $(X,x_0) = (K,k_0)$. The functor C of [2] induces a map

$$\mathrm{Hom}_{\mathcal{O}}(\underline{\pi},\underline{\pi}) \to F_1(K,k_0;K)^G$$

of the discrete space of natural self-transformations of $\underline{\pi}$ into the space of based equivariant self-maps of K which, by Proposition 2.1, is a weak homotopy equivalence. In the following we shall not distinguish between a natural self-transformation $\alpha: \underline{\pi} \to \underline{\pi}$ and the induced G-map $\alpha: K(\underline{\pi},1) \to K(\underline{\pi},1)$.

Now to the space of free G-maps into K.

For each subgroup $I \leq H$, the natural projection $p_{I,H}: G/I \to G/H$ is in $\mathcal{O}$ and hence $\underline{\pi}(p_{I,H}): \underline{\pi}(G/I) \leftarrow \underline{\pi}(G/H)$ is defined. If $\varphi = (\varphi(H/I))_{I \leq H}$ is a morphism of $\mathcal{O}_H$-groups and $\zeta \in \underline{\pi}(G/H)$, let ζ_φ be the family whose Ith member is $\varphi(H/I)$ conjugated by $\underline{\pi}(p_{I,H})(\zeta)$. As is easily checked, $\zeta\varphi$ is again a natural transformation, and so we have a group action

$$\underline{\pi}(G/H) \times \mathrm{Hom}_{\mathcal{O}_H}(\underline{\pi}_1(X)|H, \underline{\pi}|H) \to \mathrm{Hom}_{\mathcal{O}_H}(\underline{\pi}_1(X)|H, \underline{\pi}|H)$$

of $\underline{\pi}(G/H)$ on the set of natural transformation of $\underline{\pi}_1(X)|H$ into $\underline{\pi}|H$. Let $\underline{\pi}(G/H)_u$ be the stabilizer w.r.t. this action of the natural transformation $\underline{\pi}_1(u) = (\pi_1(u^I))_{I \leq H}$ induced by the G-map u.

THEOREM 2.2. *The homotopy groups of* $F_u(X;K)^H$ *are*

$$\pi_i F_u(X;K)^H = \begin{cases} \mathrm{Hom}_{\mathcal{O}_H}(\underline{\pi}_1(X) \mid H\,,\, \underline{\pi} \mid H)/\underline{\pi}(G/H) & i=0 \\ \underline{\pi}(G/H)_u & i=1 \\ 0 & i \geq 2 \end{cases}$$

PROOF: By Proposition 2.1, the homotopy sequence for the evaluation fibration

$$F_u(X,x_0;K)^H \to F_u(X;K)^H \to K^H$$

degenerates to the exact sequence of groups and sets

$$1 \to \pi_1 \to \underline{\pi}(G/H) \xrightarrow{\partial} \mathrm{Hom}_{\mathcal{O}_H}(\underline{\pi}_1(X) \mid H, \underline{\pi} \mid H) \to \pi_0 \to 1$$

where the group $\underline{\pi}(G/H)$ acts on the set of $\mathcal{O}_H$-group morphisms as above. □

Recall, finally, that $(X;A)$ denotes a $G-CW$ pair. Let $i: A \to X$ be the inclusion (assuming $x_0 \in A^G \neq \emptyset$), $\underline{\pi}_1(i)|H : \underline{\pi}_1(A)|H \to \underline{\pi}_1(X)|H$ the induced morphism of $\mathcal{O}_H$-groups, and

$$i^*: \mathrm{Hom}_{\mathcal{O}_H}(\underline{\pi}_1(X) \mid H, \underline{\pi} \mid H) \to \mathrm{Hom}_{\mathcal{O}_H}(\underline{\pi}_1(A) \mid H, \underline{\pi} \mid H)$$

pre-composition with $\underline{\pi}_1(i) \mid H$. Then, of course,

$$(i^*)^{-1}(\underline{\pi}_1(u) \mid A) = \{\varphi \in \mathrm{Hom}_{\mathcal{O}_H} \mid \underline{\pi}_1(i) = \underline{\pi}_1(u \mid A) \mid H\}$$

is the set of extensions to $\underline{\pi}_1(X) \mid H$ of the $\mathcal{O}_H$-group morphism $\underline{\pi}_1(u|A)|H$.

COROLLARY 2.3. *The homotopy groups of* $F_u(X,A;K)^H$, *where* $A^G \neq \emptyset$, *are*

$$\pi_i F_u(X,A;K)^H = \begin{cases} (i^*)^{-1}(\underline{\pi}_1(u) \mid H) & i=0 \\ 0 & i>0 \end{cases}$$

PROOF: The homotopy sequence for the restriction fibration

$$F_u(X,A;K)^H \to F_u(X,*;K)^H \to F_u(A,*;K)^H$$

reduces to the exact sequence

$$1 \to \pi_0 F_u(X,A;K)^H \to \mathrm{Hom}_{\mathcal{O}_H}(\underline{\pi}_1(X) \mid H, \underline{\pi} \mid H) \xrightarrow{i^*} \mathrm{Hom}_{\mathcal{O}_H}(\underline{\pi}_1(A) \mid H, \underline{\pi} \mid H)$$

of sets. □

For the sake of clarity of exposition, the results of the following sections will not, as in this section, be formulated for an arbitrary subgroup $H \leq G$ but only for the full group G.

3. Bredon cohomology of $\mathcal{O}$-groups.

This chapters main theme is the definition of Bredon cohomology of an $\mathcal{O}$-group $\underline{\pi}$ with coefficients in a $\underline{\pi}$-module $\underline{A}$.

Let, accordingly, $\underline{\pi}$ be an $\mathcal{O}$-group and $\underline{A}$ a $\underline{\pi}$-module ([11], Definition 3.1). This means that $\underline{A}$ is an abelian $\mathcal{O}$-group equipped with a natural transformation $\underline{\pi} \times \underline{A} \to \underline{A}$ giving $\underline{A}(G/H)$ a $\underline{\pi}(G/H)$-module structure for each $H \leq G$. The $\underline{\pi}$-modules together with the abelian groups $\mathrm{Hom}_{\underline{\pi}}(\underline{B}, \underline{A})$ of natural transformations $\underline{B} \to \underline{A}$ commuting with the $\underline{\pi}$-actions form an abelian category, $\underline{\pi}$-Mod.

EXAMPLE 3.1: Define an abelian $\mathcal{O}$-group $U(\underline{\pi})$ as follows: $U(\underline{\pi})(G/K)$ is the free $\underline{\pi}(G/K)$-module on the set of all morphisms f in $\mathcal{O}$ with domain G/K ; $U(\underline{\pi})(g)$, for $g \in \mathcal{O}(G/H\ ,\ G/K)$, is the abelian group homomorphism provided by

$$U(\underline{\pi})(g)(\eta f) = \underline{\pi}(g)(\eta)(f \circ g)\ ,\ \eta \in \underline{\pi}(G/K)\,.$$

Then $U(\underline{\pi})$ is a projective $\underline{\pi}$-module and $\mathrm{Hom}_{\underline{\pi}}(U(\underline{\pi})\ ,\ \underline{A}) \neq 0$ whenever $\underline{A} \neq 0$, cf. ([1], pp. I-23 - I-24); that is, $U(\underline{\pi})$ is a projective generator of $\underline{\pi}$-Mod.

The category $C(\underline{\pi} - \mathrm{Mod})$ of $\underline{\pi}$-Mod-complexes is again an abelian category ([7], p. 259). If $\underline{B}_* \in C(\underline{\pi} - \mathrm{Mod})$ is such a chain complex, we let

$$H^q_G(\underline{B}_*, \underline{A})$$

denote the $(-q)$-th homology group of the dual complex $\mathrm{Hom}_{\underline{\pi}}(\underline{B}_*, \underline{A})$.

Note that $\underline{\pi}$-Mod has enough injectives ([4], Theoreme 1.10.1) since it has projective generators and clearly satisfies Grothendieck's axiom AB5. Moreover, homology provides a functor $H_q : C(\underline{\pi} - \mathrm{Mod}) \to \underline{\pi} - \mathrm{Mod}$. In this situation, standard homological algebra yields a universal coefficient spectral sequence.

$$E_2^{pq} = \mathrm{Ext}^p_{\underline{\pi}}(H_q(\underline{B}_*)\ ,\ \underline{A}) \Rightarrow H^{p+q}_G(\underline{B}_*\ ,\ \underline{A})$$

where $\mathrm{Ext}^p_{\underline{\pi}}(H_q(\underline{B}_*), \cdot)$ is the pth right derived functor of $\mathrm{Hom}_{\underline{\pi}}(H_q(\underline{B}_*), \cdot) : \underline{\pi} - \mathrm{Mod} \to \mathcal{G}^{ab}$.

EXAMPLE 3.2: Let X be a G-connected $G - CW$-complex with $\underline{\pi}$ as fundamental $\mathcal{O}$-group and let $\Gamma_*(\tilde{X})$ be the $\underline{\pi}$-Mod-complex defined as follows: For each $H \leq G$, let $\tilde{X}^H$ be the universal covering complex of X^H (equipped with a base point over $x_0 \in X^G$), and for $g \in \mathcal{O}(G/H\ ,\ G/K) = (G/K)^H$, let $\tilde{g} : X^H \leftarrow X^K$ be the unique (based) lift of $g : X^H \leftarrow X^K$. Then $\Gamma_*(\tilde{X})$ takes G/H to the $\underline{\pi}(G/H)$-module complex $\Gamma_*(\tilde{X}^H)$ and g to $\tilde{g}_* : \Gamma_*(\tilde{X}^H) \leftarrow \Gamma_*(\tilde{X}^K)$.

The group

$$H^n_G(X\ ;\ \underline{A}) := H^n_G(\Gamma_*(\tilde{X})\ ,\ \underline{A})$$

is the nth Bredon cohomology group of X with local coefficients in $\underline{A}$ [11]. Obstructions to sectioning G-fibrations and G-homotoping one G-section into another lie in such groups; see [11].

In analogy with the non-equivariant case, $H^*_G(K(\underline{\pi}, 1), \underline{A})$ can be computed from a bar resolution.

EXAMPLE 3.3: For any group π, let $B_\bullet(\pi)$ be its normalized bar resolution. Composition of $\underline{\pi} : \mathcal{O} \to \mathcal{G}$ and $B_\bullet$ is a $\underline{\pi}$-Mod-complex, denoted $B_\bullet(\underline{\pi})$. This construction is natural in the sense that any $\mathcal{O}$-group endomorphism α of $\underline{\pi}$ induces an α-transformation $B_\bullet(\alpha) : B_\bullet(\underline{\pi}) \to B_\bullet(\underline{\pi})$ of $\underline{\pi}$-Mod-complexes.

An α-transformation, for $\alpha \in \mathrm{Hom}_{\mathcal{O}}(\underline{\pi}, \underline{\pi})$, is an element of $\mathrm{Hom}_{\underline{\pi}}(\underline{A}, \alpha^*\underline{A})$, where $\alpha^*\underline{A}$ is the abelian $\mathcal{O}$-group $\underline{A}$ with $\underline{\pi}$-structure

$$\underline{\pi} \times \underline{A} \stackrel{\alpha\times 1}{\to} \underline{\pi} \times \underline{A} \to \underline{A}$$

pulled back along α. Alternatively, $\varphi \in \mathrm{Hom}_{\mathcal{O}}(\underline{A}, \underline{A})$ is an α-transformation iff

$$\varphi(G/H)(\zeta a) = \alpha(G/H)(\zeta)\varphi(G/H)(a)$$

for all $H \leq G$, $\zeta \in \underline{\pi}(G/H)$, and $a \in \underline{A}(G/H)$.

DEFINITION 3.4. *The nth Bredon cohomology group of $\underline{\pi}$ with coefficients in $\underline{A}$ is*

$$H^n_G(\underline{\pi}, \underline{A}) := H^n_G(B_\bullet(\underline{\pi}), \underline{A}).$$

For $\alpha \in \mathrm{Hom}_{\mathcal{O}}(\underline{\pi}, \underline{\pi})$,

$$\alpha^* : H^n_G(\underline{\pi}, \alpha^*\underline{A}) \leftarrow H^n_G(\underline{\pi}, \underline{A})$$

is the map induced by the chain map $\mathrm{Hom}_{\underline{\pi}}(B_\bullet(\alpha)\ ,\ 1)$, and for $\varphi \in \mathrm{Hom}_{\underline{\pi}}(\underline{A}\ ,\ \alpha^*\underline{A})$,

$$\varphi_\bullet : H^n_G(\underline{\pi}\ ,\ \underline{A}) \to H^n_G(\underline{\pi}\ ,\ \alpha^*\underline{A})$$

is the coefficient group homomorphism induced by $\mathrm{Hom}_{\underline{\pi}}(1, \varphi)$.

As in the non-equivariant case, one has

PROPOSITION 3.5. *There exists an isomorphism*

$$H^*(\eta) : H^*_G(K(\underline{\pi}, 1); \underline{A}) \leftarrow H^*_G(\underline{\pi}\ ,\ \underline{A})\ .$$

PROOF: A feature of the functorial construction [2] of the Eilenberg-MacLane G-space $K(\underline{\pi}, 1)$ is the existence of cellular homotopy equivalences

$$\eta(G/H) : K(\underline{\pi}, 1)^H \to K(\underline{\pi}(G/H), 1)\ ,$$

natural in H, where the range is constructed simplicially with $B_\bullet(\underline{\pi}(G/H))$ as the cellular chain complex of its universal covering space. Take

$$\eta_\bullet : \Gamma_\bullet(\tilde{K}(\underline{\pi}, 1)) \to B_\bullet(\underline{\pi})$$

to be the family $\eta_\bullet = (\eta(G/H)_\bullet)$ of induced chain homotopy equivalences

$$\eta(G/H)_\bullet : \Gamma_\bullet(\tilde{K}(\underline{\pi}, 1)^H) \to B_\bullet(\underline{\pi}(G/H))\ .$$

Then η_* induces an isomorphism of the E_2-level of the universal coefficient spectral sequences so that the limit

$$H^*(\eta) = H(\eta_*) : H^*_G(K(\underline{\pi},1)\,;\,\underline{A}) \leftarrow H^*_G(\underline{\pi},\underline{A})$$

is also an isomorphism ([6], Theorem 3.2). □

Now note that the set

$$\Sigma^G := \{(\alpha,\varphi) \in \mathrm{Hom}_{\mathcal{O}}(\underline{\pi},\underline{\pi}) \times \mathrm{Hom}_{\mathcal{O}}(\underline{A},\underline{A}) \mid \varphi \in \mathrm{Hom}_{\underline{\pi}}(\underline{A},\alpha^*\underline{A})\}$$

is monoid under composition of natural transformations and that for $(\alpha,\varphi) \in \Sigma^G$ there are induced homomorphisms

$$H^*_G(\underline{\pi},\alpha^*\underline{A}) \overset{\alpha^*}{\leftarrow} H^*_G(\underline{\pi},\underline{A}) \overset{\varphi_*}{\rightarrow} H^*_G(\underline{\pi},\alpha^*\underline{A})$$

on cohomology. For any group cohomology class $\theta \in H^{n+1}_G(\underline{\pi},\underline{A}), n \geq 2$,

$$\Sigma^G_\theta := \{(\alpha,\varphi) \in \Sigma^G \mid \alpha^*\theta = \varphi_*\theta\}$$

is a submonoid of Σ^G. Σ^G_θ has a topological significance to be formulated in Proposition 3.6 below. However, we must first introduce the G-space K_θ which is going to play a very central role in the next chapter. In the course, we also recall some facts from [11]: There exists a sectioned G-fibration ([11], Lemma 3.2)

$$K(\underline{A},n+1) \rightarrow L(\underline{\pi},\underline{A},n+1) \underset{\check{k}}{\overset{\hat{k}}{\rightleftarrows}} K(\underline{\pi},1)$$

which classifies Bredon cohomology with local coefficients in the sense that the primary difference ([16], VI ; [11]) with $\check{k}$ induces a bijection ([11], Theorem 3.3)

$$\pi_0 F_{\hat{k}}(K;L,K)^G \rightarrow H^{n+1}_G(\underline{\pi},\underline{A})$$

between vertical G-homotopy classes of G-sections of $\hat{k} : L \rightarrow K$ and the group $H^{n+1}_G(\underline{\pi},\underline{A})$. ($L$ is short for $L(\underline{\pi},\underline{A},n+1)$ and K is short for $K(\underline{\pi},1)$.)

Let now (also) $\theta : K \rightarrow L$ be a G-section corresponding to the cohomology class $\theta \in H^{n+1}_G(\underline{\pi},\underline{A})$ in this manner and form the pull back

$$\begin{array}{ccc} K_\theta & \longrightarrow & \overline{P}L \\ \big\downarrow & & \big\downarrow{\scriptstyle e_1} \\ K & \xrightarrow{\ \theta\ } & L \end{array}$$

along θ of the equivariant path fibration over and under K [7] given by

$$\overline{P}L = \{u : I \rightarrow L \mid u(0) = \check{k}\hat{k}u(0)\,,\ \hat{k}u(I) = \{\hat{k}u(0)\}\}$$

and $e_1(u) = u(1)$, $I = [0,1]$. The element $(x,u) \in K_\theta \subset K \times \overline{PL}$ iff $\theta(x) = u(1)$, and

$$g(x,u) = (gx, gu), \quad g \in G,$$

defines a G-action on K_θ making all maps in the pull back diagram into G-maps.

Now suppose that $f : K_\theta \to K_\theta$ is a based fibre G-map, i.e. that f is a based G-map such that the diagram

$$
\begin{array}{ccc}
K_\theta & \xrightarrow{\ f\ } & K_\theta \\
{\scriptstyle p_\theta}\downarrow & & \downarrow{\scriptstyle p_\theta} \\
K & \longrightarrow & K
\end{array}
$$

commutes for some G-self-map of $(K,*)$. By taking homotopy, we obtain a pair $(\underline{\pi}_1(f), \underline{\pi}_n(f)) \in \mathrm{Hom}_{\mathcal{O}}(\underline{\pi}, \underline{\pi}) \times \mathrm{Hom}_{\mathcal{O}}(\underline{A}, \underline{A})$ of natural self-transformations of $\mathcal{O}$-groups.

PROPOSITION 3.6. $(\underline{\pi}_1(f), \underline{\pi}_n(f)) \in \Sigma_\theta^G$.

PROOF: Naturality of the $\underline{\pi}_1(K_\theta) = \underline{\pi}$-module structure on $\underline{\pi}_n(K_\theta) = \underline{A}$ forces $\varphi := \underline{\pi}_n(f)$ to be an $\alpha := \underline{\pi}_1(f)$-transformation.

Since the fibre of p_θ is $(n-1)$-G-connected, there exists, for any $\underline{\pi}$-module $\underline{M}$, an equivariant transgression homomorphism ([1], Definition 4.2)

$$\tau_G : \mathrm{Hom}_{\underline{\pi}}(\underline{A}, \underline{M}) \to H_G^{n+1}(\underline{\pi}, \underline{M})$$

given by $\tau_G(\psi) = \psi_*\theta$ for $\psi \in \mathrm{Hom}_{\underline{\pi}}(\underline{A}, \underline{M})$. Naturality of the transgression implies commutativity of the diagram

$$
\begin{array}{ccc}
\mathrm{Hom}_{\underline{\pi}}(\underline{A}, \alpha^*\underline{A}) & \xleftarrow{\ \varphi_*\ } & \mathrm{Hom}_{\underline{\pi}}(\underline{A}, \underline{A}) \\
{\scriptstyle \tau_G}\downarrow & & \downarrow{\scriptstyle \tau_G} \\
H_G^{n+1}(\underline{\pi}, \alpha^*\underline{A}) & \xleftarrow{\ \alpha^*\ } & H_G^{n+1}(\underline{\pi}, \underline{A}) \ .
\end{array}
$$

In particular,

$$\alpha^*\theta = \alpha^*\tau_G(1) = \tau_G(\varphi_*1) = \tau_G(\varphi) = \varphi^*\theta \ ,$$

that is, $(\alpha, \varphi) \in \Sigma_\theta^G$. $\square$

Finally we shall construct a spectral sequence whose E_2-level carries complete information on the group of based G-homotopy classes of equivariant homotopy self-equivalences of K_θ.

Let E_θ^G denote the group of invertible elements in Σ_θ^G and consider the action

$$\diamond : E_\theta^G \times \mathrm{Hom}_{\underline{\pi}}(B_*(\underline{\pi}), \underline{A}) \to \mathrm{Hom}_{\underline{\pi}}(B_*(\underline{\pi}), \underline{A})$$

of E_θ^G on the cochain complex $\mathrm{Hom}_{\underline{\pi}}(B_*(\underline{\pi}), \underline{A})$ given by

$$(\alpha, \varphi) \diamond \underline{c} = \varphi \circ \underline{c} \circ B_*(\alpha^{-1}), \quad (\alpha, \varphi) \in E_\theta^G \ ,$$

for all cochains $\underline{c}$. The first filtration of the associated bicomplex

$$M^\circ = \mathrm{Hom}^\circ_{E^G_\theta}(B_\bullet(E^G_\theta), \mathrm{Hom}_{\underline{\pi}}(B_\bullet(\underline{\pi}), \underline{A}))$$

of E^G_θ-linear homomorphisms generates a spectral sequence

$$(\circ) \qquad E_2^{pq} = H^p(E^G_\theta, H^q_G(\underline{\pi}, \underline{A})) \Rightarrow H^{p+q}(\mathrm{Tot}\,(M^\circ))$$

and in particular a differential

$$d_2^{0,n+1} : H^{n+1}_G(\underline{\pi}, \underline{A})^{E^G_\theta} \to H^2(E^G_\theta, H^n_G(\underline{\pi}, \underline{A}))$$

which can be applied to θ.

PROPOSITION 3.6. $H^*(Tot\,(M^\circ)) = H^*_G(\underline{\pi} \rtimes E^G_\theta, \underline{A})$.

Above, $\underline{\pi} \rtimes E^G_\theta$ is the $\mathcal{O}$-group that takes the object G/H to the semi-direct product $\underline{\pi}(G/H) \rtimes E^G_\theta$ w.r.t. the action

$$E^G_\theta \hookrightarrow \mathrm{Aut}_{\mathcal{O}}(\underline{\pi}) \times \mathrm{Aut}_{\mathcal{O}}(\underline{A}) \overset{pr_1}{\to} \mathrm{Aut}_{\mathcal{O}}(\underline{\pi}) \overset{\text{eval.}}{\to} \mathrm{Aut}\,(\underline{\pi}(G/H))$$

and the morphism $g \in \mathcal{O}(G/H; G/K)$ to $\underline{\pi}(g) \rtimes 1$; $\underline{A}$ is equipped with the $\underline{\pi} \rtimes E^G_\theta$-module structure provided by all the natural maps

$$(\underline{\pi}(G/H) \rtimes E^G_\theta) \times \underline{A}(G/H) \to \underline{A}(G/H)$$
$$((\eta, (\alpha, \varphi)), a) \to \eta\varphi(a)$$

combined.

PROOF OF PROPOSITION 3.6. (SKETCH): (1) Extend the action $\circ$ to an action

$$\circ : E^G_\theta \times \mathrm{Hom}_{\underline{\pi}}(B_\bullet(\underline{\pi} \rtimes E), \underline{A}) \to \mathrm{Hom}_{\underline{\pi}}(B_\bullet(\underline{\pi} \rtimes E), \underline{A})$$

by putting

$$e \circ \underline{c} = e \circ \underline{c} \circ B_\bullet(\overline{(1, e^{-1})}), \quad e \in E^G_\theta,$$

where $\overline{(1, e^{-1})}$ is conjugation by $(1, e^{-1})$. Show ([8], p. 348) that

$$\mathrm{Hom}_{\underline{\pi}}(B_\bullet(\underline{\pi}), \underline{A}) \overset{\text{res}}{\leftarrow} \mathrm{Hom}_{\underline{\pi}}(B_\bullet(\underline{\pi} \rtimes E^G_\theta), \underline{A})$$

is an E^G_θ-linear quasi-isomorphism.

(2) Define a new action

$$\bullet : E^G_\theta \times \mathrm{Hom}_{\underline{\pi}}(B_\bullet(\underline{\pi} \rtimes E^G_\theta), \underline{A}) \to \mathrm{Hom}_{\underline{\pi}}(B_\bullet(\underline{\pi} \rtimes E^G_\theta), \underline{A})$$

by

$$(e \bullet \underline{c})(b) = e\underline{c}((1, e^{-1})b)\,, \quad b \in B_\bullet(\underline{\pi} \rtimes E^G_\theta)$$

and show that

$$\overline{M}^\circ \cong \overline{M}^\bullet.$$

where $\overline{M}^\cdot = \mathrm{Hom}_{E^G_{\mathcal{O}}}(B_\bullet(E^G_{\mathcal{O}}), \mathrm{Hom}_{\underline{\pi}}(B_\bullet(\underline{\pi} \rtimes E^G_{\mathcal{O}}), \underline{A})$ for $\cdot \in \{\circ, \bullet\}$.

(3) Note that

$$\overline{M}^\bullet = \mathrm{Hom}_{\underline{\pi} \rtimes E^G_\theta}(B_\bullet(E) \otimes B_\bullet(\underline{\pi} \rtimes E^G_\theta), \underline{A})$$

by adjointness. Proceed as in the proof of the Lyndon spectral sequence found in MacLane [8]. □

4. Equivariant Homotopy Equivalences.

This section contains the computation of all homotopy groups of the H-space $F_1(K_\theta, *; K_\theta)^G$ of based, equivariant self-maps of K_θ. The most delicate one is the monoid of path-components. See [12] for the non-equivariant case.

In the following main theorem, $\sigma(K_\theta, *)^G := \pi_0 F_1(K_\theta, *; K_\theta)^G$ is the monoid of based G-homotopy classes of based self G-maps of K_θ. Its units form the group $\varepsilon(K_\theta, *)^G$ of G-homotopy classes of G-self-homotopy equivalences of $(K_\theta, *)$.

THEOREM 4.1. *For any $i > 0$ and any based self G-map f of K_θ,*

$$\pi_i(F_1(K_\theta, *; K_\theta)^G, f) \cong \overline{H}_G^{n-i}(\underline{\pi}, \underline{\pi}_1(f)^* \underline{A}) .$$

For $i = 0$, there exists a short exact sequence of monoids

$$0 \to H_G^n(\underline{\pi}, \underline{A}) \to \sigma(K_\theta, *)^G \to \Sigma_\theta^G \to 1$$

whose invertible elements constitute a short exact sequence of groups

$$0 \to H_G^n(\underline{\pi}, \underline{A}) \to \varepsilon(K_\theta, *)^G \to E_\theta^G \to 1$$

in the congruence class $d_2^{0,n+1}(\theta) \in H^2(E_\theta^G, H_G^n(\underline{\pi}, \underline{A}))$.

Theorem 4.1 follows from Lemma 4.1 - 4.4 below.

LEMMA 4.1. *For any $i > 0$ and any f,*

$$\pi_i(F_1(K_\theta, *; K_\theta)^G, f) \cong \overline{H}_G^{n-i}(\underline{\pi}, \underline{\pi}_1(f)^* \underline{A})$$

PROOF: Consider the homotopy sequence for the fibration

$$F_f(K_\theta, *; K_\theta, K) \to F_1(K_\theta, *; K_\theta) \xrightarrow{p_\theta} F_{p_\theta}(K_\theta, *; K) ,$$

defined by post-composition with p_θ, and use Proposition 2.1 and ([11]), Theorem 5.1). □

The rest of this paper concerns the $i = 0$ case.

We have a pull back diagram

$$\begin{array}{ccc} \mathcal{F}_1(K_\theta, *; K_\theta)^G & \longrightarrow & F_1(K_\theta, *; K_\theta)^G \\ \downarrow & & \downarrow{\scriptstyle p_\theta} \\ F_1(K, *; K)^G & \xrightarrow{\overline{p}_\theta} & F_{p_\theta}(K_\theta, *; K)^G \end{array}$$

where $\mathcal{F}_1(K_\theta, *; K_\theta)^G$ is the subspace of based fibre G-maps of the fibration $p_\theta : K_\theta \to K$ into itself, and $\overline{p}_\theta$ is defined as pre-composition with p_θ. By Proposition 2.1, $\overline{p}_\theta$ is a weak homotopy equivalence and so is then the upper horizontal inclusion map. In particular,

$$\sigma(K_\theta, *)^G = \pi_0 \mathcal{F}_1(K_\theta, *; K_\theta)^G$$

so that Proposition 3.6 implies the existence of a homomorphism

$$(\underline{\pi}_1, \underline{\pi}_n) : \sigma(K_\theta, *)^G \to \Sigma_\theta^G$$

into the monoid Σ_θ^G.

LEMMA 4.2. *$(\underline{\pi}_1, \underline{\pi}_n)$ is surjective.*

LEMMA 4.3. *The kernel of $(\underline{\pi}_1, \underline{\pi}_n)$ is (isomorphic to) $H^n_G(\underline{\pi}, \underline{A})$.*

PROOF OF LEMMA 4.2: For any element $(\alpha, \varphi) \in \Sigma^G$ there exists ([11]), Lemma 3.4) an essentially unique G-map $\varphi_\alpha : L \to L$ such that hte diagram

$$\begin{array}{ccc} L & \xrightarrow{\varphi_\alpha} & L \\ {\scriptstyle k}\downarrow\uparrow{\scriptstyle k} & & {\scriptstyle k}\uparrow\downarrow{\scriptstyle k} \\ K & \xrightarrow{\alpha} & K \end{array}$$

commutes and $\underline{\pi}_n(\varphi_\alpha) = \varphi$. By the definition of Σ^G_θ, the G-maps $\varphi_\alpha\theta$ and $\theta\alpha$ are equivariantly and vertically homotopic lifts of $\alpha : K \to K$; let

$$G(\alpha, \varphi) : I \times K \to L$$

be a homotopy of $G(\alpha,\varphi)_0 = \varphi_\alpha\theta$ to $G(\alpha,\varphi)_1 = \theta\alpha$. (Note for later use that we can assume that $\theta = k$ on the n-skeleton K_n of K and that $G(\alpha, \varphi)$ is stationary on K_{n-1}.) Define

$$\overline{G}(\alpha, \varphi) : K_\theta \to K_\theta$$

to be the based fibre G-map that takes $(x, u) \in K_\theta$ to

$$\overline{G}(\alpha,\varphi)(x,u) = (\alpha(x), G(\alpha,\varphi)(\cdot, x) \cdot \varphi_\alpha u)\,.$$

Since $\varphi_\alpha u(1) = \varphi_\alpha\theta(x) = G(\alpha,\varphi)(0,x)$, the two paths $t \to \varphi_\alpha u(t)$ and $t \to G(\alpha,\varphi)(t,x)$ have a well defined product, here denoted by a dot. By construction, $(\underline{\pi}_1, \underline{\pi}_n)(\overline{G}(\alpha,\varphi)) = (\alpha, \varphi)$. □

PROOF OF LEMMA 4.3: The kernel of $(\underline{\pi}_1, \underline{\pi}_n)$ consists of fibre G-maps $f : (K_\theta, *) \to (K_\theta, *)$ over K with $\underline{\pi}_n(f) = 1$. Associate to any such f the primary difference ([16], VI; [11])

$$\delta^n_G(f, 1) \in H^n_G(K_\theta, \underline{A})$$

of f and the identity. It is not hard to see that the sequence

$$0 \to H^n_G(\underline{\pi}, \underline{A}) \xrightarrow{p_\theta^*} H^n_G(K_\theta; \underline{A}) \xrightarrow{i^*} \mathrm{Hom}_{\mathcal{O}}(\underline{A}, \underline{A})$$

is exact. Then

$$i^*\delta^n_G(f,1) = \underline{\pi}_n(f) - 1 = 1 - 1 = 0$$

shows that $\delta^n_G(f,1)$ actually belongs to the subgroup $H^n_G(\underline{\pi}, \underline{A})$ of $H^n_G(K_\theta; \underline{A})$.

If $g : (K_\theta, *) \to (K_\theta, *)$ is another G-map representing an element of the kernel, the difference of $f \circ g$ and 1 is

$$\begin{aligned} \delta^n_G(f \circ g, 1) &= \delta^n_G(f \circ g, f) + \delta^n_G(f, 1) \\ &= \underline{\pi}_n(f)_* \delta^n_G(g, 1) + \delta^n_G(f, 1) \\ &= \delta^n_G(g, 1) + \delta^n_G(f, 1)\,. \end{aligned}$$

Hence $f \to \delta^n_G(f,1)$ is an injective homomorphism of Ker $(\underline{\pi}_1, \underline{\pi}_n)$ into $H^n_G(\underline{\pi}, \underline{A})$. It remains to show that it is also surjective.

Consider the space

$$F^G_{\theta p_2} := F_{\theta p_2}((I;\dot{I}) \times (K, K_{n-1}); L, K)^G$$

of equivariant and vertical self-homotopy equivalences of K_θ. We define maps, $< \cdot >$ and $\bar{\cdot}$, such that the diagram

$$\begin{array}{ccc} & \overset{<\cdot>}{} & H^n_G(\underline{\pi}, \underline{A}) \\ \pi_0(F^G_{\theta p_2}) & \nearrow \atop \nwarrow & \uparrow \quad \delta^n_G(1,\cdot) \\ & \bar{\cdot} & \ker(\underline{\pi}_1, \underline{\pi}_n) \end{array}$$

commutes with $< \cdot >$ surjective. In fact, for $h \in F^G_{\theta p_2}$, define $\overline{h} : K_\theta \to K_\theta$ to be the G-map over K that takes $(x,u) \in K_\theta$ to

$$\overline{h}(x,u) = (x, h(\cdot, x) \cdot u)$$

and define $< h > \in \mathrm{Hom}_{\underline{\pi}}(B_n(\underline{\pi}), \underline{A})$ to be the (homology class of the) cocycle that takes $\underline{b} \in B_n(\underline{\pi})$ to

$$< h > (\underline{b}) = d^{n+1}_G(\theta p_2, h)(i \otimes \underline{b}) \,.$$

Here, $d^{n+1}_G(\theta p_2, h) \in \Gamma^{n+1}_G((I,\dot{I}) \times (K, K_{n+1}); \underline{A})$ is the difference cochain ([16], VI;[11]), in this case a cocycle ([16], Theorem VI. 5.6.(3)), of the constant homotopy θp_2 and $h; i \in \Gamma_1(I;\dot{I})$ is the standard 1-cell.

Now $\delta^n_G(1,\overline{h}) = < h >$ and by obstruction theory, $< \cdot >$ is a bijection into $H^n_G(K, K_{n-1}; \underline{A})$ and hence a surjection onto $H^n_G(K;\underline{A}) = H^n_G(\underline{\pi}, \underline{A})$. □

We shall continue to use the notation from the proofs of Lemma 4.2 and 4.3.

Let us now focus on the self G-maps of K_θ that are homotopy equivalences. If $(\alpha, \varphi) \in E^G_\theta, (\alpha^{-1}\varphi^{-1})$ exists and

$$\overline{G}(\alpha^{-1}, \varphi^{-1}) \circ \overline{G}(\alpha, \varphi) \in \ \ker\,(\underline{\pi}_1, \underline{\pi}_n) \subset \sigma(K_\theta, *)^G$$

is a G-equivalence since it is of the form $\overline{h}$ for some self-homotopy h of θ and such maps have homotopy inverses defined by the inverse homotopies. But then also $\overline{G}(\alpha, \varphi)$ is invertible in the monoid $\sigma(K_\theta, *)^G$ for purely algebraic reasons. Together with Lemma 4.2 and 4.3 these remarks assure the existence of a short exact sequence of groups

$$0 \to H^n_G(\underline{\pi}, \underline{A}) \to \varepsilon(K_\theta, *)^G \to E^G_\theta \to 1$$

and since

$$\alpha^* \delta^n_G(f\overline{h}f^{-1}, 1) = \alpha^* \delta^n_G(f\overline{h}f^{-1}, ff^{-1}) = \delta^n_G(f\overline{h}, f) = \varphi_* \delta^n_G(\overline{h}, 1)$$

for any lift f of $(\alpha, \varphi) \in E^G_\theta$ and any self-homotopy h, this sequence realizes the action of E^G_θ on $H^n_G(\underline{\pi}, \underline{A})$. Its congruence class can thus be viewed as an element of the second cohomology group of E^G_θ with coefficients in $H^n_G(\underline{\pi}, \underline{A})$.

Lemma 4.4. *The congruence class of the above extension of $H^n_G(\underline{\pi}, \underline{A})$ by E^G_θ is $\pm d_2^{0,n+1}(\theta)$.*

Proof: For $(\alpha,\varphi) \in E^G_\theta, H(\alpha,\varphi) := G(\alpha,\varphi)\circ(1\times\alpha^{-1})$ is a vertical G-homotopy (rel. K_{n-1}) of $\varphi_\alpha\theta\alpha^{-1}$ to θ. Since θp_2, where p_2 is the projection onto the second factor, and $H(\alpha,\varphi)$ agree on the n-skeleton $(I\times K)_n$ they define together with the constant homotopy a difference cochain ([16], p. 296;[11])

$$d_G^{n+1}(\theta p_2, H(\alpha,\varphi)) \in \Gamma_G^{n+1}(I\times K;\underline{A})$$

which, since both maps are extendable, actually is an $(n+1)$-cocycle ([16], Theorem VI. 5.6. (3)). Repeating the computations of ([16], Theorem V. 5.6') one finds, with $\underline{c}\in B_{n+1}(\underline{\pi})$, that

$$\begin{aligned} 0 &= d_G^{n+1}(\theta p_2, H(\alpha,\varphi))\underline{\partial}(i\times\underline{c}) \\ &= d_G^{n+1}(\theta p_2, H(\alpha,\varphi))(1\times\underline{c} - 0\times\underline{c} - i\times\underline{\partial c}) \\ &= (d_G^{n+1}(\bar{k},\theta) - d_G^{n+1}(\bar{k},\varphi_\alpha\theta\alpha^{-1}) + <\alpha,\varphi>\underline{\partial})(\underline{c}) \\ &= (d_G^{n+1}(\bar{k},\theta) - (\alpha,\varphi)d_G^{n+1}(\bar{k},\theta) + <\alpha,\varphi>\underline{\partial})(\underline{c}) \end{aligned}$$

where $<\alpha,\varphi>\in \mathrm{Hom}_{\underline{\pi}}(B_n(\underline{\pi}),\underline{A})$ is defined as the n-cochain with value

$$<\alpha,\varphi>(\underline{b}) = -d_G^{n+1}(\theta p_2, H(\alpha,\varphi))(i\times\underline{b})$$

on any chain $\underline{b}\in B_n(\underline{\pi})$. Hence the cochain

$$(\alpha,\varphi)\cdot<\beta,\psi> - <\alpha\beta,\varphi\psi> + <\alpha,\varphi>$$

is always an n-cocycle for any pair of elements $(\alpha,\varphi),(\beta,\psi)\in E^G_\theta$. By taking each generator $[\alpha,\varphi \mid \beta,\psi]$ to the cohomology class of this cocycle we get an E^G_θ-map

$$B_2(E^G_\theta)\to H^n_G(\underline{\pi},\underline{A})$$

which is a cocycle representing $\pm d_2^{0,n+1}(\theta)$.

Next we shall construct a factor set ([8], IV.4) for the group extension of the lemma. Suppose $(\alpha,\varphi),(\beta,\psi)\in E^G_\theta$. Let

$$h[\alpha,\varphi\mid\beta,\psi](t,x) = \begin{cases} H(\alpha\beta,\varphi\psi)(1-3t,x) & 0\le t\le \frac{1}{3} \\ \varphi_\alpha H(\beta,\psi)(3t-1,\alpha^{-1}x) & \frac{1}{3}\le t\le\frac{2}{3} \\ H(\alpha\varphi)(3t-2,x) & \frac{2}{3}\le t\le 1 \end{cases}$$

for $(t,x)\in I\times K$. Then $h[\alpha,\varphi\mid\beta,\psi]$ is a vertical and equivariant self-homotopy (rel. K_{n-1}) of θ,

$$\overline{G}(\alpha,\varphi)\circ\overline{G}(\beta,\psi)\simeq \overline{h}[\alpha,\varphi\mid\beta,\psi]\circ\overline{G}(\alpha\beta,\varphi\psi)\,,$$

and for $\underline{b}\in B_n(\underline{\pi})$,

$$\begin{aligned} <h[\alpha,\varphi\mid\beta,\psi]>(\underline{b}) &= d_G^{n+1}(\theta p_2, h[\alpha,\varphi\mid\beta,\psi])(i\times\underline{b}) \\ &= ((\alpha,\varphi)<\beta,\psi> - <\alpha\beta,\varphi\psi> + <\alpha,\varphi>)(\underline{b})\,. \end{aligned}$$

Consequently, the factor set

$$\begin{aligned} B_2(E^G_\theta) &\to H^n_G(\underline{\pi},\underline{A}) \\ [\alpha,\varphi\mid\beta,\psi] &\to \delta^n_G(1,\overline{h}[\alpha,\varphi\mid\beta,\psi]) \end{aligned}$$

agrees with the 2-cocycle representing $\pm\, d_2^{0,n+1}(\theta)$ □

COROLLARY 4.5. *The group* $\varepsilon(L,*)^G$ *is isomorphic to the semi-direct product* $H^n_G(\underline{\pi},\underline{A}) \rtimes E^G_0$.

PROOF: $L = K_0$ □

REMARK 4.6. *For* $H \leq G$, *the group* $\varepsilon(K_\theta,*)^H$ *of* H*-homotopy classes of* H*-equivariant self maps of* $(K_\theta,*)$ *fits into a short exact sequence*

$$0 \to H^n_H(\underline{\pi}|H,\underline{A}|H) \to \varepsilon(K_\theta,*)^H \to E^H_\theta \to 1$$

where E^H_θ *is the group of pairs* $(\alpha,\varphi) \in \mathrm{Aut}_{\mathcal{O}_H}(\underline{\pi} \mid H) \times \mathrm{Aut}_{\mathcal{O}_H}(\underline{A} \mid H)$ *such that* φ *is an* α*-transformation. This group extension is classified by* $d_2^{0,n+1}(\theta)$ *where*

$$d_2^{0,n+1} : H^n_H(\underline{\pi}|H,\underline{A}|H)^{E^H_\theta} \to H^2(E^H_\theta, H^n_H(\underline{\pi}|H,\underline{A}|H))$$

comes from the first filtration of the bicomplex

$$\mathrm{Hom}_{E^H_\theta}(B_*(E^H_\theta), \mathrm{Hom}_{\underline{\pi}|H}(B_*(\underline{\pi} \mid H), \underline{A} \mid H)) .$$

The assertions of Remark 4.6 follow by repeating the above argument making only some obvious and purely formal modifications in the appropriate places.

In the same spirit, one can, if $\underline{\pi}$ is an abelian $\mathcal{O}_G$-group, $\underline{A}$ a trivial $\underline{\pi}$-module, $\theta \in H^{n+1}_G(K(\underline{\pi},m),\underline{A}), n > m > 1$, compute $\varepsilon(K_\theta,*)^H, H \leq G$, for the mapping fibre K_θ of $\theta : K(\underline{\pi},m) \to K(\underline{A},n+1)$. The obvious generalizations of Theorem 4.1 and Remark 4.6 hold.

REFERENCES

1. G.E. Bredon, *Equivariant Cohomology Theories*, Lecture Notes in Mathematics **34** (1967), Springer-Verlag, Berlin-New York.
2. A.D. Elmendorf, *Systems of Fixed Point Sets*, Trans.Amer.Math.Soc **277** (1983), 275–284.
3. D.H. Gottlieb, *Covering transformations an universal fibration*, Illinois J. Math **13** (1969), 432–43
4. A. Grothendieck, *Sur quelques Points d'Algèbre Homologique*, Tohoku Math. J. **9** (1957), 119–221.
5. V.L. Hansen, *Spaces of maps into Eilenberg-MacLane spaces*, Canad. J. Math. **XXXIII** (1981), 782–785.
6. J. McCleary, *User's Guide to Spectral Sequences*, Mathematics Lecture Series **12** (1985), Publish or Perish, Wilmington.
7. J.F. McClendon, *Obstruction Theory in Fiber Spaces*, Math. Z. **120** (1971), 1–17.
8. S. MacLane, *Homologie. Third Corrected Printing*, Die Grundlehren der mathematischen Wissenschaften **114** (1975), Springer-Verlag, Berlin-Heidelberg-New York.
9. K. Maruyama, *A Remark on The Group of Self-homotopy Equivalences*, Mem. Fac. Sci. Kyushu Univ.Ser. A **41** (1987), 81–84.
10. J.M. Møller, *Spaces of sections of Eilenberg-MacLane fibrations*, Pacific J. Math **130** (1987), 171–186.
11. J.M. Møller, *On Equivariant Function Spaces*, Preprint (1987).
12. J.M.Møller, *Homotopy Equivalences of Group Cohomology Spaces*, Preprint (1988).
13. W. Shih, *On the group* $\varepsilon(X)$ *of homotopy equivalence maps*, Bull. Amer. Math. Soc **492** (1964), 361–365.

14. R.M. Switzer, *Counting elements in homotopy sets*, Math. Z. 178 (1981), 527–554.
15. K. Tsukiyama, *Self-homotopy-equivalences of a space with two non-vanishing homotopy groups*, Proc. Amer. Math. Soc 79 (1980), 134–138.
16. G.W. Whitehead, *Elements of Homotopy Theory*, Graduate Texts in Mathematics 61 (1978), Springer-Verlag, Berlin-Heidelberg-New York.

ON SKELETON PRESERVING HOMOTOPY SELF-EQUIVALENCES OF CW COMPLEXES

John W. Rutter
Department of Pure Mathematics,
University of Liverpool,
Liverpool L69 3BX, England.

1 Introduction

A cellular map which is a homotopy self-equivalence of a CW complex X need not induce homotopy self-equivalences of the skeleta $\{X_n\}$: it evidently does so of course if X has cells in no two consecutive dimensions. For Postnikov and simplicial set decompositions a homotopy self-equivalence does induce homotopy self-equivalences on the n-th stages, and this has led to various general results on the group $\mathcal{E}(X)$ of based homotopy classes of based homotopy self-equivalences of X being proved using these latter decompositions rather than CW complexes. Here we consider the subgroup $\bar{\mathcal{E}}(X)$ of $\mathcal{E}(X)$ consisting of those classes which can be represented by cellular maps $f:X \to X$ inducing homotopy equivalences $f_r = f|X_r : X_r \to X_r$ of each r–skeleton and investigate the circumstances in which $\bar{\mathcal{E}}(X)$ has finite index in $\mathcal{E}(X)$. Further we draw conclusions regarding the commensurability of $\bar{\mathcal{E}}(X)$ with an arithmetic group.

Condition A The CW complex X is countable and satisfies (i) $X_1 = \{pt\}$, (ii) for finitely many dimensions $r_1, \ldots, r_k$ there are finitely many cells of each dimension r_i $(1 \le i \le k)$, and (iii) for all other dimensions r $(r \ge 2)$ the r-cells are attached by maps into X_{r-2}.

Certainly any simply-connected finite complex X with $X_1 = \{pt\}$ satisfies condition A.

Theorem B Let X be a complex satisfying condition A, then the group $\bar{\mathcal{E}}(X)$ has finite index in $\mathcal{E}(X)$.

Using the result of Wilkerson and Sullivan (theorem B of [9] and 10.3 of [7]), we then have the following theorem.

Theorem C Let X be a simply-connected finite complex with $X_1 = \{pt\}$, then the group $\bar{\mathcal{E}}(X)$ is commensurable with an arithmetic group, and can therefore be finitely presented.

Remark 1.1 Let $X = S^n \cup_q e^{n+1}$ $(q \ge 2,\ n \ge 2)$ and consider the commutative diagram

---

This paper is in final form and no version of it will be submitted for publication elsewhere.

$$\begin{array}{ccccc} S^n & \xrightarrow{q} & S^n & \to & X \\ r\downarrow & & r\downarrow & & \bar{r}\downarrow \\ S^n & \to & S^n & \to & X \end{array}$$

where r for example is the $(n-1)$-st suspension of the self-map $e^{i\theta} \rightsquigarrow e^{ir\theta}$ of S^1 and $\bar{r}$ is the mapping cone on the two maps of degree r $(r \in \mathbf{Z})$. Then $\bar{r}$ is a homotopy equivalence if, and only if, the highest common factor $(r,q)=1$. If $q=5$, therefore, there are homotopy equivalences of X which cannot be represented by a cellular map restricting to a homotopy equivalence of $X_n = S^n$, so that $\bar{\mathcal{E}}(X)$ is properly contained in $\mathcal{E}(X)$. The group $\bar{\mathcal{E}}(S^n \cup_q e^{n+1})$ is calculated in example 2.8 of [6].

In §2 we prove a result of independent interest, namely that, up to cellular homotopy type (see §2), X_n may be obtained from X_{n-1} by first adding n-cells which give no homology in dimension n, then adding n-cells all of which give free homology and whose boundaries in the simply-connected case lie in X_{n-2}. In §3 we show that the subgroup of $\mathcal{E}(X_n)$, whose elements can be represented by those cellular maps $f_n: X_n \to X_n$ for which $f_{n-1}: X_{n-1} \to X_{n-1}$ is a homotopy equivalence, has finite index in $\mathcal{E}(X_n)$ provided that $X = X_n$ is simply-connected and either i) there are finitely many n–cells, or ii) the boundaries of all the n-cells are in X_{n-2}. Finally in §4 we prove theorem B by an induction argument. All maps and homotopies are base-point preserving.

2 The canonical form for a complex

Let A be an (n-1)-dimensional $(n \geq 2)$ path-connected CW complex with one zero cell, and let $\vee_{\alpha \in T} S_\alpha^{n-1}$ be the 1-point union of copies of the (n-1)-sphere, and $h: \vee_{\alpha \in T} S_\alpha^{n-1} \to A$ a based attaching map. In the case where $n = 2$, suppose further that $\operatorname{im} H_1(h) \subset H_1(A)$ is finitely generated. We now modify the attaching map by a homotopy equivalence E of $\vee_{\alpha \in T} S_\alpha^{n-1}$ thus obtaining a canonical form for the attaching map for the n-cells.

Lemma 2.1 There is a homotopy equivalence E of $\vee_{\alpha \in T} S_\alpha^{n-1}$ such that the mapping cone $C_{h'}$, of $h' = h \circ E$, has the form $C_{h'} = A \cup B \cup C$, where $h' = h \circ E$, $B = \cup_{\alpha \in U} e_\alpha^n$, $C = \cup_{\alpha \in V} e_\alpha^n$, $H_n(A \cup B) = 0$ and the inclusion and boundary induce isomorphisms $H_n(A \cup C) \cong H_n(A \cup B \cup C)$ and $H_n(A \cup C) \cong H_{n-1}(\vee_{\alpha \in V} S_\alpha^{n-1})$. Furthermore if A is simply-connected, $A_1 = \{pt\}$ and $n \geq 3$, then modifying h' by a homotopy we can assume that $h'(\vee_{\alpha \in V} S_\alpha^{n-1}) \subset A_{n-2}$. In each of these cases C_h and $C_{h'}$ have the same homotopy type rel A.

Remark 2.2 $H_n(A \cup B) = 0$ is equivalent to $H_{n-1}(\vee_{\alpha\in U} S_\alpha^{n-1}) \to H_{n-1}(A) \to H_{n-1}(A \cup B)$ being short exact; and $H_n(A \cup C) \cong H_{n-1}(\vee_{\alpha\in V} S_\alpha^{n-1})$ is equivalent to $H_{n-1}(A) \cong H_{n-1}(A \cup C)$.

We say that the n-cells of a complex X (with $X_1 = \{pt\}$ in the simply-connected case) are **in canonical form** if $X_n = X_{n-1} \cup B \cup C$ where B and C satisfy the conditions given in lemma 2.1, and we say that X is **in canonical form** if its n-cells are in canonical form for each $n \geq 2$. By applying the procedure of lemma 2.1 to successive dimensions, we can replace X by a complex Y in canonical form. We note that various related normal forms are well known: see for example Chang ([1]) and Hilton (p 53 of [3]). The canonical form given here is specific to the proof of theorem B.

Definition Two complexes X and Y are **of the same cellular homotopy type** if there are cellular maps $f: X \to Y$, $g: Y \to X$, and homotopies $H: g \circ f \sim 1_X$ and $K: f \circ g \sim 1_Y$ such that H_t and K_t are cellular maps for each t.

The following theorem is now immediate.

Theorem 2.3 Let X be a countable complex which is either simply-connected and $X_1 = \{pt\}$, or for which $\operatorname{im} H_1(h) \subset H_1(X_1)$ is finitely generated. Then X has the cellular homotopy type of a complex in canonical form.

Remark 2.4 Arising from the canonical decomposition of lemma 2.1, the mapping cone $C_{h'}$ may be the (based) sum of a complex with a contractible complex. A contractible complex will not in general have the same cellular homotopy type as a point. For example the canonical decomposition arising from $S^3 = e^0 \cup e^2 \cup_1 e^3 \cup_1 e^3$ is $(e^0 \cup e^2 \cup_1 e^3) \vee (e^0 \cup e^3)$, which does not have the same cellular homotopy type as $e^0 \cup e^3$. However, if Y is contractible, then $\mathfrak{S}(Y) = \bar{\mathfrak{S}}(Y) = (pt)$, so that, as far as theorems B and C are concerned, contractible summands may be discarded.

Proof of Lemma 2.1 Let G be the kernel of $\sigma: H_{n-1}(\vee_{\alpha\in T} S_\alpha^{n-1}) \to H_{n-1}(A)$. Since $H_{n-1}(A)$ is free abelian, we prove that there is an automorphism of $H_{n-1}(\vee_{\alpha\in T} S_\alpha^{n-1})$ induced by a homotopy equivalence E such that $T = U \cup V$ say and $H_{n-1}(\vee_{\alpha\in V} S_\alpha^{n-1})$ is mapped onto G by E_*. In case $n \geq 3$ the proof is clear. Let $n = 2$ and suppose that T is finite. Then $H_1(\vee_{\alpha\in T} S_\alpha^1) = G \oplus \theta(\operatorname{im}\sigma)$ is generated by $\{\psi_\alpha\}$ where ψ_α is the image of the generator of $\pi_1(S_\alpha^1)$. A new basis consisting of a basis for G and a basis for $\theta(\operatorname{im}\sigma)$ determines a change of basis matrix, and, writing this as a product of elementary matrices, the result follows from 7.3.4 of [2]. Now consider the general case. Since $\operatorname{im}\sigma$ is finitely generated $\theta(\operatorname{im}\sigma)$ is contained in a subgroup L generated by finitely many of the ψ_α, say $\psi_{\alpha_1}, \ldots, \psi_{\alpha_s}$. Choose a new basis $\omega_{\alpha_1}, \ldots, \omega_{\alpha_s}$ for L as above, where $\omega_{\alpha_1},\dots,\omega_{\alpha_t}$ is a basis for $\theta(\operatorname{im}\sigma)$. Also, for $\alpha\neq\alpha_r$ $(1\le r\le s)$, $\theta\sigma(\psi_\alpha)=\Sigma a_r\omega_{\alpha_r}$ $(1\le r\le t)$ say: choose $\omega_\alpha=\psi_\alpha-\theta\sigma(\psi_\alpha)=\psi_\alpha-\Sigma a_r\omega_{\alpha_r}$. The change of basis for L determines a homotopy equivalence as before, and the further change of basis determines a second homotopy equivalence: E is the composite. Now let A be simply-connected, $n\ge3$, and $A_1=\{pt\}$. Consider the following diagram

$$\begin{array}{ccccc} & & \pi_{n-1}(A_{n-2}) & & \\ & & \downarrow & & \\ \pi_{n-1}(\vee_{\alpha\in T}S_\alpha^{n-1}) & \to & \pi_{n-1}(A) & \to & \pi_{n-1}(A,A_{n-2}) \\ & & \downarrow & & \\ H_{n-1}(\vee_{\alpha\in T}S_\alpha^{n-1}) & \xrightarrow{(hE)_*} & H_{n-1}(A) & \to & H_{n-1}(A,A_{n-2}) \end{array}$$

The vertical sequence is exact (§1 and §10 of [8]), and the sequence $\pi_{n-1}(A_{n-2})\to\pi_{n-1}(A)\xrightarrow{\rho}\pi_{n-1}(A,A_{n-2})$ is exact. Let $k_\alpha : S_\alpha^{n-1}\to A$ $(\alpha\in V)$ be an attaching map which is a component of $h\circ E$. Then $\rho[k_\alpha]=0$, and therefore the class of k_α can be represented by a map $S_\alpha^{n-1}\to A_{n-2}$.

3 Classes giving equivalences on the (n – 1) – skeleton

We now recall some notation used in [5]. Given a map $f:X\to Y$, we consider classes u and $\tilde{u}$ which make the following diagram commutative.

$$\begin{array}{ccc} X & \xrightarrow{f} & Y \\ u\downarrow & & \downarrow\tilde{u} \\ X & \to & Y \end{array}$$

We define

$$\mathcal{R}_f(X)=\{u\in\mathcal{E}(X):\exists\,\bar{u}\in\mathcal{E}(Y).\,f^*(\bar{u})=f_*(u)\},$$
$$\mathcal{L}_f(Y)=\{\bar{u}\in\mathcal{E}(Y):\exists\,u\in\mathcal{E}(X).\,f_*(u)=f^*(\bar{u})\},$$
$$\mathcal{R}^1_f(X)=\{u\in\mathcal{E}(X):f_*(u)=f^*(1_Y)\}$$
$$\mathcal{L}^1_f(Y)=\{\bar{u}\in\mathcal{E}(Y):f^*(\bar{u})=f_*(1_X)\}.$$

The normal subgroups $\mathcal{R}^1_f(X)$ and $\mathcal{L}^1_f(Y)$ correspond to the indeterminacy involved in determining u from $\tilde{u}$ and vice-versa. We have the basic result:

($\mathcal{L}$–$\mathcal{R}$ duality theorem 2.1 of [5]) $\mathcal{L}_f(Y)/\mathcal{L}_f^1(Y)\cong\mathcal{R}_f(X)/\mathcal{R}_f^1(X)$.

Consider the cofibre sequence

$$X_{n-1} \xrightarrow{i_n} X_n \to \vee S_\alpha^n \to \dots .$$

Then $\mathcal{E}_{i_n}(X_n)$ consists of those classes of $\mathcal{E}(X_n)$ which can be represented by a cellular map whose restriction to X_{n-1} is a homotopy equivalence.

In this section we prove the theorems.

Theorem 3.1 Let X_{n-1} $(n \geq 3)$ be simply-connected and let X_n be obtained by adding finitely many n-cells, then $\mathcal{E}_{i_n}(X_n)$ has finite index in $\mathcal{E}(X_n)$.

Theorem 3.2 Let X_n $(n \geq 3)$ be simply-connected and let $H_{n-1}(X_n)$ be free abelian, then $\mathcal{E}_{i_n}(X_n) = \mathcal{E}(X_n)$. In particular this is true if X_n is obtained from X_{n-1} by adding n–cells with boundaries in X_{n-2}.

Corollary Let X be a finite dimensional and $H_*(X)$ free abelian, then $\bar{\mathcal{E}}(X) = \mathcal{E}(X)$.

Proof of theorem 3.2 Since in this case $[A \cup B, SK] = 0$, we have $\mathcal{E}_{i_{A \cup B}}(A \cup B \cup C) = \mathcal{E}(A \cup B \cup C)$ as in lemma 3.4. Using the notation of §2, let $K = K_{n-1} = \vee_{\alpha \in V} S_\alpha^{n-1}$ and $J = J_{n-1} = \vee_{\alpha \in U} S_\alpha^{n-1}$; and let K_{n-2} and J_{n-2} be similarly defined for the complex A. Let f be a homotopy self-equivalence of $A \cup B$ which restricts to a map $g : A \to A$, and $Sh = f/g : J \to J$. Then there is the commutative diagram of split short exact sequences of free abelian groups:

$$\begin{array}{ccccc}
H_{n-1}(J_{n-1}) & \rightleftarrows & H_{n-1}(A) & \to & H_{n-1}(A \cup B) \\
\downarrow h_* & & \downarrow g_* & & \cong \downarrow f_* \\
H_{n-1}(J_{n-1}) & \rightleftarrows & H_{n-1}(A) & \to & H_{n-1}(A \cup B)
\end{array}$$

A splitting is given by a composite function $A \to SK_{n-2} \xrightarrow{s} J_{n-1}$ by lemma 2.1. Using the coaction define a new function

$$\bar{g} : A \to A \vee SK_{n-2} \xrightarrow{(g,\,(1-h)s)} A \vee J_{n-1} \to A .$$

Then $i_{A \cup B} \circ g \sim i_{A \cup B} \circ \bar{g}$ extends to $f \sim f$ and $h^* = id : H_{n-1}(J_{n-1}) \to H_{n-1}(J_{n-1})$. Thus $\bar{g}$ is a homology equivalence and therefore a homotopy equivalence.

We prove theorem 3.1 below by first proving it for the canonical form $A \cup B \cup C$ for X. Since the homotopy equivalence $X \to A \cup B \cup C$ is rel A, it induces isomorphisms making the following diagram commutative and the result follows.

$$\begin{array}{ccc} \mathcal{L}_{i_n}(X_n) & \cong & \mathcal{L}_{i_A}(A\cup B\cup C) \\ \downarrow & & \downarrow \\ \mathcal{E}(X_n) & \cong & \mathcal{E}(A\cup B\cup C) \end{array}$$

Consider the cofibre sequence

$$A\cup B \xrightarrow{i_{A\cup B}} A\cup B\cup C \xrightarrow{p_C} SK$$

Define $\mathcal{E}_0(A\cup B\cup C) = \{\alpha \in \mathcal{E}(A\cup B\cup C) : p_C \circ \alpha \circ i_{A\cup B} = 0\}$. Let α be represented by $f{:}A\cup B\cup C \to A\cup B\cup C$. Since $p_C \circ f \circ i_{A\cup B} \sim *$, there is a map $\bar{f}{:}SK \to SK$ satisfying $\bar{f} \circ p_C \sim p_C \circ f$: furthermore, using the isomorphism $[X,S^n] \cong H^n(X)$ for $\dim X \leq n$, it follows that the class of $\bar{f}$ is unique since, by lemma 2.1, $H^{n-1}(A\cup B)$ is finite. By lemma 2.1 $H_n(\bar{f})$ is an isomorphism, and thus $\bar{f}$ is a homotopy equivalence. It now follows, by proposition 4.4 of [4], that there exists a homotopy equivalence f_1 making the following diagram homotopy commutative.

$$\begin{array}{ccccc} A\cup B & \to & A\cup B\cup C & \to & SK \\ f_1\downarrow & & f\downarrow & & \bar{f}\downarrow \\ A\cup B & \to & A\cup B\cup C & \to & SK \end{array}$$

Futhermore we have easily that

$$\mathcal{E}_0(A\cup B\cup C) = \mathcal{L}_{i_{A\cup B}}(A\cup B\cup C) = \mathcal{R}_{p_C}(A\cup B\cup C).$$

Remark 3.3 If A is 1-dimensional, then a map $f_1 : A \to A$ extending to a homotopy equivalence $f : A\cup C \to A\cup C$ need not itself be a homotopy equivalence. For example let $A = S^1 \vee S^1$ and $h = \iota_1\iota_2\iota_1^{-1}\iota_2^{-1} : S^1 \to S^1 \vee S^1$, then $A \subset A\cup C$ is equivalent to $S^1 \vee S^1 \subset S^1 \times S^1$. Choose $f_1 = (\iota_2\iota_1\iota_2^{-1}, \iota_1\iota_2\iota_1^{-1})$, then f_1 is not a homotopy equivalence, though f and $\bar{f}$ are homotopic to the identity maps. Note that, in this case, there is no map $g{:}S^1 \to S^1$ for which f is the mapping cone on (f_1, g).

Lemma 3.4 Let A be simply-connected and $(n-1)$-dimensional $(n \geq 3)$, then the set of cosets $\mathcal{E}(A\cup B\cup C)/\mathcal{E}_0(A\cup B\cup C)$ is bijective with a subgroup of $\hom([SK,SK];[A\cup B,SK])$. In particular if B and C have finitely many cells, then $\mathcal{E}_0(A\cup B\cup C)$ has finite index in $\mathcal{E}(A\cup B\cup C)$.

Proof The quotient map $A\cup B\cup C \to A\cup B\cup C/A^{n-2}$ induces the diagram

$$\begin{array}{ccccccc} 0 \leftarrow H^n(A\cup B) & \leftarrow & H^n(A\cup B\cup C) & \leftarrow & H^n(SK) \leftarrow 0 \\ \uparrow\cong & & \uparrow\cong & & \uparrow\cong \\ H^n(A\cup B/A^{n-2}) & \leftarrow & H^n((A\cup B\cup C/A^{n-2})\vee SK) & \leftarrow & H^n(SK) \end{array}$$

Clearly there is a canonical splitting $H^n(A\cup B)\to H^n(A\cup B\cup C)$ onto the torsion subgroup. The inclusion $SK=\vee S^n\subset \prod S^n$ determines abelian group structures for which the following sequence is split exact by the previous remarks

$$0\to [SK,SK]\to [A\cup B\cup C,SK] \underset{\kappa}{\rightleftarrows} [A\cup B,SK]\to 0.$$

Also $\kappa([A\cup B,SK])$ is the torsion subgroup of $[A\cup B\cup C,SK]$. Thus given $\beta\in[A\cup B\cup C,A\cup B\cup C]$ there is the decomposition $p_C\circ\beta=\hat{\beta}\circ p_C+\kappa(\beta')$, where $\hat{\beta}\in[SK,SK]$ and $\beta'=p_C\circ\beta\circ i_{A\cup B}$. Given that $\beta\in\mathcal{E}(A\cup B\cup C)$, then $\hat{\beta}\in\mathcal{E}(SK)$, since $H_n(p_C)$ is an isomorphism by lemma 2.1. Suppose next that $\beta,\gamma\in[A\cup B\cup C,A\cup B\cup C]$, then

$$p_C\circ\beta\circ\gamma=\hat{\beta}\circ p_C\circ\gamma+\kappa(\beta')\circ\gamma=\hat{\beta}\circ\hat{\gamma}\circ p_C+\hat{\beta}\circ\kappa(\gamma')+\kappa(\beta')\circ\gamma .$$

Thus $(\beta\circ\gamma)\hat{}=\hat{\beta}\circ\hat{\gamma}$, and hence $g:\mathcal{E}(A\cup B\cup C)\to\mathcal{E}(SK)$, given by $g(\beta)=\hat{\beta}$ is a homomorphism. Also $p_C\circ\beta\circ\gamma\circ i_{A\cup B}=(\beta\circ\gamma)'=\hat{\beta}\circ\gamma'+\gamma_b(\beta')$, where γ_b is the automorphism of $[A\cup B,SK]$ induced by the automorphism γ^* of $[A\cup B\cup C,SK]$: thus $\gamma_b(x)=(i_{A\cup B})^*\gamma^*\kappa(x)$. Now let $F=[SK,SK]$ and $T=[A\cup B,SK]$, then F is free abelian, and T is a torsion group. Thus there is a split exact sequence

$$0\to \hom(F,T)\to \operatorname{aut}(F\oplus T)\rightleftarrows \operatorname{aut}F\times\operatorname{aut}T\to 1$$

and the elements of $\operatorname{aut}(F\oplus T)$, operating on the right, may be written in matrix form $\begin{pmatrix} f & \tau\\ 0 & t\end{pmatrix}$ with $t\in\operatorname{aut}T$, $f\in\operatorname{aut}F$ and $\tau\in\hom(F,T)$, and with the multiplication

$$\begin{pmatrix} f & g\\ 0 & t\end{pmatrix}\begin{pmatrix} f' & \tau'\\ 0 & t'\end{pmatrix} = \begin{pmatrix} f\circ f' & f\circ\tau'+\tau\circ t'\\ 0 & t\circ t'\end{pmatrix}$$

Right composition yields the homomorphism $\phi:(A\cup B\cup C)\to\operatorname{aut}(F\oplus T)$ and

$$\phi(\beta) = \begin{pmatrix} \hat{\beta}^* & (\beta')^*\\ 0 & \beta_b\end{pmatrix}$$

where

$$\begin{pmatrix} \hat{\beta}^* & (\beta')^* \\ 0 & \beta_b \end{pmatrix}\begin{pmatrix} \hat{\gamma}^* & (\gamma')^* \\ 0 & \gamma_b \end{pmatrix} = \begin{pmatrix} \hat{\beta}^* \circ \hat{\gamma}^* & \hat{\beta}^* \circ (\gamma')^* + (\beta')^* \circ \gamma_b \\ 0 & \beta_b \circ \gamma_b \end{pmatrix}$$

Since $\mathcal{E}_0(A \cup B \cup C) = \{\beta : \beta' = 0\}$, we have the pullback diagram

$$\begin{array}{ccc} \mathcal{E}_0(A \cup B \cup C) & \rightarrow & \mathcal{E}(A \cup B \cup C) \\ \downarrow & & \downarrow \phi \\ \text{aut } F \times \text{aut } T & \rightarrow & \text{aut}(F \oplus T) \end{array}$$

The cosets of aut $F \times$ aut T in aut $(F \oplus T)$ are indexed by hom (F,T), and therefore the cosets of $\mathcal{E}_0(A \cup B \cup C)$ in $\mathcal{E}(A \cup B \cup C)$ are indexed by $\text{hom}(F,T) \cap \phi(\mathcal{E}(A \cup B \cup C))$.

Next consider the cofibre sequence

$$A \xrightarrow{i_A} A \cup B \xrightarrow{p_B} SJ.$$

Lemma 3.5 Let A be simply-connected and $(n-1)$-dimensional $(n \geq 3)$, and let $A \cup B$ have finitely many cells, then $\mathcal{L}_{i_A}(A \cup B)$ has finite index in $\mathcal{E}(A \cup B)$.

Proof By proposition 4.4 of [4] we have $\mathcal{L}_{i_A}(A \cup B) = \mathcal{R}_{p_B}(A \cup B)$. The isotropy group of $p_B \in (A \cup B, SJ)$ under the action of $\mathcal{E}(A \cup B)$ is $\mathcal{R}^1_{p_B}(A \cup B)$. Since $(A \cup B, SJ)$ is finite, $\mathcal{R}^1_{p_B}(A \cup B)$ has finite index in $\mathcal{E}(A \cup B)$, and the result follows.

Consider the cofibre sequence

$$A \xrightarrow{i_A} A \cup B \cup C \xrightarrow{p_B} S(J \vee K).$$

Lemma 3.6 Let A be simply-connected and $(n-1)$-dimensional $(n \geq 3)$, and let $A \cup B \cup C$ have finitely many n-cells, then $\mathcal{L}_{i_A}(A \cup B \cup C)$ has finite index in $\mathcal{E}(A \cup B \cup C)$.

Proof Consider the pullback diagram

$$\begin{array}{ccc} P & \rightarrow & \mathcal{R}_{i_{A \cup B}}(A \cup B) \cap \mathcal{L}_{i_A}(A \cup B) / \mathcal{R}^1_{i_{A \cup B}}(A \cup B) \cap \mathcal{L}_{i_A}(A \cup B) \\ \downarrow & & \downarrow \chi \\ \mathcal{L}_{i_{A \cup B}}(A \cup B \cup C) & \rightarrow & \mathcal{R}_{i_{A \cup B}}(A \cup B) / \mathcal{R}^1_{i_{A \cup B}}(A \cup B) \end{array}$$

The lower horizontal map is given by $\mathcal{L}$-$\mathcal{R}$ duality (theorem 2.1 of [5]), and χ is the monomorphism induced by the inclusion $\mathfrak{R}_{i_{A\cup B}}(A\cup B)\cap\mathfrak{L}_{i_A}(A\cup B)\subset\mathfrak{R}_{i_{A\cup B}}(A\cup B)$. Now $\mathfrak{R}_{i_{A\cup B}}(A\cup B)\cap\mathfrak{L}_{i_A}(A\cup B)$ has finite index in $\mathfrak{R}_{i_{A\cup B}}(A\cup B)$ by lemma 3.5, and therefore χ is a monomorphism onto a subgroup of finite index by the first isomorphism theorem. Hence $P\rightarrow\mathfrak{L}_{i_{A\cup B}}(A\cup B\cup C)$ is a monomorphism onto a subgroup of finite index: identifying P with its image, we have, by lemma 3.4, that P has finite index in $\mathcal{E}(A\cup B\cup C)$. Also it is evident from the construction of P that $P=\mathfrak{L}_{i_A}(A\cup B\cup C)$, and the lemma follows.

4 Proof of theorem B

We now assume that X is a simply-connected finite complex, and prove theorem B in this case by a finite induction. The general case of theorem B then follows by modifying this proof using theorem 3.2.

Let X be n–dimensional and let $i_r: X_{r-1}\rightarrow X_r$ be the inclusion of the (r–1)-skeleton $(3\leq r\leq n)$. Recall that $X_1=pt$.

Let $\mathfrak{F}_r\subset\mathcal{E}(X_r)$ $(2\leq r\leq n)$ consist of those classes α which can be represented by a map $f_r: X_r\rightarrow X_r$ such that f_r extends to $f_n: X_n\rightarrow X_n$ and $f_n|X_s: X_s\rightarrow X_s$ is a homotopy equivalence $(r\leq s\leq n)$. Also let $\mathfrak{F}^1_r\subset\mathfrak{F}_r$ consist of those classes for which $i_{r+1}\circ\alpha=i_{r+1}$: this condition is equivalent to there being an extension f_n such that $f_n|X_s$ is homotopic to the identity $(r+1\leq s\leq n)$. Consider the diagram

$$\begin{array}{ccc} \mathfrak{F}_r\cap\mathfrak{L}_{i_r}(X_r) & \xrightarrow{i_r^*} & (X_{r-1}, X_r) \\ & & \uparrow i_{r*} \\ & & \mathfrak{F}_{r-1} \end{array}$$

Given u in $\mathfrak{F}_r\cap\mathfrak{L}_{i_r}(X_r)$, choose $\bar{u}$ in $\mathfrak{F}_{r-1}$ with $i_r\circ\bar{u}=u\circ i_r$, and define the product

$$i_r^*(u)\,.\,i_r^*(v)=i_r^*(u\circ v)=i_{r*}(\bar{u}\circ\bar{v})=i_{r*}(\bar{u})\,.\,i_{r*}(\bar{v})$$

in $i_r^*(\mathfrak{F}_r\cap\mathfrak{L}_{i_r}(X_r))=i_{r*}(\mathfrak{F}_{r-1})$. The multiplication on this latter set is well defined and gives the set a group structure (cf. §2 of [5]).

The following result is now immediate.

Lemma 4.1 The subgroups $\mathfrak{F}_r\cap\mathfrak{L}^1_{i_r}(X_r)$ and $\mathfrak{F}^1_{r-1}$ are normal in $\mathfrak{F}_r\cap\mathfrak{L}_{i_r}(X_r)$ and $\mathfrak{F}_{r-1}$ respectively, and the following sequence of homomorphisms is exact

$$1 \to \mathfrak{F}_r \cap \mathfrak{L}^1_{i_r}(X_r) \to \mathfrak{F}_r \cap \mathfrak{L}_{i_r}(X_r) \to \mathfrak{F}_{r-1} / \mathfrak{F}^1_{r-1} \to 1.$$

Next let $\mathcal{E}_r$ $(2 \le r \le n)$ be the subgroup of $\mathcal{E}(X)$ consisting of those classes which can be represented by maps f such that $f|X_s : X_s \to X_s$ is a homotopy equivalence $(r \le s \le n)$. Define $\theta : \mathcal{E}_r \to [X_r, X_{r+1}]$ by $\theta([f]) = [i_{r+1} \circ f_r]$. As before define a group structure on $i_{r+1*}(\mathfrak{F}_r)$ by $(i_{r+1})_*(\bar{u}).(i_{r+1})_*(\bar{v}) = (i_{r+1})_*(\bar{u} \circ \bar{v})$, then θ is a homomorphism onto $(i_{r+1})_*(\mathfrak{F}_r)$ with kernel $\mathcal{E}^1_r$ consisting of those classes which can be represented by cellular maps f with $f|X_r = 1 : X_r \to X_r$. We then have the obvious lemma.

Lemma 4.2 The following sequence of homorphisms is exact

$$1 \to \mathcal{E}^1_r \to \mathcal{E}_r \to \mathfrak{F}_r / \mathfrak{F}^1_r \to 1 .$$

Proof of theorem B Let X be n-dimensional $(n \ge 3)$. By lemma 3.6 $\mathcal{E}_{n-1} = \mathfrak{L}_{i_n}(X_n)$ has finite index in $\mathcal{E}_n = \mathcal{E}(X)$. Let $2 \le r \le n-2$: we show that $\mathcal{E}_r$ has finite index in $\mathcal{E}_{r+1}$. The following diagram is a pullback

$$\begin{array}{ccc} \mathcal{E}_r & \to & \mathfrak{F}_{r+1} \cap \mathfrak{L}_{i_{r+1}} / \mathfrak{F}^1_{r+1} \cap \mathfrak{L}_{i_{r+1}} \\ \downarrow & & \downarrow \\ \mathcal{E}_{r+1} & \to & \mathfrak{F}_{r+1} / \mathfrak{F}^1_{r+1} \end{array}$$

Now $\mathfrak{F}_{r+1} \cap \mathfrak{L}_{i_{r+1}}$ has finite index in $\mathfrak{F}_{r+1}$ by lemma 3.6, and therefore $\mathcal{E}_r$ has finite index in $\mathcal{E}_{r+1}$ as in the proof of lemma 3.6.

References

1 S.C. Chang, Homotopy invariants and continuous mappings, Proc. Royal Soc. A 202 (1950) 253-263.
2 M. Hall Jnr., The theory of groups, Macmillan, New York (1959).
3 P.J. Hilton, Homotopy theory and duality, Gordon and Breach, 1965.
4 J.W. Rutter, Maps and equivalences into equalizing fibrations and from coequalizing cofibrations, Math. Z. 122 (1971) 125-141.
5 J.W. Rutter, Self equivalences and principal morphisms, Proc. London Math. Soc., 20 (1970) 644-58.
6 J.W. Rutter, The group of homotopy self-equivalences of CW complexes, Math. Proc. Cambridge Phil. Soc. 93 (1983) 275-293.
7 D. Sullivan, Infinitesimal computations in topology, Publ. Math. I.H.E.S. 47 (1977) 269 - 331.
8 J.H.C. Whitehead, A certain exact sequence, Ann. Math., 52 (1950) 51 - 110.
9 C.W. Wilkerson, Applications of minimal simplicial groups, Topology 15 (1976) 111-130.

Self-Homotopy Equivalences and Highly Connected Poincaré Complexes

Kohhei Yamaguchi

Department of Mathematics, The University of Electro-Communications,
1-5-1, Chofugaoka, Chofu, Tokyo 182, Japan.

§0. Introduction.

For a based space X, let $E(X)$ be the group consisting of all homotopy classes of based self-homotopy equivalences of X with the multiplication induced from the composition of maps. The group $E(X)$ is a natural one to study and plays an important role in the homotopy type classification problem. The homotopy types of (n-1)-connected 2n dimensional Poincaré complexes were already classified by J. Milnor and C.T.C. Wall [14] and in this paper, we would like to study the homotopy types of (n-2)-connected 2n dimensional Poincaré complexes. Since the case n = 3 was treated by C.T.C. Wall [15], A.V. Zubr [19] and the author [18], we mainly consider the case n greater than 3. For this purpose, we determine the group $E(X)$ here in detail for some elementary spaces, the r-times bouquet of the k-fold suspensions of 2 dimensional complex projective spaces,

(0.1) $X = \Sigma^{k-1}K_r$, where we put $K_r = \overset{r}{\vee}\Sigma CP^2$.

In particular, the case r = 1 was already determined by S. Oka [9] and he determined the ring structure of $[\Sigma^k K_1, \Sigma^k K_1]$.

Then our main results are as follows:

Theorem A. If $k \geq 1$, the composite of homomorphisms

$\Sigma^{k-1} \circ \Xi: E(\Sigma^k K_r) \longrightarrow Inv(M(r))$ is an isomorphism of groups, where for a ring R, let Inv(R) denote the group consisting of all unit elements of R.

(Here Σ^{k-1} denotes the (k-1)-fold suspension homomorphism, and the homomorphism Ξ and the ring $M(r)$ are defined in §3.)

On the other hand, if k = 0, the homotopy set $[K_r, K_r]$ does not become a ring and it seems difficult to determine explicitely the group structure of $E(K_r)$.

This paper is in final form and no version of it will be submitted for publication elsewhere.

Theorem B. The sequence

$$1 \longrightarrow 1 + \mathrm{Ker}\Sigma \longrightarrow \mathcal{E}(K_r) \xrightarrow{\Xi_\circ \Sigma} \mathrm{Inv}(M(r)) \longrightarrow 1$$

is exact as a multiplicative group, where we put

(0.2) $1 + \mathrm{Ker}\Sigma = \{1+\mu\colon \mu \text{ is an element of } \mathrm{Ker}\Sigma\}$,

1 denotes the identity map of K_r and $\mathrm{Ker}\Sigma$ is given by (2.8).

Remark 0.3. The group structure of $1 + \mathrm{Ker}\Sigma$ is induced from the composition of maps. However, we could not determine its multiplication, explicitely.

(0.4) $(1+\mu)_\circ(1+\gamma) \equiv 1 + \mu + \gamma + \mu_\circ\gamma \pmod{\mathrm{Ker}\Sigma}$ for $\mu, \gamma \in \mathrm{Ker}\Sigma$.

Theorem C. Let M and N be Poincaré complexes which satisfy the condition (4.8) and are of type I or II. If $n \geq 4$ and the cohomology rings $H^*(M,Z)$ and $H^*(N,Z)$ are isomorphic as graded rings, then the suspensions ΣM and ΣN are of the same homotopy type.

Remark 0.5. (H. Ishimoto, [3]) If M has the type I and N has the type II, the cohomology rings $H^*(M,Z/2)$ and $H^*(N,Z/2)$ are not isomorphic as A_2-modules, where A_p denotes the mod p Steenrod algebra.
Thus, under the same assumptions as Theorem C, if M and N have different types, they are not of the same homotopy type.

Conjecture D. Under the same assumptions as Theorem C, if M and N have the same type X (X = I or II), then they are of the same homotopy type.

This paper is organized as follows:
In §1, we construct the generators of $\pi_*(K_r)$ and determine the homotopy group $\pi_*(K_r)$. In §2, we determine the additive structure of the homotopy set $[K_r,K_r]$. In §3, we investigate its multiplicative structure and give the proofs of theorems A and B. In §4, we define the types of Poincaré complexes and prove Theorem C. In §5, we consider the Poincaré complexes of type O.

The author would like to take this opportunity to sincerely thank Professors H. Ishimoto and S. Sasao for their advice and good suggestions.

§1. Homotopy Groups.

We work only with pointed spaces and basepoint-preserving maps or homotopies. We denote by $[X,Y]$ the set of based homotopy classes of maps from X to Y and will not distinguish between a map and its homotopy class. Let $Z\{x\}$ (resp. $Z/m\{x\}$) be the infinite additive cyclic group (resp. additive cyclic group of order m) with the generator x. Let η_2 (resp. ω) be the Hopf map (resp. the Blakers-Massey map) contained in $\pi_3(S^2)$ (resp. $\pi_6(S^3)$). We put $\eta_n = E^{n-2}\eta_2$, $\eta_n^2 = \eta_n\circ\eta_{n+1}$, and $\eta_n^3 = \eta_n\circ\eta_{n+1}\circ\eta_{n+2}$ for $n \geq 2$, where E^m denotes the m-fold suspension homomorphism.

Then one easily checks:

<u>Lemma 1.1.</u> (H. Toda, [13]) (1) $\pi_{n+1}(S^n) = Z/2\{\eta_n\}$ $(n \geq 3)$, $\pi_3(S^2) = Z\{\eta_2\}$, $\pi_{n+2}(S^n) = Z/2\{\eta_n^2\}$ $(n \geq 2)$, $\pi_6(S^3) = Z/12\{\omega\}$, $\pi_7(S^4) = Z\{\nu_4\} \oplus Z/12\{E\omega\}$.

(2) $\eta_3\circ E\omega = 0$, $\eta_3\circ\nu_4 = \omega\circ\eta_6$.

Let $C(f)$ denote the mapping cone of the map f and id_n be the identity map of S^n. Consider the cofiber sequence

$$(1.2) \quad S^4 \xrightarrow{\eta_3} S^3 \xrightarrow{i} \Sigma CP^2 \xrightarrow{p} S^5 \xrightarrow{\eta_4} S^4.$$

Since the order of η_3 is two, there is a coextension of $2id_4$, $\ell \in \pi_5(\Sigma CP^2)$ satisfying the condition

$$(1.3) \quad p\circ\ell = 2id_5.$$

Let $[f,g]$ (resp. $[f,g]_r$) be the Whitehead product (resp. relative Whitehead product) of the maps f and g, and α be the characteristic map of the top cell of ΣCP^2. Then using the Blakers-Massey Theorem and Theorem 2.1 in [4], one easily checks:

<u>Lemma 1.4</u> (1) <u>The pair</u> $(\Sigma CP^2, S^3)$ <u>is 4-connected,</u> $\pi_5(\Sigma CP^2, S^3) = Z\{\alpha\}$, <u>and</u> $\pi_6(\Sigma CP^2, S^3) = \alpha_*\pi_6(D^5, S^4) \cong Z/2$.

(2) $\pi_7(\Sigma CP^2, S^3) = Z\{[\alpha, id_3]_r\} \oplus \alpha_*\pi_7(D^5, S^4) \cong Z \oplus Z/2$.

(3) $\pi_8(\Sigma CP^2, S^3) = Z/2\{[\alpha, \eta_3]_r\} \oplus \alpha_*\pi_8(D^5, S^4) \cong Z/2 \oplus Z \oplus Z/12$.

Let $j: \Sigma CP^2 \longrightarrow (\Sigma CP^2, S^3)$ be the inclusion map. Since SU(3) is the total space of S^3-bundle over S^5, there is a fiber sequence

(1.5) ξ: $S^3 = SU(2) \longrightarrow SU(3) \longrightarrow SU(3)/SU(2) = S^5$.

It follows from the ring structure of $H^*(SU(3), Z/2)$ that the 7-skeleton of SU(3) is homotopy equivalent to ΣCP^2. Then one easily checks:

Lemma 1.6. There exists some element β in $\pi_7(\Sigma CP^2)$ satisfying the following conditions:

(1) $SU(3) = C(\beta)$.

(2) $j_\circ \beta = [\alpha, id_3]_r$.

(3) $p_\circ \beta = 0$.

Proof. The assertions (1) and (2) follow immediately from (3.3) in [7] and (5.1) in [8]. To prove (3), it suffices only to show $E(p_\circ \beta) = 0$. Let $c = c(\beta) \in \pi_4(SO(4))$ be the characteristic element of ξ. From (3.1) in [6], we have $E\beta = Ei_\circ J(c)$, where J denotes the J-homomorphism. Since $p_\circ i = 0$, $E(p_\circ \beta) = E(p_\circ i)_\circ J(c) = 0$. Q.E.D.

Proposition 1.7. (1) $\pi_k(\Sigma CP^2) = 0$ for $k = 1, 2, 4$.

(2) $\pi_3(\Sigma CP^2) = Z\{i\}$.

(3) $\pi_5(\Sigma CP^2) = Z\{\ell\}$.

(4) $\pi_6(\Sigma CP^2) = Z/6\{i_\circ \omega\}$.

(5) $\pi_7(\Sigma CP^2) = Z\{\beta\}$.

Proof. Consider the homotopy exact sequence of the pair $(\Sigma CP^2, S^3)$. It is well-known that

(1.8) $\partial([f,g]_r) + [\partial(f), g] = 0$, where ∂ denotes the boundary homomorphism of the homotopy exact sequence.

Since $\partial(\alpha) = \eta_3$, the above results follow from (1.1), (1.4), (1.6) and (1.8) This completes the proof. Q.E.D.

Remark 1.9. In general, there are two possibilities of the choice of the coextension ℓ. However, it is easy to see that $i_*(\eta_3^2) = 0$. Thus the coextension ℓ is uniquely determined.

Let $j_t : \Sigma CP^2 \longrightarrow K_r$, and $p_t : K_r \longrightarrow \Sigma CP^2$ be the inclusion map and projection map to the t-th factor. Similarly, let

$$i_t : S^3 \longrightarrow \overset{r}{\vee} S^3 \quad (\text{resp. } i'_t : S^5 \longrightarrow \overset{r}{\vee} S^5)$$

and

$$\pi_t : \overset{r}{\vee} S^3 \longrightarrow S^3 \quad (\text{resp. } \pi'_t : \overset{r}{\vee} S^5 \longrightarrow S^5)$$

denote the inclusion map and projection map to the t-th factor, respectively.

Then it follows from (1.7) and the Hilton-Milnor Theorem that we easily have:

Proposition 1.10. (1) $\pi_3(K_r) = \sum_{t=1}^{r} Z\{j_t \circ i\}$.

(2) $\pi_k(K_r) = 0$ for $k = 1, 2, 4$.

(3) $\pi_5(K_r) = (\sum_{t=1}^{r} Z\{j_t \circ \ell\}) \oplus (\sum_{t \neq s} Z\{[j_t \circ i, j_s \circ i]\})$.

(4) $\pi_7(K_r) = (\sum_{t=1}^{r} Z\{j_t \circ \beta\}) \oplus (\sum_{t \neq s} Z\{[j_t \circ i, j_s \circ \ell]\}) \oplus$

$(\sum_{t \neq s, s \neq m, m \neq t} Z\{[j_s \circ i, [j_t \circ i, j_m \circ i]]\})$.

§2. The Homotopy Set $[K_r, K_r]$.

Since K_r is a suspension space, the homotopy set $[K_r, K_r]$ has the group structure induced from the track addition. Consider the cofiber sequence

$$(2.1) \quad \overset{r}{\vee} S^4 \xrightarrow{\vee \eta_3} \overset{r}{\vee} S^3 \xrightarrow{\vee i} K_r \xrightarrow{\vee p} \overset{r}{\vee} S^5 \xrightarrow{\vee \eta_4} \vee S^4.$$

Since $[\vee S^4, K_r] = 0$, we have the following exact sequence:

$$0 \longrightarrow [\overset{r}{\vee} S^5, K_r] \xrightarrow{(\vee p)^*} [K_r, K_r] \xrightarrow{(\vee i)^*} [\overset{r}{\vee} S^3, K_r] \longrightarrow 0,$$

(2.2) where $[\overset{r}{\vee} S^5, K_r] = \sum_{t=1}^{r} \pi_5(K_r) \circ \pi'_t$ and

$[\overset{r}{\vee} S^3, K_r] = \sum_{t=1}^{r} \pi_3(K_r) \circ \pi_t$.

Remark 2.3. Since K_r is a single suspension space, the group structure of $[K_r, K_r]$ is not necessarily abelian. Thus it seems difficult to solve the extension problem of (2.2).

Definition 2.4. We define the elements of $[K_r, K_r]$, σ_{ts}, λ_{ts} and μ^m_{ts} as follows:

$\sigma_{ts} = j_t \circ p_s$, $\lambda_{ts} = j_t \circ \ell \circ p \circ p_s$ for $1 \leq t, s \leq r$ and

$\mu^m_{ts} = [j_t \circ i, j_s \circ i] \circ p \circ p_m$ for $1 \leq t, s, m \leq r$; $t \neq s$.

Lemma 2.5. (1) $\sigma_{ts} \circ \sigma_{mn} = \delta_{sm} \sigma_{tn}$,

(2) $\sigma_{ts} \circ \lambda_{mn} = \lambda_{ts} \circ \sigma_{mn} = \delta_{sm} \lambda_{tn}$, and

(3) $\lambda_{ts} \circ \lambda_{mn} = (2\delta_{sm}) \lambda_{tn}$, where δ_{sm} denotes the Kronecker delta.

Proof. This follows immediately from (1.2), (1.3) and (2.4). Q.E.D.

Lemma 2.6. The group $[K_r, K_r]$ is generated by the following elements: σ_{ts}, λ_{ts} for $1 \leq t, s \leq r$, and μ^m_{ts} for $1 \leq t, s, m \leq r$; $t \neq s$.

Proof. From (1.10), (2.2) and (2.4) the result follows. Q.E.D.

On the other hand, if $k \geq 1$, the space $\Sigma^k K_r$ is a double suspension space and $[\Sigma^k K_r, \Sigma^k K_r]$ has the abelian group structure. Thus similar calculations show the following results:

Theorem 2.7. If $k \geq 1$, there is an isomorphism of abelian groups

$[\Sigma^k K_r, \Sigma^k K_r] = (\sum_{1 \leq t, s \leq r} Z\{\Sigma^k \sigma_{ts}\}) \oplus (\sum_{1 \leq t, s \leq r} Z\{\Sigma^k \lambda_{ts}\})$.

In particular, the suspension homomorphism

$\Sigma: [\Sigma^k K_r, \Sigma^k K_r] \longrightarrow [\Sigma^{k+1} K_r, \Sigma^{k+1} K_r]$ is an isomorphism.

Corollary 2.8. If $k \geq 1$, the iterated suspension homomorphism

$\Sigma^k: [K_r, K_r] \longrightarrow [\Sigma^k K_r, \Sigma^k K_r]$ is an epimorphism and

(2.9) $\mathrm{Ker}\Sigma^k = \sum_{1 \leq t, s, m \leq r; t \neq s} Z\{\mu^m_{ts}\}$.

Proof. This is obtained by combining (2.6) and (2.7). Q.E.D.

Remark 2.10. If $r = 1$, (2.7) was already obtained by S. Oka [9].

The homotopy set $[X, X]$ has the multiplicative structure defined by the composition of maps and in the next section we consider it for $X = \Sigma^k K_r$. ($k \geq 0$)

§3. The Multiplicative Structure.

First, we need the following:

Definition 3.1. For a positive integer m, let Mat(m,Z) (resp. GL(m,Z)) be the ring (resp. group) consisting of all (m,m)-matrices (resp. (m,m)-matrices which have their inverse matrices) whose entries are integers.
For elements of Mat(r,Z), A and B, we define the element of Mat(2r,Z), [A,B], by the following:

$$(3.2)\quad [A,B] = \begin{pmatrix} A & O \\ O & A+2B \end{pmatrix}$$

Let M(r) be the set of all matrices of the form (3.2). Then one easily checks

Lemma 3.3. (1) $[A,B] + [C,D] = [A+C, B+D]$.

(2) $m[A,B] = [mA, mB]$.

(3) $[A,B][C,D] = [AC, AD+BC+2BD]$.

(4) If the matrix $[A,B]$ is contained in $GL(2r,Z)$, then A and $A+2B$ are contained in $GL(r,Z)$ and

$$[A,B]^{-1} = [A^{-1}, -(A+2B)^{-1}BA^{-1}].$$

Thus, the set M(r) is closed under the addition and multiplication. In particular, M(r) is a subring of Mat(2r,Z).

Definition 3.4. We define the homomorphism of abelian groups

$\Xi : [\Sigma K_r, \Sigma K_r] \longrightarrow M(r)$ by the following:

(3.5) $\Xi(f) = [A,B]$ for $f \in [\Sigma K_r, \Sigma K_r]$,

where if the map f is the following form

(3.6) $f = \sum_{1\le t,s\le r}\{a_{ts}(\Sigma\sigma_{ts}) + b_{ts}(\Sigma\lambda_{ts})\}$, then we put

(3.7) $A = (a_{ts})$ and $B = (b_{ts})$.

Remark 3.8. Since the space $\Sigma^k K_r$ is a stable complex for $k \ge 1$, the homotopy set $[\Sigma^k K_r, \Sigma^k K_r]$ has a ring structure whose addition and multiplication are the track addition and the map-composition, respectively.

To prove Theorem A, it suffices only to show the following result:

Theorem A'. If $k \ge 1$, the composite of the homomorphisms $\Sigma^{k-1} \circ \Xi : [\Sigma^k K_r, \Sigma^k K_r] \longrightarrow M(r)$ is an isomorphism of rings.

Proof. From (2.7), (3.5), (3.6) and (3.7), one may easily verify the following:

(3.9) (1) The homomorphism $\Sigma^{k-1}\circ\Xi$ is an isomorphism of abelian groups.

(2) $\Xi(1) = E$, where we denote by 1 and E, the identity map of ΣK_r and the identity matrix of $Mat(2r,Z)$.

It suffices to show that Ξ is a multiplicative homomorphism. Let f and g be the elements of $[\Sigma K_r, \Sigma K_r]$ of the forms:

$$f = \sum_{1\le t,s\le r}\{a_{ts}(\Sigma\sigma_{ts}) + b_{ts}(\Sigma\lambda_{ts})\},$$

$$g = \sum_{1\le t,s\le r}\{c_{ts}(\Sigma\sigma_{ts}) + d_{ts}(\Sigma\lambda_{ts})\}.$$

We put $A = (a_{ts})$, $B = (b_{ts})$, $C = (c_{ts})$ and $D = (d_{ts})$. Then, we have $\Xi(f) = [A,B]$ and $\Xi(g) = [C,D]$.

On the other hand, since two maps f and g are suspension elements, using (2.5) an easy calculation shows the following:

$$f\circ g = \sum_{1\le t,s\le r}\{x_{ts}(\Sigma\sigma_{ts}) + y_{ts}(\Sigma\lambda_{ts})\}, \text{ where we put}$$

$$x_{ts} = \sum_{1\le m\le r} a_{tm}c_{ms} \text{ and } y_{ts} = \sum_{1\le m\le r}\{a_{tm}d_{ms} + b_{tm}(c_{ms}+2d_{ms})\}.$$

Hence, using (3.3) we have

$$\Xi(f\circ g) = [AC, AD+BC+2BD] = [A,B][C,D] = \Xi(f)\Xi(g).$$ Q.E.D.

Proof of Theorem B. This follows immediately from (2.6), (2.8), (2.9) and Theorem A'. Q.E.D.

§4. Homotopy Types of Poincaré Complexes.

In this section we consider the classification problem of homotopy types of (n-2)-connected 2n dimensional Poincaré complexes. First, we need the following:

Definition 4.1. Let M be an (n-2)-connected 2n dimensional Poincaré complex $(n \geq 3)$ with torsion free homology group $H_*(M,Z)$ and we put $r = \mathrm{rank}(H_{n-1}(M,Z))$.

Let $\phi: H^{n-1}(M,Z/2) \times H^{n-1}(M,Z/2) \longrightarrow Z/2$ be the symmetric bilinear form defined by

(4.2) $\phi(x,y) = \langle Sq^2(x)\cdot y, [M]_2\rangle$ for $x, y \in H^{n-1}(M,Z/2)$,

where $[M]_2$ denotes the mod 2 fundamental class of $H_{2n}(M,Z/2)$ and $\langle\ ,\ \rangle$ the Kronecker product.

Remark 4.3. Using the Poincaré duality, it is easy to see that ϕ is a symmetric bilinear form.

Definition 4.4. ([3], page 212) Under the same assumptions as above, we say that M is of type O if the Steenrod square $Sq^2: H^{n-1}(M,Z/2) \longrightarrow H^{n+1}(M,Z/2)$ is trivial (that is, $\mathrm{rank}\phi = 0$).

We call M of type I if there is an element $x \in H^{n-1}(M,Z/2)$ such that $\phi(x,x) = 1$ and $\mathrm{rank}\phi = r$; M is of type II if $\phi(y,y) = 0$ for any element $y \in H^{n-1}(M,Z/2)$ and $\mathrm{rank}\phi = r$.

Similarly, if $0 < \mathrm{rank}\phi < r$, M is of type (O+I) if there is an element $x \in H^{n-1}(M,Z/2)$ satisfying the condition $\phi(x,x) = 1$, and M is of type (O+II) if $\phi(y,y) = 0$ for any element $y \in H^{n-1}(M,Z/2)$.

In particular, M belongs to some type and its type is uniquely determined.

However, in this paper we only consider the cases of type O, I and II and in a subsequent paper we will study the cases of type (O+I) and (O+II).

Lemma 4.5. ([3], page 212) Let M be an (n-2)-connected 2n dimensional closed smooth manifold with the condition $n \geq 4$.

Then if $n \equiv 0,4,6,7 \pmod 8$, M is (n-1)-parallelizable.

In particular, if M is a π-manifold or an almost parallelizable manifold, it is (n-1)-parallelizable.

Theorem 4.6. (H. Ishimoto, [3]: Theorem 3)

Let M be an (n-2)-connected 2n dimensional closed smooth manifold with torsion free homology group $H_*(M, Z)$ and be (n-1)-parallelizable. ($n \geq 4$)

(1) Then we have the decomposition $M = M_1 \# M_2$ (up to diffeomorphism), where # denotes the connected sum, M_1 is an (n-2)-connected 2n dimensional closed smooth manifold with the conditions $H_n(M_1, Z) = 0$ and (n-1)-parallelizable, and M_2 is an (n-1)-connected 2n dimensional closed smooth manifold.

(2) The manifold M_2 is always unique up to diffeomorphism mod θ_{2n}.

Here, two 2n dimensional closed smooth manifolds N and N' are called diffeomorphic mod θ_{2n} if there is a 2n dimensional homotopy sphere H such that N and N'#H are diffeomorphic, where θ_{2n} is the group of 2n dimensional homotopy spheres.

Since the homotopy types of (n-1)-connected 2n dimensional Poincaré complexes were classified by J. Milnor and C.T.C. Wall [14], we would like to consider the following:

Problem 4.7. Classify the homotopy types of M_1 under the above conditions.

So from now on we will handle the classification problem of homotopy types of Poincaré complexes which satisfy the following conditions:

(4.8) (1) M is an (n-2)-connected 2n dimensional Poincaré complex with torsion free homology group $H_*(M, Z)$. ($n \geq 3$)

(2) $H_n(M, Z) = 0$ and $r = \operatorname{rank}(H_{n-1}(M, Z)) \geq 1$.

Here we recall

Lemma 4.9. (H. Ishimoto, [3])

Let M be a Poincaré complex which satisfies the condition (4.8). If M is of type I or II and $n \geq 4$, there is an element $\gamma \in \pi_{2n-1}(\Sigma^{n-4} K_r)$ such that M and the mapping cone $C(\gamma)$ are of the same homotopy type.

Proof. This follows from the argument of the proof of Lemma 10.3 given in [3]. Q.E.D.

To study the homotopy types of mapping cones, the following result is usefull

Lemma 4.10. Let K be a simply connected m dimensional CW complex and f, g be two elements of $\pi_n(K)$, where $m + 1 \leq n$. Then the mapping cones C(f) and C(g) are of the same homotopy type if, and only if, there exists an element θ of $E(K)$ such that $g = \pm \theta \circ f$.

Proof. This is well-known and the proof is obvious. Q.E.D.

Now our result is as follows:

Theorem 4.11. Let f and g be two elements of $\pi_{2n-1}(\Sigma^{n-4}K_r)$ and $n \geq 4$. If the cohomology rings $H^*(C(f), Z)$ and $H^*(C(g), Z)$ are isomorphic as graded rings, then the suspensions of the mapping cones $\Sigma C(f)$ and $\Sigma C(g)$ are of the same homotopy type.

Proof The case $n = 4$ belongs to the metastable range and the case $n \geq 5$ to the stable range. So it suffices only to show the case $n = 4$. Similar argument shows the case $n \geq 5$. Using (1.10), we may assume the following:

(4.12) $f \equiv \sum_{t=1}^{r} a_t (j_t \circ \beta) \bmod W$, $g \equiv \sum_{t=1}^{r} b_t (j_t \circ \beta) \bmod W$, where we put

$$W = (\sum_{t \neq s} Z\{[j_t \circ i, j_s \circ i]\}) + (\sum_{t \neq s, s \neq m, m \neq t} Z\{[j_t \circ i, [j_s \circ i, j_m \circ i]]\}).$$

Since the suspension of any Whitehead product is trivial, $\Sigma(W) = \{0\}$. Hence, $\Sigma f = \sum_{t=1}^{r} a_t (\Sigma(j_t \circ \beta))$ and $\Sigma g = \sum_{t=1}^{r} b_t (\Sigma(j_t \circ \beta))$. Since $H^*(C(f), Z)$ and $H^*(C(g), Z)$ are isomorphic as cohomology rings, using (3.3) in [5] there is an element of $GL(r, Z)$, $D = (d_{ts})$, such that

(4.13) $^t(a_1, a_2, \ldots, a_r) = D \cdot {}^t(b_1, b_2, \ldots, b_r)$.

We put $\theta = \sum_{1 \leq t, s \leq r} d_{ts} (\Sigma \sigma_{ts})$. Since the matrix $[D, O]$ is contained in $GL(2r, Z)$, using Theorem A, the map θ is a self-homotopy equivalence of ΣK_r.

On the other hand, using (4.12) and (4.13), an easy calculation shows that

(4.14) $\theta \circ \Sigma f = \Sigma g$.

Thus, it follows from (4.10) that $\Sigma C(f)$ and $\Sigma C(g)$ have the same homotopy type. Q.E.D.

Proof of Theorem C. This easily follows from (4.9) and (4.11). Q.E.D.

§5. Poincaré Complexes of Type O.

In this section we consider the homotopy types of (n-2)-connected 2n dimensional Poincaré complexes which satisfy the condition (4.8) and have type O for n = 3, and we will show that the analogous assertion to Theorem C does not hold for type O case.

For s = 1 or 2, let $i_s : S^m \longrightarrow S^2 \vee S^4$ (m = 2 or 4) be the inclusion map to the first cordinate or the second cordinate, respectively.
Let M_m (m = 1. 2, 3) be simply connected six dimensional Poincaré complexes of type O defined by the following:

$$(5.1)\quad M_1 = C([i_1, i_2]) = S^2 \times S^4,\ M_2 = C([i_1, i_2] + i_1 \circ \eta_2^3),$$
$$M_3 = C([i_1, i_2] + i_2 \circ \eta_4),\ \text{where}$$
$$\pi_5(S^2 \vee S^4) = Z\{[i_1, i_2]\} \oplus Z/2\{i_1 \circ \eta_2^3\} \oplus Z/2\{i_2 \circ \eta_4\}.$$

Then it is easy to prove:

Lemma 5.2 ([18]) (1) $H^*(M_m, Z) = E(x_2, x_4)$, where dim x_k = k. (m=1,2,3)
(2) $H^*(M_m, Z/p)$ (s = 1,2,3) are all isomorphic as A_p-modules for any odd prime p.
(3) $H^*(M_1, Z/2)$ and $H^*(M_2, Z/2)$ are isomorphic as A_2-modules.
(4) $H^*(M_m, Z/2)$ (m = 1 or 2) and $H^*(M_3, Z/2)$ are not isomorphic as A_2-modules.
(5) M_m (m = 1,2,3) have the same type O.

Proof Except for (3) and (4), the assertions are obvious; the statements (3) and (4) easily follow from the Hopf invariant one problem. Q.E.D.

Proposition 5.3. The Poincaré complexes M_m (s = 1,2,3) have all different homotopy types. So the analogous result to Theorem C does not hold for the type O case.

Proof. Using (5.2) (4), M_3 and M_m (s = 1 or 2) are of the different homotopy types. Since $\Sigma M_2 = C(\Sigma i_2 \circ \eta_3^3)$ and $\Sigma M_1 = S^3 \vee S^5 \vee S^7$, the homotopy groups $\pi_6(\Sigma M_1)$ and $\pi_6(\Sigma M_2)$ are not isomorphic. Q.E.D.

Remark 5.4. ([18], (6.14)) The Poincaré complexes M_m (m = 1,2,3) have the homotopy types of closed smooth manifolds.

References.

[1] W.D. Barcus and M.G. Barratt: On the homotopy classification of a fixed map, Trans. Amer. Math. Soc., 88 (1958), 57-74.

[2] H. Ishimoto: On the structure of (n-2)-connected 2n dimensional π—manifolds, Publ. RIMS. Kyoto Univ., 5 (1969), 65-77.

[3] ________: On the classification of (n-2)-connected 2n-manifolds with torsion free homology groups, ibid., 9 (1973), 211-260.

[4] I.M. James: On the homotopy groups of certain pairs and triads, Quart. J. Math. Oxford, 5 (1954), 260-270.

[5] ________: Note on cup-products, Proc. Amer. Math. Soc., 8 (1957), 374-383.

[6] ________: On sphere bundles over spheres, Comment. Math. Helv., 35 (1961), 126-135.

[7] I.M. James and J.H.C. Whitehead: The homotopy theory of sphere bundles over spheres (I), Proc. London Math. Soc., 4 (1954), 196-218.

[8] ________________: The homotopy theory of sphere bundles over spheres (II), ibid., 5 (1955), 148-166.

[9] S. Oka: Groups of self-equivalences of certain complexes, Hiroshima Math. J., 2 (1972), 285-293.

[10] S. Oka, N. Sawashita and M. Sugawara: On the group of self-equivalences of a mapping cone, ibid., 4 (1974), 9-23.

[11] J.W. Rutter: The group of homotopy self-equivalence classes of CW complexes, Math. Proc. Camb. Phil. Soc., 93 (1983), 275-293.

[12] A.J. Sieradski: Twisted self-homotopy equivalences, Pacific J. Math., 34 (1970), 782-802.

[13] H. Toda: Composition methods in homotopy groups of spheres, Ann. Math. Studies, 49 (1962), Princeton Univ. Press.

[14] C.T.C. Wall: Classification of (n-1)-connected 2n-manifolds, Ann. of Math., 75 (1962), 163-189.

[15] ________: Classification problem in differential topology V: On certain 6-manifolds, Inventiones Math., 1 (1966), 355-374.

[16] ________: Poincaré complexes I, Ann. of Math., 86 (1967), 213-245.

[17] K. Yamaguchi: On the self-homotopy equivalences of the wedge of certain complexes, Kodai Math. J., 6 (1983), 346-375.

[18] ________: The group of self-homotopy equivalences of S^2-bundles over S^4, II; Applications, ibid., 10 (1987), 1-9.

[19] A.V. Zubr: Classification of simply connected six dimensional spinor manifolds, Math. USSR. Izvestja, 9 (1975), 793-812.

THE GROUP OF SELF-HOMOTOPY EQUIVALENCES – A SURVEY

Martin Arkowitz
Department of Mathematics and Computer Science
Dartmouth College, Hanover, NH 03755

§1 Introduction

If X is an object in a category $\mathcal{C}$, then let Eq(X) denote the set of morphisms $f : X \to X$ which are equivalences, i.e., there is a morphism $g : X \to X$ with $f \circ g = 1_X$ and $g \circ f = 1_X$. Composition of morphisms gives Eq(X) the structure of a group, called the group of self equivalences of X, with the identity morphism the unit of the group. For many categories the group of equivalences is a familiar and well-studied object. For example, if $\mathcal{C}$ is the category of groups, Eq(G) is Aut G, the automorphism group of G, if $\mathcal{C}$ is the category of topological spaces, Eq(X) is Homeo(X), the homeomorphism group of X, if $\mathcal{C}$ is the category of C^∞-manifolds, Eq(M) is Diffeo(M), the diffeomorphism group of M, if $\mathcal{C}$ is the category of Riemannian manifolds, Eq(M) is Isom(M), the group of isometries of M, and so on. The category frequently studied in algebraic topology is $\mathcal{T}_h$, the category whose objects are topological spaces with base point and whose morphisms are based *homotopy classes* of based maps. For this category the group of equivalences of X is denoted $\mathcal{E}(X)$ and called the *group (of homotopy classes) of self-homotopy equivalences* of X. Thus $\mathcal{E}(X)$ is the analogue for the homotopy category of the automorphism group of a group, the homeomorphism group of a space, etc., and therefore can be regarded as a kind of homotopy symmetry group of a space. In this paper we give a survey of known results on $\mathcal{E}(X)$.

Clearly $\mathcal{E}(X)$ is an invariant of the homotopy type of X in that spaces of the same homotopy type have isomorphic self-homotopy equivalence groups. But $\mathcal{E}(X)$ is not functorial since maps (or homotopy classes) $X \to Y$ do not naturally induce homomorphisms between $\mathcal{E}(X)$ and $\mathcal{E}(Y)$. However, if $T : \mathcal{T}_h \to \mathcal{C}$ is a covariant functor, then T induces a homomorphism $T_* : \mathcal{E}(X) \to Eq(TX)$ in the obvious way. Many of the results on $\mathcal{E}(X)$ are based on this simple observation. For a particular functor T, one determines properties of T_* and then attempts to determine $\mathcal{E}(X)$. Examples are $T(X) = \sum_n \pi_n(X)$ (homotopy groups), $T(X) = \sum_n H_n(X; G)$ (homology groups with coefficients), $T(X) = X^n$ (nth Postnikov section), $T(X) = X_l$ (localization at a set of primes l), and $T(X) = \Sigma X$ (reduced suspension). Many of these will be illustrated in later sections.

We now indicate the organization of the paper and give a summary of the contents. In §2 we present some of the early history of the subject and outline the methods that have been used to study $\mathcal{E}(X)$. We discuss general structure theorems for the group of self-homotopy equivalences. These give conditions for $\mathcal{E}(X)$ to be finite, solvable, Hopfian, and so on. The deepest of these results is the Sullivan-Wilkerson Theorem which states that the group $\mathcal{E}(X)$ is finitely presented if X is a l-connected, finite CW-complex. We also consider spaces with few cells and spaces with few homotopy groups and show how the general methods yield detailed information on $\mathcal{E}(X)$ in these cases. We conclude the section by describing the group of self-homotopy equivalences of a product and of a wedge of two spaces.

Section 3 deals with several variations and generalizations of the group of self-homotopy equivalences. We begin by discussing the group $\mathcal{E}(X)_f$ of free self-homotopy equivalences of a space X, and we relate this group to $\mathcal{E}(X)$. Next we treat the group of fibre homotopy classes of self fibre homotopy equivalences of a fibration. This is a generalization of $\mathcal{E}(X)$, and many of the methods and results on $\mathcal{E}(X)$ carry over to this group. We then briefly survey the group of G-equivalent homotopy classes of G-equivariant homotopy equivalences of a G-space and the group $\mathcal{E}_H(X)$ of homotopy classes of self H-homotopy equivalences of an H-space X. After that we consider the space $E(X)_f$ of free self-homotopy equivalences of X. The path-components or 0th homotopy group of $E(X)_f$ is just $\mathcal{E}(X)_f$. We describe a few results on the higher homotopy groups of this space. We also consider the space of free fibre homotopy equivalences of a fibration to itself. There are then two short subsections, one on the relation between localization and $\mathcal{E}(X)$ and the other on the group of stable self-homotopy equivalences of X, i.e., $\mathcal{E}(\Sigma^i X)$ for large i. We conclude this section with a discussion of properties of certain naturally defined subgroups of $\mathcal{E}(X)$ such as the subgroup consisting of all homotopy equivalences which induce the identity on homotopy (or homology) groups.

Section 4 deals with computations and examples. We present known results on the calculation of the group of self-homotopy equivalences for pseudo-projective spaces, generalized lens spaces, Moore spaces, twisted Eilenberg-MacLane spaces, and projective spaces. By stating results or referring to the literature, we see that $\mathcal{E}(X)$ has been determined for all 1-connected, rank two, finite H-complexes X (with one exception). Similarly we observe that $\mathcal{E}_H(X)$ has been computed for all 1-connected, rank two, finite H-complexes X. The dependency of $\mathcal{E}_H(X)$ on the multiplication of X in these cases is also discussed. The last subsection of §4 is concerned with examples of spaces whose group of self-homotopy equivalences has certain given properties. We present examples of non-simply connected, finite complexes X with $\mathcal{E}(X)$ infinitely generated. We also take up the realizability question: For a given group Π, when does there exist a space (or finite complex) X with $\mathcal{E}(X) \approx \Pi$?

The last section deals with several applications of the group of self-homotopy equivalences to other parts of algebraic topology. We give an interpretation of $\mathcal{E}(X)_f$ as equivalence classes of Hurewicz fibrations over the circle with fibre of the homotopy type of X. We then consider the action of $\mathcal{E}(X) \times \mathcal{E}(Y)$ on the homotopy set $[X, Y]$. For an H-space X this induces an action of $\mathcal{E}(X)$ on the set of multiplications of X, and we state some results on the quotient set. In the next subsection we examine the non-uniqueness of the Postnikov invariant $k^{n+1} \in H^{n+2}(X^n; \pi_{n+1})$ for a Postnikov decomposition X^n of a space X. The equivalence class of k^{n+1} in $H^{n+2}(X^n; \pi_{n+1})/(\mathcal{E}(X^n) \times \mathrm{Aut}\, \pi_{n+1})$ is the element that uniquely determines X^{n+1}. We then consider the set of homotopy types which have the same n-type as a space X for all n, and relate this set to $\lim^1$ of the groups $\mathcal{E}(X^n)$. We conclude the paper with a discussion of the homotopy action of a group G on a space X which is defined to be a homomorphism $G \to \mathcal{E}(X)_f$. We are interested, among other things, in when the action is equivalent to a topological action of G on a homotopically equivalent space Y. We briefly describe a connection between a special case of this question and a well-known problem of Steenrod.

This paper is intended as a survey and review of results on the group of self-homotopy equivalences and related topics. For the most part, our presentation is chronological and we cover the period from the beginning to the day before the Montréal Conference[1]. Most of our references are listed in the "List of Papers on or Relevant to Groups of Self-Homotopy Equivalences" which appear in these notes. The references that do not appear on this list are under the heading "References" at the end of the paper.

We are grateful to the University of Arizona for their hospitality during the time that this work was done. We would like to thank Kouzou Tsukiyama for several helpful comments.

§2 The Group $\mathcal{E}(X)$ of Self-Homotopy Equivalences of X

(a) Early History. To the best of our knowledge, the first appearance in print of results on the group of self-homotopy equivalences was in 1958 in a paper of Barcus-Barratt. The entire discussion, occupying less than one page, was an application of their results on homotopy classification. Barcus and Barratt considered a 1-connected CW-complex K of dimension $< q$ to which a (q+1)-cell is attached by a map α to form the space $X = K \cup_\alpha e^{q+1}$. They established the following three term exact sequence

$$(2.1) \qquad i_* (\pi_{q+1}(K)) \to \mathcal{E}(X) \to \begin{cases} \mathcal{E}(K) & \text{if } 2\alpha \neq 0 \\ \mathcal{E}(K) \oplus \mathcal{E}(S^{q+1}) & \text{if } 2\alpha = 0 \end{cases}$$

where $i_* : \pi_{q+1}(K) \to \pi_{q+1}(X)$ is induced by the inclusion $i : K \to X$. This sequence allows one to start an inductive procedure for determining $\mathcal{E}(X)$, at least for certain cell complexes X.

In 1964 four papers appeared dealing with the group of self-homotopy equivalences ([Ka, D, 1964], [Sh, 1964], [AC, 1964_1], [AC, 1964_2]). These papers all used the Postnikov system X^n of a space X to obtain information on $\mathcal{E}(X)$. Postnikov systems are well-suited to this study since a homotopy equivalence $X^n \to X^n$ induces a homotopy equivalence $X^{n-1} \to X^{n-1}$. D. Kahn [1964] proved that the following

---

[1] Since this paper was written, the book, Algebraic Homotopy by H. J. Baues (Cambridge Studies in Advanced Math. 15, Cambridge University Press), has appeared. Interspersed throughout this book are a number of interesting and important results which bear on the subject of this survey. There are generalizations and new proofs of many of the theorems stated here and several computations and computational methods for the group of self-homotopy equivalences.

sequence is exact

$$(2.2) \qquad \mathcal{E}_{\mathcal{F}}(X^n) \to \mathcal{E}(X^n) \xrightarrow{\rho} \mathcal{E}(X^{n-1})$$

where ρ is the restriction homomorphism and $\mathcal{E}_{\mathcal{F}}(X^n)$ is the group of fibre homotopy classes of fibre homotopy equivalences $X^n \to X^n$ of the fibration $X^n \to X^{n-1}$. He showed that $\mathcal{E}_{\mathcal{F}}(X^n)$ is in one-one correspondence with a subset of $H^n(X^n; \pi_n(X))$ and identified Image ρ. Shih [1964] considered a filtration $\mathcal{F}_m$ of $\mathcal{E}(X)$ defined by $\mathcal{F}_m =$ Kernel $(\mathcal{E}(X) \to \mathcal{E}(X^{m-1}))$. He proved that there exists a spectral sequence $E_r^{p,q}$ of groups and sets with distinguished element that converges to the associated graded group of the filtration. Shih then described $E_1^{p,q}$ in terms of cohomology. Arkowitz-Curjel [1964_1] considered the subgroup $\mathcal{E}_{\#}(X)$ of $\mathcal{E}(X)$ consisting of the self equivalences which induce the identity homomorphism on homotopy groups, i.e., the kernel of $\mathcal{E}(X) \to \sum_n \text{Aut}\ \pi_n$, where $\pi_n = \pi_n(X)$. They proved that the sequence

$$(2.3) \qquad T^n(X; \pi_n) \to \mathcal{E}_{\#}(X^n) \to \mathcal{E}_{\#}(X^{n-1})$$

is exact, where $T^n(X; \pi_n)$ is the kernel of $H^n(X; \pi_n) \to \text{Hom}\ (\pi_n, \pi_n)$.

In 1966 Nomura extended and generalized these methods to induced fibrations. For Postnikov systems this yielded the exact sequence

$$(2.4) \qquad 1 \to \text{Iso}(1) \to \text{Im}\ p^* \to \mathcal{E}(X^n) \xrightarrow{\rho'} \mathcal{E}(X^{n-1}) \times \text{Aut}\ \pi_n,$$

where $p^* : H^n(X^{n-1}; \pi_n) \to H^n(X^n; \pi_n)$ is induced by the fibre map $p : X^n \to X^{n-1}$ and Iso(1) is the isotropy subgroup of the identity class $1 \in [X^n, X^n]$ under the action of $H^n(X^n; \pi_n)$ on $[X^n, X^n]$. Furthermore, Image ρ' consists of all (α, β), $\alpha \in \mathcal{E}(X^{n-1})$, $\beta \in \text{Aut}\ \pi_n$, which are compatible with the k-invariant $k^n \in H^{n+1}(X^{n-1}; \pi_n)$, $\alpha^* k^n = \beta_* k^n$.

It is easily shown that, for a finite-dimensional CW-complex X with Postnikov system X^n, $\mathcal{E}(X) \approx \mathcal{E}(X^n)$ for $n \geq \dim X$ [AC, 1964_2, Lemma 5.1]. Therefore the use of Postnikov systems to determine $\mathcal{E}(X)$ is available either for spaces with a finite number of (non-trivial) homotopy groups or for finite-dimensional CW-complexes which admit a Postnikov decomposition.

Generalizations and refinements of the Barcus-Barratt approach, which we shall call the *mapping cone method*, were given by Kudo-Tsuchida [1967], Rutter [1970, 1983], and Oka-Sawashita-Sugawara [1974]. Rutter [1988_2] expanded this method to the homology decomposition of a space. Extensions and refinements of the Postnikov system approach, which we shall call the *induced fibration method*, were given by Rutter [1970] and Oka-Sawashita-Sugawara [1974]. In general, the mapping cone method has been useful for computations of $\mathcal{E}(X)$ for spaces X with a few cells, and the induced fibration method has been useful for obtaining general results on $\mathcal{E}(X)$.

(b) General Properties of $\mathcal{E}(X)$. We first show why it is frequently necessary to put restrictions on a space X to ensure that the group $\mathcal{E}(X)$ has certain properties. Let Π be any finite group. Then Π can be regarded as a subgroup of the symmetric group S_n on n letters, for some n. We choose a non-contractible space Y and form the space X $= Y \times \cdots \times Y$ (n factors). By permuting factors, each element of S_n acts on X as a homotopy equivalence. Thus an arbitrary finite group Π can be imbedded in $\mathcal{E}(X)$ as a subgroup, for some space X. One may even assume that the space X is a member of a nice class of spaces such as l-connected, finite CW-complexes. Therefore the groups $\mathcal{E}(X)$ will not have a group-theoretic property which is inherited by subgroups and which not all finite groups have. For example, it is not true that $\mathcal{E}(X)$ is a solvable group for all l-connected, finite CW-complexes X.

Unless otherwise stated we assume in §2 that X has the homotopy type of a l-connected CW-complex of finite type which is either a finite CW-complex or else has a finite number of homotopy groups. We let dim X denote the dimension of the CW-complex in the former case and the dimension of the last homotopy group in the latter.

The first question that arises is whether or not the groups $\mathcal{E}(X)$ are trivial. It is easy to show that $\mathcal{E}(S^n) = \mathbb{Z}_2$ and $\mathcal{E}(K(\Pi, n)) =$ Aut Π. More generally, D. Kahn proved [1976] that for large i, the additive inverse of the identity map of the i-fold suspension of X, $-1 : \Sigma^i X \to \Sigma^i X$, is an element of order two in $\mathcal{E}(\Sigma^i X)$, where X is a non-contractible, finite CW-complex. From this we see that $\mathcal{E}(\Sigma X) \neq 1$ since $\Sigma^{i-1}(-1)$ $= -1$. Kahn also proved that $\mathcal{E}(X) \neq 1$ for any space X with two non-vanishing homotopy groups in dimensions n and k, $n < k < 2n - 1$ [Ka, D, 1979].

Regarding the structure of $\mathcal{E}(X)$, we have the following result.

(2.5) Proposition. [*AC*, 1964$_1$] *Let rank* $\pi_i(X) \leq 1$ *for all* $i \leq \dim X$. *Then*
(1) $\mathcal{E}(X)$ *and all its subgroups are finitely generated and*
(2) $\mathcal{E}(X)$ *is a finite group if and only if* $Hom(\operatorname{coker} h_i, \pi_i(X))$ *is finite for all* $i \leq \dim X$, *where* $h_i : \pi_i(X) \to H_i(X)$ *is the Hurewicz homomorphism.*
(3) *If for all* $i \leq \dim X$, *all primes p, and all positive integers k, the number of times the cyclic group* $\mathbb{Z}_{p^k}$ *occurs in the decomposition of the finitely generated abelian group* $\pi_i(X)$ *is* ≤ 2 *for* $p = 2$ *or* 3 *and* ≤ 1 *for other p, then* $\mathcal{E}(X)$ *is a solvable group.*

Further results on $\mathcal{E}(X)$ were obtained by limiting the class of spaces X. P. Kahn considered closed, oriented, C^∞, $(n-1)$-connected, 2n-manifolds M, $n \geq 2$, and used the mapping cone approach to obtain an extended exact sequence for $\mathcal{E}(M)$. This yielded

(2.6) Proposition. [*Ka, P*, 1966, 1969] (1) $\mathcal{E}(M)$ *is finitely generated.*
(2) *If n is even and the intersection pairing* $H_n(M) \otimes H_n(M) \to \mathbb{Z}$ *is either definite or has rank 2 and index 0, then* $\mathcal{E}(M)$ *is finite. Otherwise* $\mathcal{E}(M)$ *is infinite.*

Arkowitz and Curjel obtained results on the group of self-homotopy equivalences of an H-space. They proved that $\mathcal{E}(Y)$ is finitely generated for associative H-spaces Y [AC, 1964$_1$]. This was later generalized by Sunday.

(2.7) Proposition. [*Su*, 1973] *If Y is an H-space, then* $\mathcal{E}(Y)$ *is finitely presented.*

Arkowitz and Curjel also investigated the question of the finiteness of $\mathcal{E}(Y)$ for *H_0-spaces* Y, i.e., spaces with $H^*(Y; \mathbb{Q}) = \Lambda(x_{n_1}, \ldots, x_{n_r})$, an exterior algebra on generators x_{n_i} of odd degree n_i. Let $\rho(\Pi)$ denote the rank of the not-necessarily-abelian group Π and let $\beta_i(Y)$ be the ith Betti number of Y.

(2.8) Proposition. [*AC*, 1964$_2$] (1) *If rank* $\pi_i(Y) > 1$ *for some* $i \leq \dim Y$, *then* $\mathcal{E}(Y)$ *contains a free group on* ≥ 2 *generators.*
(2) $\mathcal{E}(Y)$ *has finite rank if and only if all* n_i *are distinct. In this case,*

$$\rho(\mathcal{E}(Y)) = \sum_{i=1}^{r} (\beta_{n_i}(Y) - 1).$$

Sasao [1983] proved (1) under the assumption that Y is an associative H-space which is not necessarily simply-connected. Maruyama [1989] obtained an expression for $\rho(\mathcal{E}_{\#}(Y))$ similar to the one in (2) for any finite H_0-complex Y.

Since finite groups are those of zero rank we have

(2.9) *Corollary.* *$\mathcal{E}(Y)$ is finite if and only if $\beta_{n_i}(Y) = 1$ for $i = 1, 2, \ldots, r$.*

D. Kahn [1976] generalized this corollary to spaces Y with $H^*(Y; Q)$ a free-commutative algebra on generators of dimension n_i, $i = 1, 2, \ldots, r$.

We note that several of the previous results assert that under certain conditions $\mathcal{E}(X)$ is finitely generated or finitely presented for a l-connected, finite CW-complex X. This led to the conjecture that for such X, $\mathcal{E}(X)$ is always finitely presented. This was independently proved by Wilkerson [1976] and Sullivan [1977].

(2.10) *Sullivan-Wilkerson Theorem. If X is a simply-connected space which is either a finite CW-complex or a space with finitely many homotopy groups, then the group $\mathcal{E}(X)$ is finitely presented.*

We briefly indicate what is involved in the proof. Two groups G and G′ are *commensurable* if there is a finite sequence of homomorphisms $G \to G_1 \leftarrow G_2 \to \ldots \leftarrow G'$ each of which has finite kernel and image of finite index. Finite presentation of a group is preserved by the commensurability relation. The main steps in the proof are to show (1) If $X_{\mathbf{Q}}$ is the *rationalization* of X, then $\mathcal{E}(X_{\mathbf{Q}})$ is a linear algebraic group over $\mathbf{Q}$, say $\mathcal{E}(X_{\mathbf{Q}}) \subseteq GL(V)$ where V is a finite dimensional vector space over $\mathbf{Q}$. (2) $\mathcal{E}(X)$ is commensurable with an *arithmetic subgroup* of $\mathcal{E}(X_{\mathbf{Q}})$, i.e., a subgroup of the matrix group $\mathcal{E}(X_{\mathbf{Q}})$ which carries an integral lattice of V isomorphically onto itself. Since arithmetic groups are finitely presented [Bo, 1962, p. 14], the result follows. The finite subgroups of an arithmetic group are also known to have a finite number of conjugacy classes. Thus

(2.11) *Proposition.* [*Wi*, 1976] *$\mathcal{E}(X)$ has a finite number of conjugacy classes of finite subgroups. Therefore, there exists an integer $N(X)$ such that every element in $\mathcal{E}(X)$ of finite order has order $\leq N(X)$.*

The Sullivan-Wilkerson Theorem holds more generally for nilpotent spaces instead of l-connected spaces. In fact, Dror-Dwyer-Kan [1981] have shown that it holds for *virtually nilpotent spaces.* These include all nilpotent spaces and all spaces with finite fundamental group.

There are many examples of non-simply connected, finite CW-complexes X with $\mathcal{E}(X)$ infinitely generated. See §4(c) for details.

Another general result on $\mathcal{E}(X)$ concerns residual finiteness. Recall that a group Π is *residually finite* if for every $x \neq 1$ in Π there is a normal subgroup K of finite index with $x \notin K$. Finitely generated, nilpotent groups have the property that their automorphism group is residually finite. The homotopy analogue of this was proved by Roitberg [1985].

(2.12) ***Proposition.*** *If X is a nilpotent, finite complex, then $\mathcal{E}(X)$ is residually finite.*

A finitely generated, residually finite group is known to be *Hopfian*, i.e., not isomorphic to a proper quotient of itself. From Roitberg's result and the Wilkerson-Sullivan Theorem it follows that $\mathcal{E}(X)$ is Hopfian if X is a nilpotent, finite CW-complex. This is also a consequence of a theorem of Sunday [1973].

We conclude this subsection by stating an important result regarding $\mathcal{E}_{\#}(X)$, the group of self-homotopy equivalences that induce the identity on $\pi_i(X)$, $i \leq \dim X$. Previously it was shown that $\mathcal{E}_{\#}(X)$ is a solvable group [AC, 1964_1]. Dror-Zabrodsky [1979] proved the following generalization.

(2.13) ***Proposition.*** *$\mathcal{E}_{\#}(X)$ is a nilpotent group.*

(c) Spaces with Few Homotopy Groups and Spaces with Few Cells. Let X be a space with two homotopy groups, $\pi_i = \pi_i(X)$, $i = n, m$, and $1 < n < m$. Let $k \in H^{m+1}(\pi_n, n; \pi_m)$ be the k-invariant and let $R \subseteq \mathrm{Aut}\,\pi_n \oplus \mathrm{Aut}\,\pi_m$ be $\{(\alpha, \beta), \alpha \in \mathrm{Aut}\,\pi_n, \beta \in \mathrm{Aut}\,\pi_m, \alpha^*(k) = \beta_*(k)\}$, where $\alpha^*, \beta_* : H^{m+1}(\pi_n, n; \pi_m) \to H^{m+1}(\pi_n, n; \pi_m)$ are induced by α and β. Then the induced fibration method yields the exact sequence

$$0 \to H^m(\pi_n, n; \pi_m) \to \mathcal{E}(X) \to R \to 1. \tag{2.14}$$

This was proved by Shih [1964] and Nomura [1966]. For $n = 1$, the exact sequence (2.14) was obtained by Didierjean [1985] and Yamanashita [1986] with the cohomology group now standing for cohomology with local coefficients determined by the action of π_1 on π_m. If X is an n-dimensional complex with $\pi_i(X) = 0$ for $1 < i < n$, the sequence (2.14) or a similar one was derived by Schellenberg [1973_2], Smallen [1974], Dyer [1976_2], Tsukiyama [1980_1], and Maruyama [1987]. Rutter [1970] generalized (2.14) to induced fibrations with conditions on the vanishing of the homotopy groups of the base and fibre. For a space X with three non-trivial homotopy groups, Didierjean [1985] established a five term exact sequence with middle group $\mathcal{E}_{\#}(X)$. This provided information on $\mathcal{E}_{\#}(X)$ for any simply-connected, 4-dimensional complex X. Rutter [1970] obtained some results on $\mathcal{E}(X)$ of a space X with three non-vanishing homotopy groups and non-trivial k-invariant the cup product map.

For spaces X with two (positive-dimensional) cells, written $X = S^n \cup_\alpha e^{n+k}$, $k \geq 2$, the Barcus-Barratt result (2.1) yields a three-term exact sequence for $\mathcal{E}(X)$. This does not however determine the group. Further information was given by Oka [1972] in the case α is a double suspension with α and its desuspension having the same finite order. Oka-Sawashita-Sugawara [1974] considered $X = S^n \cup_\alpha e^{n+k}$, $k \geq 2$, when α is a suspension. They derived the following exact sequence which splits for $2\alpha \neq 0$

$$(2.15) \qquad 1 \to \begin{Bmatrix} H & \text{if } 2\alpha \neq 0 \\ D(H) & \text{if } 2\alpha = 0 \end{Bmatrix} \to \mathcal{E}(X) \to \mathbb{Z}_2 \to 1,$$

where $H = \pi_{n+k}(S^n)/(\alpha_*\pi_{n+k}(S^{n+k-1}) + (\Sigma\alpha)^*\pi_{n+1}(S^n))$ and D(H) is the split extension of H by $\mathbb{Z}_2$ with $\mathbb{Z}_2$ acting on H by $(-1)h = -h$. Rutter [1978] obtained the same sequence for $2\alpha = 0$ when the desuspension of α has order two. In this case he gave the action of $\mathbb{Z}_2$ on D(H) explicitly and proved that the sequence splits. Oka-Sawashita-Sugawara [1974] determined $\mathcal{E}(X)$ for several different standard elements $\alpha \in \pi_{n+k-1}(S^n)$ and also deduced partial results for $\mathcal{E}(X)$ in the case α is not a suspension.

Computation of $\mathcal{E}(X)$ for a space X with three cells was carried out by several people when X is the total space of a fibration with spherical base and fibre. For a trivial fibration, X is a product of spheres, and so subsection (d) below applies. For principal S^3-bundles which are rank two H-spaces, see §4(a). We briefly discuss the other cases. Mimura-Sawashita [1984] considered principal S^3-bundles over S^n, $n \geq 5$, and determined the group of self-homotopy equivalences up to extension. Sasao [1985] considered the same bundles, $n \geq 6$, and described the group up to two extensions. Nomura [1983] investigated $\mathcal{E}(V_{n,2})$ and $\mathcal{E}(W_{n,2})$ for real and complex Stiefel varieties $V_{n,2} = O(n)/O(n-2)$ and $W_{n,2} = U(n)/U(n-2)$, and obtained the groups either exactly or up to extension. Sasao [1984] examined S^m-bundles over S^n with $3 < m + 1 < n < 2m - 2$ and determined the groups up to two extensions. Yamaguchi [1986] studied S^2-bundles over S^4 and made some explicit calculations. For an S^m-bundle over S^n and an odd prime p such that m and n are odd, $3 < m + 1 < n$, and the inclusion $S^m \to X$ localized at p is an H-map, Mimura and Sawashita [1986] proved that $\mathcal{E}(X)$ is a finite group with unique p-Sylow subgroup that is the semi-direct product of p-primary components $\pi_{m+n}(X; p)$ and $\pi_n(S^m; p)$.

Rutter [1988_2] computed (up to extension) the group of self-homotopy equivalences of the cone on a Moore space of type $(\mathbb{Z}_{2^s}, n-1)$ attached to an (n-1)-sphere. Thus he determined (up to extension) $\mathcal{E}(X)$ for certain spaces $X = S^{n-1} \vee S^n \cup e^{n+1}$ with three cells in successive dimensions.

P. Kahn [1966] in his work on $(n-1)$-connected, 2n-manifolds studied $\mathcal{E}(X)$ for $X = (S^n \vee \ldots \vee S^n) \cup_\alpha e^{2n}$. Finally, Rutter [1983] considered the mapping cone C_h of a map $h : \vee S^{n-1} \to A$, where A is an m-dimensional CW-complex, $n > m \geq 1$, and obtained several exact sequences useful for computing $\mathcal{E}(C_h)$.

(d) Cartesian Products and Wedges of Spaces. It was early recognized that the group of self-homotopy equivalences of a cartesian product is usually more complicated than the group of self-homotopy equivalences of its factors (see the first paragraph of (a)). The first non-trivial cartesian products considered were products of two spheres. P. Kahn [1966] computed $\mathcal{E}(S^n \times S^n)$ and Sieradski [1970] calculated $\mathcal{E}(S^m \times S^n)$ for $m,n \in \{1, 3, 7\}$ as a special case of his results on $\mathcal{E}(A \times B)$ for H-spaces A and B. Metzler and Zimmerman [1971] determined $\mathcal{E}(S^3 \times S^3)$ using the quaternions. In 1975, Sawashita, using mapping cone methods, derived for $m > n \geq 2$, the short exact sequence

$$(2.16) \qquad 0 \to H \to \mathcal{E}(S^m \times S^n) \to G \to 1$$

where H is a factor group of $\pi_{m+n}(S^m) \oplus \pi_{m+n}(S^n)$ and G a subgroup of $\mathcal{E}(S^m \vee S^n)$. Although the extension is not always known, many cases were treated separately, and a great deal of information obtained. Another approach was given in [MS, 1986].

For the cartesian product of arbitrary spaces X and Y, Ando and Yamaguchi [1982] gave the following split exact sequence under the assumptions $[Y, X] = 0$ and $[X \wedge Y, X] = 0$

$$(2.17) \qquad 1 \to \mathrm{Inv}[X, Y^Y] \to \mathcal{E}(X \times Y) \to \mathcal{E}(X) \times \mathcal{E}(Y) \to 1$$

where Y^Y is the (H-)space of maps $Y \to Y$ and $\mathrm{Inv}[X, Y^Y]$ is the group of invertible elements in the monoid $[X, Y^Y]$. This had previously been done for $X = K(\Pi, 1)$ by Sasao-Ando [1982]. Yamanashita [1985_2] obtained a result similar to (2.17) which expresses $\mathcal{E}(X \times Y)$ as a semi-direct product under different hypotheses (see (3.9)).

The results on the group of self-homotopy equivalences of a wedge of two spaces are more sporadic. Sieradski [1970] briefly considered $\mathcal{E}(X \vee Y)$ when X and Y are coH-spaces. Rutter [1988_2] calculated $\mathcal{E}(X \vee Y)$ when X and Y are Moore spaces. Oka-Sawashita-Sugawara [1974] examined $\mathcal{E}(Y \vee \Sigma X)$ for X $(m-2)$-connected and $\dim Y \leq m-1$ using mapping cone methods for a trivial attaching map. Maruyama-Mimura [1984] obtained results on $\mathcal{E}(KP^2 \vee S^m)$ where KP^2 denotes the complex, quaternionic, or Cayley projective plane. They gave extensive calculations for $2 \leq m \leq 16$. Yamaguchi [1983_1] considered $\mathcal{E}(X \vee \ldots \vee X)$ where X is a certain two or three cell complex. Finally, we remark that Frank-Kahn [1977] investigated $\mathcal{E}(S^1 \vee S^n \vee S^{2n-1})$ and showed that it is not finitely generated (see §4(c)).

§3 Variations on $\mathcal{E}(X)$

In this section we discuss several generalizations and refinements of the group of self-homotopy equivalences. Our standing assumption is that all spaces are of the homotopy type of connected CW-complexes of finite type. We begin with the simplest variation, the group of free or unbased self-homotopy equivalences.

(a) Free Homotopy Equivalences. If X is a space with base point, one can ignore the base point (or work with spaces without a chosen base point), and consider the collection of *homotopy classes of free maps* $X \to X$ *which are homotopy equivalences*. This forms a group under composition of homotopy classes which is denoted $\mathcal{E}(X)_f$ (the subscript "f" signifies free). For a based space X there is an obvious homomorphism $\mathcal{E}(X) \to \mathcal{E}(X)_f$ which is an epimorphism since X is connected. If X is 1-connected or if X is an H-space, then clearly $\mathcal{E}(X) \approx \mathcal{E}(X)_f$. However, it has been shown by Becker and Gottlieb [1973] that $\mathcal{E}(\mathbb{R}P^{2n})_f = 0$. Thus $\mathcal{E}(X) \not\approx \mathcal{E}(X)_f$ for $X = \mathbb{R}P^{2n}$ (see §4(a)). There have been a few computations in the literature of $\mathcal{E}(X)_f$ such as for $X = S^1 \cup_q e^2$ (q a map of degree q) by Olum [1965] and for $X = S^2 \times \mathbb{R}P^2$ by Matumoto [1979]. Unbased homotopy equivalences also appear in connection with the group of self fibre homotopy equivalences, the group of equivariant self-homotopy equivalences, and the space of self-homotopy equivalences, all of which are discussed below.

(b) Fibre Homotopy Equivalences. Let $p : E \to B$ be a Serre or Hurewicz fibration with fibre $F = p^{-1}(*)$. We examine the set of fibre homotopy classes of fibre homotopy equivalences $E \to E$. With composition of fibre homotopy classes this set becomes a group denoted $\mathcal{E}_{\mathcal{F}}(E)$ and called the *group of self fibre homotopy equivalences* of the fibration $p : E \to B$. One can also work with free fibre homotopy classes of free fibre homotopy equivalences $E \to E$ to obtain a group $(\mathcal{E}_{\mathcal{F}}(E))_f$. By considering the fibration $X \to *$, we observe that $\mathcal{E}_{\mathcal{F}}(X) \approx \mathcal{E}(X)$, so that the group of self fibre homotopy equivalences is a generalization of the group of self-homotopy equivalences. Many of the results for $\mathcal{E}(X)$ have a counterpart for $\mathcal{E}_{\mathcal{F}}(E)$. Nomura studied the fibration $\Omega Z \to E \xrightarrow{p} B$ induced by a map $\theta : B \to Z$. If B is m-connected and ΩZ has homotopy groups only in dimensions n + 1 to m + n + 1, then the following sequence was proved exact [No, 1965]

$$(3.1) \qquad 1 \to \operatorname{Im} p^* \to \mathcal{E}_{\mathcal{F}}(E) \to \{\Omega\alpha \in \mathcal{E}(\Omega Z), \alpha\theta = \theta\} \to 1,$$

where $p^* : [B, \Omega Z] \to [E, \Omega Z]$. A similar sequence was obtained by Rutter [1970].

Tsukiyama [1982] established an exact sequence for $(\mathcal{E}_{\mathcal{F}}(E))_f$ for a Hurewicz fibration over a sphere and Sasao [1982] established one for certain sphere bundles over spheres. Many computations of $(\mathcal{E}_{\mathcal{F}}(E))_f$ were given in these papers when E is a fibration over a sphere with fibre a sphere.

The Sullivan-Wilkerson Theorem was extended to $\mathcal{E}_{\mathcal{F}}(E)$ by Scheerer [1980].

(3.2) ***Proposition.*** *If $E \to B$ is a Serre fibration such that E and B are nilpotent spaces each of the homotopy type of a finite CW-complex or each having finitely many homotopy groups, then $\mathcal{E}_{\mathcal{F}}(E)$ and $(\mathcal{E}_{\mathcal{F}}(E))_f$ are finitely presented.*

The proof is an adaptation of the method of Sullivan-Wilkerson and consists of showing $\mathcal{E}_{\mathcal{F}}(E)$ commensurable with an arithmetic subgroup of a linear algebraic group over $\mathbf{Q}$. Scheerer also sketched a proof of the analogous result for cofibrations.

In a somewhat different direction, James investigated the subgroup $(\mathcal{E}'_{\mathcal{F}}(E))_f$ of $(\mathcal{E}_{\mathcal{F}}(E))_f$ consisting of homotopy classes of those fibre homotopy equivalences that are homotopic to the identity on each fibre. Under the assumption that B has finite Lusternik-Schnirelmann category and the fibres are compact, he proved that the group $(\mathcal{E}'_{\mathcal{F}}(E))_f$ is $(\mathcal{E}_{\mathcal{F}}(E))_f$-nilpotent of class $\leq$ cat B [Ja, 1979]. In 1982, Meiwes was able to weaken some of the hypotheses.

Gottlieb considered a Hurewicz fibration $p : E \to B$ with fibre F. If $E_\infty \to B_\infty$ is the universal fibration for Hurewicz fibrations with fibre of the homotopy type of F, then p is classified by a map $k : B \to B_\infty$. Let $M(B, B_\infty)_f$ denote the space of free maps from B to B_∞ with the compact-open topology.

(3.3) ***Proposition.*** *[Go, 1968]* $(\mathcal{E}_{\mathcal{F}}(E))_f \approx \pi_1(M(B, B_\infty)_f, k)$.

See (e) for a generalization.

<u>(c)</u> <u>Equivariant</u> <u>Homotopy</u> <u>Equivalences</u>. If G is a group (finite or topological) which acts on a space X, then one can consider the *group* $\mathcal{E}_G(X)$ *of G-equivariant homotopy classes of G-equivariant homotopy equivalences* $X \to X$ and, by using free maps, the group $(\mathcal{E}_G(X))_f$. Tsukiyama [1985] examined the situation in which the action of G on X yields a principal G-bundle $X \to X/G = B$ with classifying map $k : B \to B_G$. He obtained an exact sequence which, in the case G is compact, X is 1-connected, and $B = S^n$, becomes

$$(3.4) \quad \pi_{n+1}(X) \to \mathbf{Z}_2 \to \pi_n(G)/\langle \tilde{k}, \pi_1(G)\rangle \to \mathcal{E}_G(X) \to \begin{Bmatrix} \mathbf{Z}_2 & \text{if } 2k=0 \\ 0 & \text{if } 2k\neq 0 \end{Bmatrix} \to 0,$$

where $\langle \tilde{k}, \pi_1(G)\rangle$ denotes Samelson product of $\tilde{k} \in \pi_{n-1}(G)$ with $\pi_1(G)$. Ōshima and Tsukiyama [1986] extended this from spheres to suspensions. Many examples were worked out in [Ts, 1985] and [ŌT, 1986] with $(\mathcal{E}_G(X))_f$ computed explicitly. In addition, there are some general results in [ŌT, 1986] which give conditions for $(\mathcal{E}_G(X))_f$ to be finite or finitely presented. Matsuda [1978, 1979] investigated an orthogonal representation $G \to \mathrm{Aut}\, V$ of a finite group G into a finite-dimensional vector space V. The unit sphere $S(V) \subseteq V$ inherits a G-action. Then the order of the group $\mathcal{E}_G(S(V))$ was determined in the cases G is abelian, G is the dihedral group of order 2n, and G is the symmetric group S_n. Triantafillou [1984] considered a finite group acting on X nilpotently, i.e., so that the fixed point spaces X^H are non-empty and nilpotent for all subgroups H of G. By extending Sullivan's minimal models to equivariant spaces, she proved that $\mathcal{E}_G(X)$ is commensurable with an arithmetic subgroup of the linear algebraic group $\mathcal{E}_G(X_{\mathbf{Q}})$ and hence established an equivariant Sullivan-Wilkerson Theorem.

(3.5) ***Proposition.*** [*Tr*, 1984] *With the above hypothesis, $\mathcal{E}_G(X)$ is a finitely presented group.*

(d) H-homotopy equivalences. If X is an H-space (with a fixed multiplication μ which is usually not mentioned), then a map $X \to X$ which is an H-map and a homotopy equivalence is called an *H-homotopy equivalence*. The set $\mathcal{E}_H(X)$ of all homotopy classes of H-homotopy equivalences $X \to X$ is a subgroup of $\mathcal{E}(X)$ called the *group of self H-homotopy equivalences* of X. If we wish to show dependence on the multiplication μ we write this as $\mathcal{E}_H(X, \mu)$. The first results on $\mathcal{E}_H(X)$ were proved by Arkowitz-Curjel and D. Kahn. A weaker version of the following result of D. Kahn [1972_1] appeared in [AC, 1967].

(3.6) ***Proposition.*** *If X is a finite H-complex, then $\mathcal{E}_H(X)$ is finitely presented.*

Arkowitz-Curjel [1967] also investigated the influence of the rank of the homotopy groups on the size of $\mathcal{E}_H(X)$.

(3.7) ***Proposition.*** *Let X be a finite, associative H-complex. If rank $\pi_i(X) \leq 1$ for all i, then $\mathcal{E}_H(X)$ is finite. If rank $\pi_i(X) > 1$ for some i, then $\mathcal{E}_H(X)$ contains a non-abelian free group.*

Sawashita [1984] considered the induced fibration method applied to the fibration $\Omega Z \to E \to B$ induced from an H-map $B \to Z$. In the case of a Postnikov decomposition X^n of an H-space X this yielded the exact sequence

(3.8) $$0 \to H_n \to \mathcal{E}_H(X^n) \to G_n \to 1,$$

where $G_n \subseteq \mathcal{E}_H(X^{n-1})$ and H_n is a subquotient of $H^n(X^{n-1}; \pi_n(X))$. This sequence was used to compute the group of H-homotopy equivalences for certain rank two H-spaces (see §4).

An interesting variant of the group of self H-homotopy equivalences was introduced by Sawashita and Sugawara. They defined the group $\mathcal{H}(X) = \cap\, \mathcal{E}_H(X, \mu)$, where the intersection is taken over all multiplications μ of the H-space X. It was proved in [SS, 1986] that if X is one of the classical Lie groups U(n), SU(n), or Sp(n), then $\mathcal{H}(X)$ is a finite, nilpotent group all of whose elements induce the identity on integral cohomology. If X is one of the five exceptional Lie groups G_2, F_4, E_6, E_7, E_8, respectively, then Sawashita and Sugawara proved [1987] that any element of $\mathcal{H}(X)$ induces the identity homomorphism on $H^*(X)/\text{torsion}$, $H^*(X; \mathbb{Z}_p)$, and $H^*(X; \mathbb{Z}_{\{p\}})$ for $p > 1, 3, 3, 3, 5$, respectively.

<u>(e) The Space of Self-Homotopy Equivalences and the Space of Self Fibre Homotopy Equivalences</u>. For spaces X and Y, let M(X, Y) denote the space of based maps $X \to Y$ and $M(X, Y)_f$ the space of free maps $X \to Y$, both with the compact-open topology. Then let $E(X) \subseteq M(X, X)$ and $E(X)_f \subseteq M(X, X)_f$ be the subspaces of maps which are homotopy equivalences. Clearly $\pi_0(E(X)) = \mathcal{E}(X)$ and $\pi_0(E(X)_f) = \mathcal{E}(X)_f$. It is therefore natural to explore properties of the spaces E(X) and $E(X)_f$, such as their homotopy groups (see §5(a)). Gottlieb [1965] showed that if X is an aspherical space, then $E(X)_f$ is aspherical with $\pi_1(E(X)_f)$ isomorphic to the center of $\pi_1(X)$. McCullough [1981_2] considered the connected sum M of $r \geq 2$ aspherical n-manifolds of dimension ≥ 3 and obtained information on $\pi_i(E(M)_f)$ for $1 \leq i \leq n-2$. In particular, the groups $\pi_i(E(M)_f)$ were determined for $1 \leq i \leq n-4$ and $\pi_{n-2}(E(M)_f)$ was proved to be infinitely generated. The homomorphism $\pi_0(E(M)) = \mathcal{E}(M) \to \text{Aut}\, \pi_1(M)$ was investigated in [Mc, 1981_3] and shown to have kernel $\oplus_{i=1}^{n-1} \mathbb{Z}_2$. An expression for the image was given in [Mc, 1985].

Yamanoshita [1985_2] examined $E(X \times Y)$ and $E(X \times Y)_f$ under the hypothesis that Y is n-connected and $\pi_i(X) = 0$ for $i > n$ or $\dim X \leq n$. He obtained weak homotopy equivalences which in the based case is

$$E(X \times Y) \equiv E(X) \times E(Y) \times M(Y, E(X)_f) \times M(X, E(Y)_f). \tag{3.9}$$

For certain fibrations $F \to E \to B$, Yamanoshita also studied the space of fibre homotopy equivalences $E \to E$ which fix the fibre [Ya, 1986].

Booth, Heath, Morgan, and Piccinini [1984] considered a topological category $\mathcal{F}$ (of fibres). They then defined Dold $\mathcal{F}$-fibrations $p : E \to B$ and restricted attention to those for which a universal Dold $\mathcal{F}$-fibration $p_\infty : E_\infty \to B_\infty$ exists. Rather than give the relevant definitions, we remark that the following (with base a CW-complex) are some of the examples of Dold $\mathcal{F}$-fibrations of the above type: (1) Hurewicz fibrations with fibre of the homotopy type of a fixed space F (2) principal G-bundles for a fixed topological group G (3) n-dimensional vector bundles (4) fibre bundles with fibre F and group G, where G is a fixed topological group and F a fixed G-space. For each Dold $\mathcal{F}$-fibration $p : E \to B$ one defines the space $\mathcal{G}(p)_f$ of free, self fibre homotopy equivalences which is an associative H-space under composition. To the fibration p we assign a classifying map $k : B \to B_\infty$ and denote by $M(B, B_\infty; k)_f$ the path-component in the space of free maps $M(B, B_\infty)_f$ which contains k.

(3.10) ***Proposition.*** [*BHMP*, 1984] *There is a Dold $\mathcal{F}$-fibration with base $M(B, B_\infty; k)_f$ and fibre homeomorphic to $\mathcal{G}(p)_f$ such that the connecting map $\Omega M(B, B_\infty; k)_f \to \mathcal{G}(p)_f$ is a weak H-homotopy equivalence.*

We remark that Gottlieb [1972] essentially proved (3.10) for a principal bundle. Also a based version of (3.10) appeared in [BHMP, 1984]. A consequence of (3.10) is

(3.11) ***Corollary.*** $\pi_0(\mathcal{G}(p)_f) \approx \pi_1(M(B, B_\infty; k)_f)$.

Thus if $p : E \to B$ is any one of the four types of fibrations considered above, this corollary is just Gottlieb's result (3.3) for that fibration, namely, that $(\mathcal{E}_{\mathcal{F}}(E))_f$ is isomorphic to $\pi_1(M(B, B_\infty)_f, k)$.

(f) Stable Homotopy Equivalences. If X is a finite CW-complex and $\Sigma^i X$ is the i-fold suspension of X, then $\lim_r \mathcal{E}(\Sigma^r X) = \mathcal{E}(\Sigma^i X)$ for sufficiently large i, by the generalized Freudenthal theorem. We call this group the *group of stable self-homotopy equivalences of X* and denote it $\mathcal{E}_S(X)$. Johnston [1972] proved analogues of (2.5) for $\mathcal{E}_S(X)$ regarding finite generation, finiteness, and solvability. D. Kahn [1972_2] showed that $\mathcal{E}_S(X)$ is finitely presented and that there are only a countable number of possible groups $\mathcal{E}_S(X)$ as X ranges over all finite complexes. Some computations of $\mathcal{E}_S(X)$ were given by Sasao [1981] when X is the total space of a sphere bundle over a sphere. Sieradski [1972] computed $\mathcal{E}_S(P_q)$ for $P_q = S^1 \cup_q e^2$.

(g) Localization. We have already discussed rationalization (or localization at the empty set) in connection with the proof of the Sullivan-Wilkerson Theorem. There are some other scattered results on localization which we collect here. Let X_P denote the localization of the space X at the set of primes P. Lieberman-Smallen [1974] showed that if P and R are complementary sets of primes, then $\mathcal{E}(X)$ is the pull-back of $\mathcal{E}(X_P) \to \mathcal{E}(X_{\mathbf{Q}}) \leftarrow \mathcal{E}(X_R)$, where X is a finite, nilpotent complex. Arkowitz [1988] considered the homomorphism $J : \mathcal{E}(X_{\mathbf{Q}}) \to \mathrm{Aut}\ H^*(X_{\mathbf{Q}};\ \mathbf{Q})^{opp}$, where the latter denotes the group with opposite multiplication of the group of automorphisms of the algebra $H^*(X_{\mathbf{Q}};\ \mathbf{Q})$. He proved that if X is a formal space, then $\mathcal{E}(X_{\mathbf{Q}})$ is a semi-direct product of Aut $H^*(X_{\mathbf{Q}};\ \mathbf{Q})$ and Kernel J. He also showed $\mathcal{E}_H(\Omega X_{\mathbf{Q}}) \approx \mathrm{Aut}\ \pi_*(\Omega X_{\mathbf{Q}})$, the group of Lie algebra automorphisms of $\pi_*(\Omega X_{\mathbf{Q}})$. Glover and Homer investigated flag manifolds over **C**, i.e., homogeneous spaces $M = U(n)/U(n_1) \times \cdots \times U(n_k)$, $n = n_1 + \cdots + n_k$, $k \geq 2$, $n_1 \leq \cdots \leq n_k$. They proved [GM, 1981] that the homomorphism $J : \mathcal{E}(M_{\mathbf{Q}}) \to \mathrm{Aut}\ H^*(M_{\mathbf{Q}};\ \mathbf{Q})^{opp}$ is an isomorphism. For $k = 2$, the flag manifolds are complex Grassmannians and, in this case, Hoffman [1984] determined the structure of Aut $H^*(M_{\mathbf{Q}};\ \mathbf{Q})$. Finally, $\mathcal{E}_\#(X)$ is a nilpotent group by (2.13), and so, for any set of primes P, the localization $\mathcal{E}_\#(X)_P$ exists. Maruyama [1989] proved that $\mathcal{E}_\#(X_P) \approx \mathcal{E}_\#(X)_P$.

(h) Subgroups and Quotients of $\mathcal{E}(X)$. We have already mentioned the subgroup $\mathcal{E}_\#(X)$ of $\mathcal{E}(X)$ in §2 (see (2.13)). Some general results on $\mathcal{E}_\#(X)$ were given by Tsukiyama [1975, 1977, 1980]. A similar subgroup $\mathcal{E}_*(X)$ is defined by considering those homotopy equivalences which induce the identity on integral homology groups. Wilkerson [1976] showed that $\mathcal{E}_*(X)$ is a finite extension of a nilpotent group which is nilpotent if $H_*(X)$ is torsion-free. Zabrodsky [1985] proved that $\mathcal{E}_*(X)$ is a finite, nilpotent group if X is a nilpotent, H_0-space which is either a finite-dimensional complex or a space with finitely many homotopy groups. Oka [1981_1, 1981_2] determined (exactly or up to extension) the groups $\mathcal{E}_*(X)$ for X = Sp(2), Sp(3), SU(4), and $G_{2,b}$ (see §4) and Sasao [1984, 1985] gave an exact sequence for calculating $\mathcal{E}_*(X)$ up to extension when X is a sphere bundle over a sphere. Dror-Zabrodsky [1979] proved that any subgroup of $\mathcal{E}(X)$ which, for all $i \leq \dim X$, either acts nilpotently on $\pi_i(X)$ or acts nilpotently on $H_i(X)$ is itself a nilpotent group. Sullivan [1977] observed that any subgroup of $\mathcal{E}(X)$ which fixes a cohomology class is commensurable with an arithmetic subgroup and hence finitely presented.

Hurvitz in [1981] studied certain subgroups of a product of groups of self-homotopy equivalences. For a map $f : X \to Y$ with homotopy class $\alpha \in [X, Y]$, she defined the subgroup $\mathcal{E}(f) \subseteq \mathcal{E}(X) \times \mathcal{E}(Y)$ by $\mathcal{E}(f) = \{(\beta, \gamma) \in \mathcal{E}(X) \times \mathcal{E}(Y), \alpha\beta = \gamma\alpha\}$ and the related subgroup $\mathcal{E}_Y(f) \subseteq \mathcal{E}(X)$ by $\mathcal{E}_Y(f) = \{\beta \in \mathcal{E}(X), \alpha\beta = \alpha\}$. She proved the following generalization of (2.10) and (2.11).

(3.12) ***Proposition.*** [*Hu*, 1981] *If $f : X \to Y$ is a map between simply-connected, finite CW-complexes, then $\mathcal{E}(f)$ and $\mathcal{E}_Y(f)$ are finitely presented and have a finite number of conjugacy classes of finite subgroups.*

For non-simply-connected spaces which are low-dimensional complexes or have few homotopy groups, there has been interest in $I : \mathcal{E}(X) \to \mathrm{Aut}\, \pi_1(X)$. Olum [1965] calculated Image I and Kernel I for the space $X = P_q = S^1 \cup_q e^2$ (see §4) and Schellenberg [1973_1] did it for the 2-skeleton of $P_q \times P_r$. For a finite, 2-dimensional CW-complex with one 2-cell, Jagodia [1979] showed that I is an epimorphism with kernel equal to $H^2(\pi_1(X); \pi_2(X))$. Sieradski [1976] determined Image I for those 2-complexes whose fundamental group is a direct sum of cyclic groups. Schellenberg [1973_2] obtained an exact sequence for Kernel I when X is a complex with $\pi_i(X) = 0$ for $1 < i < \dim X$.

Finally we note that for a CW-complex X, Rutter [1983] obtained information on the subgroup of $\mathcal{E}(X)$ consisting of those self-homotopy equivalences which induce self-homotopy equivalences on all of the skeleta of X.

§4 Computations and Examples

Many of the computations and examples which we present in this section have been obtained by various mathematicians with a great amount of work. One should keep in mind that, not only is one solving a homotopy problem in calculating $\mathcal{E}(X)$, but that one is also solving a composition (of homotopy classes) problem in determining the group structure of $\mathcal{E}(X)$. This may account for the difficulty of some of the computations in the literature. Although we only state the end result of the computations and examples (and in cases when that is too complicated, simply refer to the original papers), we have tried to arrange these results into a coherent form. Some computations and examples have already been alluded to in earlier sections, and we shall usually not repeat them here.

(a) Computations of $\mathcal{E}(X)$. An early paper by Olum in 1965 demonstrated both the difficulties in computing $\mathcal{E}(X)$ and the richness of the structure of $\mathcal{E}(X)$. Olum made an exhaustive study of $\mathcal{E}(P_q)$ for the pseudo-projective plane $P_q = S^1 \cup_q e^2$, where $q : S^1 \to S^1$ is a map of degree q. He proved that there is a split, short exact sequence

$$(4.1) \qquad 1 \to \text{Kernel I} \to \mathcal{E}(P_q) \xrightarrow{I} \text{Aut}\ \pi_1(P_q) \to 1$$

and that Kernel I is isomorphic to U_q^1, the group of units in the group-ring $\mathbb{Z}[\pi_1(P_q)]$ of augmentation 1. Clearly Aut $\pi_1(P_q) = \text{Aut}\ \mathbb{Z}_q = \mathbb{Z}_q^*$, the units of $\mathbb{Z}_q$, and so $\mathcal{E}(P_q)$ can be expressed as a semi-direct product $U_q^1 \rtimes \mathbb{Z}_q^*$. Olum then analyzed U_q^1 and the operation of $\mathbb{Z}_q^*$ on U_q^1 using algebraic number theory. Since P_q is the 2-skeleton of the lens space L(q, m), he was able to calculate

$$(4.2) \qquad \mathcal{E}(L(q,m)) = \begin{cases} \mathbb{Z}_2 & \text{if } q = 2 \\ \text{subgroup of } \mathbb{Z}_q^* \text{ whose squares} = \pm 1 & \text{if } q > 2. \end{cases}$$

The results on the pseudo-projective planes were generalized by Schellenberg [1973_2] and Plotnick [1982] who determined $\mathcal{E}(X)$ for X a generalized lens space (of dimension 2n + 1) minus a point. The results on the lens spaces were extended to spaces X which are the orbit spaces of an odd-dimensional homotopy sphere under the free action of a finite group Π. Smallen [1974] calculated $\mathcal{E}(X)$ for $\Pi = \mathbb{Z}_m$ and for various groups acting on a homotopy 3-sphere. For arbitrary finite Π acting on an odd-dimensional homotopy sphere, Tsukiyama [1975] showed that $\mathcal{E}_{\#}(X) = 1$ and Plotnick [1982] determined $\mathcal{E}(X)$ as a subgroup of Aut Π.

Moore spaces $M = M(G, n)$, $n \geq 2$, are higher dimensional analogues of the pseudo-projective planes. Sieradski [1970] noted the exact sequence

$$(4.3).\qquad 0 \to \mathrm{Ext}(G, \pi_{n+1}(M)) \to \mathcal{E}(M) \xrightarrow{J} \mathrm{Aut}\, G \to 1$$

with J an isomorphism if and only if G has no 2-torsion. This was also obtained by Rutter [1983] with mapping cone methods.

Twisted Eilenberg-MacLane have also been considered. Given the action of a group Π on an abelian group A and an integer $n > 1$, a twisted Eilenberg-MacLane space $L = L(\Pi, A, n)$ is defined by (1) L has two non-trivial homotopy groups $\pi_1(L) = \Pi$ and $\pi_n(L) = A$ (2) the action of $\pi_1(L)$ on $\pi_n(L)$ is just the given action of Π on A (3) the k-invariant in $H^{n+1}(\Pi; A)$ is zero. Møller [1988] determined $\mathcal{E}(L)$ and $\mathcal{E}(L)_f$ in terms of group-theoretic invariants of Π and A.

For real projective spaces $\mathbb{R}P^n$, it is easily seen that $\mathcal{E}(\mathbb{R}P^n) = \mathbb{Z}_2$ (e.g., [BG, 1973]). Complex projective space CP^n also has its group of self-homotopy equivalences equal to $\mathbb{Z}_2$, a fact noted by P. Kahn [1969]. Kahn also considered quaternionic projective space $\mathbb{H}P^n$, $n \geq 2$, and showed that $\mathcal{E}(\mathbb{H}P^2) = \mathbb{Z}_2$ and $\mathcal{E}(\mathbb{H}P^n) = \mathcal{E}^*(\mathbb{H}P^n)$, the group of self-homotopy equivalences which induce the identity on cohomology. As a consequence of Mislin's work on the classification of self-maps of $\mathbb{H}P^\infty$ [1987], it follows that $\mathcal{E}(\mathbb{H}P^\infty) = 1$.

There has been a great deal of interest in determining the group of self-homotopy equivalences of H-spaces of low rank. The rank one H-spaces are easily dispensed with since they are S^1, S^3, S^7, $\mathbb{R}P^3$, and $\mathbb{R}P^7$, and so their groups of self-homotopy equivalences are just $\mathbb{Z}_2$. We concentrate for the most part on rank two H-spaces.

We use the following notation introduced earlier: If H is an abelian group, then D(H) denotes the split extension of H by $\mathbb{Z}_2$, where the operation of $\mathbb{Z}_2$ on H is given by $(-1)h = -h$. For a cyclic group H of order n, D(H) is just the dihedral group of order 2n. In 1974, Oka-Sawashita-Sugawara, using mapping cone methods, proved

$$(4.4) \qquad \begin{cases} \mathcal{E}(SU(3)) = D(\mathbb{Z}_{12}) \times \mathbb{Z}_2 \\ \mathcal{E}(Sp(2)) = D(\mathbb{Z}_{120}). \end{cases}$$

Sawashita [1977] reproved (4.4) as part of his study of the group of self H-homotopy equivalences of these spaces by means of Postnikov decompositions. Oka [1981_1] showed $\mathcal{E}_*(Sp(2)) = \mathbb{Z}_{120}$ and derived (4.4) for Sp(2). He also obtained an exact sequence for $\mathcal{E}_*(Sp(3))$ and expressed $\mathcal{E}_*(SU(4))$ as the product of three Sylow subgroups [1981_2].

Principal S^3-bundle over S^7 are classified by elements of $\pi_6(S^3) = \mathbb{Z}_{12}$. If $\omega \in \pi_6(S^3)$ is the canonical generator (sometimes called the Blakers-Massey element), we denote by E_k the S^3-bundle over S^7 classified by $k\omega$, k = 0, 1, ..., 11. It is known that there are only seven homotopy types of spaces E_k, and these occur for k = 0, 1, ..., 6. Furthermore, E_k is an H-space if and only if k = 0, 1, 3, 4, 5. Indeed, $E_0 = S^3 \times S^7$, $E_1 = Sp(2)$, and we shall call the H-spaces E_k, k = 3, 4, 5, the *Hilton-Roitberg-Stasheff H-spaces*. Rutter [1978] calculated $\mathcal{E}(E_k)$ for k = 0, 1, ..., 5 and $\mathcal{E}(E_6)$ up to extension. For the Hilton-Roitberg-Stasheff H-spaces this yielded

$$(4.5) \qquad \begin{cases} \mathcal{E}(E_3) = D(\mathbb{Z}_3 \times \mathbb{Z}_{120}) \\ \mathcal{E}(E_4) = \mathbb{Z}_2 \times D(\mathbb{Z}_{120}) \\ \mathcal{E}(E_5) = D(\mathbb{Z}_{120}). \end{cases}$$

The other collection of simply-connected, rank two H-spaces consists of eight homotopy types denoted $G_{2,b}$, $-2 \leq b \leq 5$. All of these are of type (3, 11), have 2-torsion in their homology, and are principal S^3-bundles over the real Stiefel manifolds $V_{7,2}$ [MNT, 1973]. Furthermore, $G_{2,0}$ is the compact, exceptional Lie group G_2. The groups $\mathcal{E}(G_{2,b})$, $-1 \leq b \leq 5$, were determined up to extension by Mimura-Sawashita [1981] and completely as a semi-direct product of known groups with explicitly defined action by Oka [1981_2]. The group $\mathcal{E}(G_{2,-2})$ was determined up to two extensions in [MS, 1981] and up to one extension in [Ok, 1981_2]. Because the results are somewhat complicated to state, we refer the reader to these two papers.

The simply-connected, rank two, finite H-complexes are $S^p \times S^q$ (p, q = 3,7), SU(3), Sp(2), E_k (k = 3,4,5), and $G_{2,b}$ (b = -2,...,5) [MNT, 1973]. The groups of equivalences $\mathcal{E}(S^p \times S^q)$ have been calculated (see §2(d)). In particular, $\mathcal{E}(S^3 \times S^7) = \mathcal{E}(E_0) = \mathbb{Z}_2 \times D(\mathbb{Z}_{15}) \times D(\mathbb{Z}_{24})$ by Rutter [1978], and $\mathcal{E}(S^3 \times S^3)$ and $\mathcal{E}(S^7 \times S^7)$ were given by Sawashita [1975, Theorem 6.4]. Thus we have the following result.

(4.6) *The group of self-homotopy equivalences for all simply-connected, rank two, finite H-complexes has been computed with the exception of* $\mathcal{E}(G_{2,-2})$ *which is determined up to extension.*

(b) Computations of $\mathcal{E}_H(X)$. There have been extensive calculations of the group of self H-homotopy equivalences for H-spaces of low rank. For rank one H-spaces, it is easily seen that

$$(4.7)\qquad \begin{array}{rl} \mathcal{E}_H(S^n) = & \left\{ \begin{array}{ll} \mathbb{Z}_2 & n = 1 \\ 1 & n = 3,7 \end{array} \right. \\ \mathcal{E}_H(\mathbb{R}P^n) = & \quad 1 \qquad\quad n = 3,7. \end{array}$$

For, if X is one of the above five H-spaces, then $\mathcal{E}(X) = \mathbb{Z}_2$. There are thus two self-homotopy equivalences represented by $1 : X \to X$, the identity map, and by $-1 : X \to X$, the negative of the identity with respect to the H-space structure on X. But -1 is an H-map if and only if X is homotopy-commutative. Since S^3, S^7, $\mathbb{R}P^3$, and $\mathbb{R}P^7$ are not homotopy-commutative and S^1 is homotopy-commutative, (4.7) follows.[2]

We turn next to rank two H-spaces. Sawashita [1976] considered the product of two spheres and showed

$$(4.8)\qquad \begin{array}{l} \mathcal{E}_H(S^3 \times S^7) = 1 \\ \mathcal{E}_H(S^p \times S^p) = \{A = (a_{ij}) \in GL(2, \mathbb{Z}),\ a_{ij} \equiv \dfrac{1+(-1)^{ij}\det A}{2} \bmod k_p\},\quad p = 3,7, \end{array}$$

where $k_3 = 24$ and $k_7 = 240$. Sawashita also determined the group of self H-homotopy equivalences for SU(3) and Sp(2)

$$(4.9)\qquad \left\{ \begin{array}{l} \mathcal{E}_H(SU(3)) = \mathbb{Z}_2 \\ \mathcal{E}_H(Sp(2)) = 1. \end{array} \right.$$

For the three Hilton-Roitberg-Stasheff H-spaces E_k, Maruyama [1981] showed

---

[2]The non-associative H-spaces S^7 and $\mathbb{R}P^7$ require a more elaborate argument.

(4.10) $$\mathcal{E}_H(E_k) = 1, \quad \text{for } k = 3, 4, 5.$$

Finally, there are the eight homotopy types $G_{2,b}$ of (3, 11) H-spaces, $-2 \le b \le 5$ Sawashita [1984] used the sequence (3.7) together with a Postnikov decomposition of $G_{2,b}$ to calculate their group of self H-homotopy equivalences:

(4.11) $$\mathcal{E}_H(G_{2,b}) = 1, \quad \text{for } -2 \le b \le 5.$$

By putting the preceding paragraphs together, we obtain the following result.

(4.12) *The group of self H-homotopy equivalences for all simply-connected, rank two H-complexes has been calculated.*

It is interesting to observe that some of the previous results hold for any multiplication, not just the standard one. Maruyama and Oka [1981] used complex K-theory to prove that $\mathcal{E}_H(E_k,\mu) = 1$ for any multiplication μ in E_k, where k = 0, 1, 3, 4, 5. Maryuama [1984] also showed, in contrast to (4.9), that there is a multiplication μ on SU(3) such that $\mathcal{E}_H(SU(3), \mu) = 1$. Since the generator of $\mathcal{E}_H(SU(3))$ in (4.9) is represented by complex conjugation $C : SU(3) \to SU(3)$, we have that $C : SU(3), \mu \to SU(3), \mu$ is not an H-map for the above multiplication μ. Finally, Oka [1986] proved that (4.11) holds more generally. He showed that $\mathcal{E}_H(G_{2,b}, \mu) = 1$ for any multiplication μ on $G_{2,b}$, $-1 \le b \le 5$.

<u>(c)</u> <u>Examples</u>. Many examples have been produced in order to show the possibilities and limitations of the groups $\mathcal{E}(X)$. The first class of examples deals with the question of finite generation of $\mathcal{E}(X)$. If Π is a group, then $\mathcal{E}(K(\Pi, 1)) = \text{Aut } \Pi$. In 1967 Lewin gave an example of a finitely presented group Π whose automorphism group is not finitely generated. For this Π, $K(\Pi, 1)$ is a space whose group of self-homotopy equivalences is not finitely generated. On the other hand, there is the Sullivan-Wilkerson Theorem which shows that $\mathcal{E}(X)$ is finitely generated for all nilpotent, finite CW-complexes X. It is therefore natural to ask if there is a finite CW-complex X, of necessity not nilpotent, such that $\mathcal{E}(X)$ is infinitely generated. The first answer to this question was the following example of Frank-Kahn [1977].

(4.13) $\mathcal{E}(S^1 \vee S^n \vee S^{2n-1})$ *is infinitely generated for all* $n > 1$.

The authors also showed that $\mathcal{E}(S^1 \vee S^n \vee S^m)$ is finitely-generated if $1 < n < m < 2n-1$. McCullough in [1980] gave examples of 4-dimensional, finite CW-complexes X which are $K(\Pi, 1)$'s with $\mathcal{E}(X) = \text{Aut } \Pi$ infinitely generated. He asked if there were 2-dimensional examples. This was answered in 1982 by Brunner and Ratcliffe who presented infinitely many such examples. In all of these two and four dimensional examples, $\mathcal{E}(X) = \text{Aut } \pi_1(X)$ is not finitely generated. Brunner-Ratcliffe [1982] then raised the question of the existence of 2-dimensional, finite complexes X with $\mathcal{E}(X)$ infinitely generated and Aut $\pi_1(X)$ finitely generated. It was shown by McCullough [1984] that these exist.

The second class of examples deal with realizability. This concerns the problem of when a given group Π can be realized as $\mathcal{E}(X)$ for some space X, $\mathcal{E}(X) \approx \Pi$. (Note that $\Pi \subseteq \mathcal{E}(X)$ for some space X by the first paragraph of §2(b).) The obvious place to start is with $\Pi = 1$ and to determine if it can be realized by a non-contractible, finite complex. D. Kahn [1976] showed that there is a space Y with two non-trivial homotopy groups $\pi_4(Y) = \mathbb{Z}$ and $\pi_7(Y) = \mathbb{Z}_3$ such that $\mathcal{E}(Y) = 1$. He then constructed a 9-dimensional, finite complex X of the same 9-type as Y such that $\mathcal{E}(X) = 1$. A similar construction in [Ka, D, 1976] starting with the space Y and adjoining homotopy group $\mathbb{Z}$ in dimension 8 and $\mathbb{Z}_5$ in dimension 15 yielded a space Z with four non-vanishing homotopy groups such that $\mathcal{E}(Z) = \mathbb{Z}$. Finally, for the cyclic groups $\mathbb{Z}_n$, Oka [1980] proved that if $n \not\equiv 0 \bmod 8$ or if $n \equiv 16 \bmod 32$, then there exists a finite complex X_n with $\mathcal{E}(X_n) = \mathbb{Z}_n$.

§5 Applications

In this section we describe several diverse applications of the group of self-homotopy equivalences to different parts of algebraic topology. We begin with an interpretation of $\mathcal{E}(X)_f$.

(a) Classification of Hurewicz Fibrations over the Circle. Stasheff's Classification Theorem [1963] established a natural one-one correspondence between equivalence classes of Hurewicz fibrations with base B and fibre of the homotopy type of F and free homotopy classes of maps of B into a classifying space B_∞. Since we work with based homotopy classes, we adopt the modifications in Allaud [1966]. We consider a Hurewicz fibration $p : E \to B$ together with a homotopy equivalence $g : F \to p^{-1}(*)$ from a fixed space F called the fibre to $p^{-1}(*)$. If $p_i : E_i \to B$, $g_i : F \to p_i^{-1}(*)$ are two of these fibrations over B with fibre F, i = 1, 2, then they are *equivalent* if there is a map $f : E_1 \to E_2$ such that $p_2 f = p_1$ and $fg_1 \simeq g_2$. Let $\mathcal{H}(B, F)$ denote the set of equivalence classes of such Hurewicz fibrations with base B and fibre F. Then the modified classification theorem asserts that there is a fibration $p_\infty : E_\infty \to B_\infty$ with homotopy equivalence $g_\infty : F \to p_\infty^{-1}(*)$ such that the mapping $[B, B_\infty] \to \mathcal{H}(B, F)$ which assigns to a (based) homotopy class $h \in [B, B_\infty]$ the equivalence class of the pull-back of p_∞ by h is a one-one correspondence. Furthermore, if $E(F)_f$ is the H-space of free self-homotopy equivalences of F, then $\pi_i(B_\infty, *) \approx \pi_{i-1}(E(F)_f, 1)$. This yields a one-one correspondence between the (i-1)st homotopy group of $E(F)_f$ and the equivalence classes of Hurewicz fibrations over an i-sphere with fibre F. By setting i = 1 we obtain the one-one correspondence,

(5.1) $$\mathcal{E}(F)_f \approx \mathcal{H}(S^1, F).$$

Thus the group of free self-homotopy equivalences of any space F is in one-one correspondence with the equivalence classes of Hurewicz fibrations over the circle with fibre F. To obtain an interpretation for $\mathcal{E}(F)$ similar to (5.1), see May [1975, Chapter 9] for the appropriate classification theorem.

<u>(b) The Operation of Groups of Self Equivalences on Homotopy Sets.</u> The groups $\mathcal{E}(X)$ and $\mathcal{E}(Y)$ each operate on the set [X, Y] as follows: If $\alpha \in \mathcal{E}(X)$, $\beta \in \mathcal{E}(Y)$, and $\phi \in [X, Y]$, then

$$\alpha \cdot \phi = \phi \circ \alpha^{-1} \text{ and } \beta \cdot \phi = \beta \circ \phi.$$

The group $\mathcal{E}(X) \times \mathcal{E}(Y)$ acts on [X, Y] by

$$(\alpha, \beta) \cdot \phi = \beta \circ \phi \circ \alpha^{-1}.$$

We denote the quotient set of [X, Y] under the action of $\mathcal{E}(X) \times \mathcal{E}(Y)$ by {X, Y}. A modification of this action is the action of $\mathcal{E}(X)$ on $[X \times X, X]$ defined by

$$\alpha \cdot \phi = \alpha \circ \phi \circ (\alpha^{-1} \times \alpha^{-1}).$$

For an H-space X, let M(X) and $M_0(X)$ denote the subsets of $[X \times X, X]$ which are, respectively, the collection of homotopy classes of H-space multiplications of X and the collection of homotopy classes of homotopy-associative, H-space multiplications of X. Clearly the action of $\mathcal{E}(X)$ on $[X \times X, X]$ induces an action of $\mathcal{E}(X)$ on $M(X)$ and on $M_0(X)$. If $\mu \in M(X)$, then the isotropy subgroup $\mathcal{E}(X)_\mu$ of μ in $\mathcal{E}(X)$ is just the group $\mathcal{E}_H(X, \mu)$ of self H-homotopy equivalences with respect to μ (§3(d)). The quotient of $M(X)$ under the action of $\mathcal{E}(X)$ is denoted by $\widetilde{M}(X)$. Two multiplications μ and ν of X are thus equivalent if there is a homotopy equivalence H-map $(X, \mu) \to (X, \nu)$. A similar definition holds for $\widetilde{M}_0(X)$. In 1968 Curjel showed that for a finite, homotopy-associative H-complex X, $\widetilde{M}_0(X)$ is always finite. He proved that $\widetilde{M}(X)$ is infinite if and only if, for some n, the nth Betti number of $X \wedge X$ times the rank of $\pi_n(X)$ is non-zero. Sawashita [1976] determined the cardinality of the sets $\widetilde{M}(S^p)$, $\widetilde{M}(S^1 \times S^p)$, $(p = 1, 3, 7)$, and $\widetilde{M}(S^3 \times S^7)$. Finally we remark that the cardinality of $\widetilde{M}(E_k)$ can be determined for the H-space $E_1 = Sp(2)$ and the three Hilton-Roitberg-Stasheff H-spaces E_3, E_4, E_5. All the ingredients for the computation are known, namely, the cardinality of the following sets: $\mathcal{E}(E_k)$ (§4(a)), $\mathcal{E}_H(E_k, \mu)$ for any multiplication μ (§4(b)), $M(E_1)$ (Mimura [1969]), and $M(E_k)$, $k = 3, 4, 5$ (Arkowitz-Murley-Shar [1975]).

(c) Postnikov Invariants. A completely different but important application of the action of $\mathcal{E}(X) \times \mathcal{E}(Y)$ on $[X, Y]$ deals with the non-uniqueness of Postnikov invariants. For convenience, all spaces will be assumed 1-connected. We introduce some terminology. A space X' is called an *n-section* if $\pi_i(X') = 0$ for $i > n$. Given a space X, the *n-type* of X is the homotopy type of any n-section X' such that there is a map $p : X \to X'$ with $p_* : \pi_i(X) \to \pi_i(X')$ an isomorphism for all $i \leq n$. Note that one can always take X^n, the nth Postnikov section of X, for X'. We consider the following problem: Fix an n-section X' with $n \geq 2$ and an abelian group G. How many homotopy types of (n+1)-sections X are there which have n-type X' and $\pi_{n+1}(X) = G$? If X has these properties, then there is a map $p : X \to X'$ such that $p_* : \pi_i(X) \to \pi_i(X')$ is an isomorphism for $i \leq n$. We can without loss of generality assume that p is a fibre map, and then it easily follows that the fibre of p is $K(G,n+1)$. But it is well-known [Hi, 1965, Theorem 7.1] that such a fibration is induced by a homotopy class $k \in [X', K(G,n+2)]$ from the path-space fibration $EK(G,n+2) \to K(G,n+2)$. Therefore the (n+1)-sections X that we seek are of the homotopy type of I_k, the (total space of the) fibration induced by some $k \in [X', K(G,n+2)]$, $X \equiv I_k$. If I_k and I_l are two fibrations induced by k and l in $[X', K(G,n+2)]$, then $I_k \equiv I_l$ if and only if there exist $\alpha \in \mathcal{E}(X')$ and $\beta \in \mathcal{E}(K(G,n+2)) = \mathrm{Aut}\ G$ such that $l\alpha = \beta k$ [Hi, 1965, Theorem 7.3]. Thus we have

(5.2) *If X' is an n-section, $n \geq 2$, and G an abelian group, then the collection of homotopy types of (n+1)-sections which have n-type X' and (n+1)st homotopy group G is in one-one correspondence with the set $\{X', K(G,n+2)\} = [X', K(G,n+2)]/(\mathcal{E}(X') \times \mathrm{Aut}\ G)$. The correspondence is obtained by assigning to an element $k \in [X', K(G,n+2)]$, the induced fibre space I_k.*

This result or a similar one is well-known. To the best of our knowledge it first appeared in a paper of Adams [1956]. For a Postnikov decomposition X^n of a space X, (5.2) asserts that the homotopy type of X^{n+1} is uniquely determined by the homotopy type of X^n, the homotopy group $\pi_{n+1} = \pi_{n+1}(X)$, and the equivalence class $\{k^{n+1}\}$ of the Postnikov invariant in $\{X^n, K(\pi_{n+1}, n+2)\} = H^{n+2}(X^n;\pi_{n+1})/(\mathcal{E}(X^n) \times \mathrm{Aut}\ \pi_{n+1})$. It would therefore be more precise to regard the invariants of a Postnikov decomposition X^n of a space X as elements $\{k^{n+1}\}$ in the set $H^{n+2}(X^n;\ \pi_{n+1})/(\mathcal{E}(X^n) \times \mathrm{Aut}\ \pi_{n+1})$.

For a different and more general discussion of this question within a simplicial setting see the paper of Dwyer-Kan-Smith [1989, §10].

There are dual results for the homology decomposition of a space, but these are complicated by the fact that a map of spaces does not always induce a map of homology n-sections. We refer to [Hi, 1965, Chapter 7] and [BC, 1959].

<u>(d)</u> <u>Spaces of the Same n-Type</u>. Another connection between the group of self-homotopy equivalences and Postnikov sections appeared in the work of Wilkerson [1976]. We begin with some definitions. Let $G_1 \xleftarrow{q_2} G_2 \longleftarrow \cdots \xleftarrow{q_n} G_n \longleftarrow \cdots$ be a sequence of groups and homomorphisms, where $e_n \in G_n$ is the identity of the group. Consider the action of the product group $\prod G_n$ on the product set $\prod G_n$ defined by

$$(\alpha_n) \cdot (\gamma_n) = (\alpha_n \cdot \gamma_n \cdot (q_{n+1}(\alpha_{n+1}))^{-1}).$$

The quotient set of this action, denoted by $\varprojlim^1 G_n$, is a set with distinguished element the equivalence class of (e_n). For a CW-complex X, let SNT(X) denote the set of homotopy types of CW-complexes Y such that X and Y have the same n-type for all n.

Then Wilkerson proved

(5.3) ***Proposition.*** *There is a one-one correspondence of sets with distinguished element*

$$SNT(X) \approx \varprojlim{}^1 \mathcal{E}(X^n).$$

By showing $\varprojlim{}^1 \mathcal{E}(X^n) = 1$, Wilkerson proved that, for 1-connected spaces X of finite type, SNT(X) consists of one element in each of the following cases (a) $\pi_i(X)$ is finite for all $i > 0$ (b) X is the rationalization of some space (c) X is the profinite completion of some space. Earlier, Gray [1966] had presented an example of a space X for which SNT(X) has more than one element.

Dror, Dwyer, and Kan gave a generalization of (5.3) in 1979. Let X^n denote the nth Postnikov section of X and $B_n = B_{E(X^n)}$ the classifying space of the H-space $E(X^n)$. By replacing the maps $B_n \to B_{n-1}$ by fibre maps and taking the limit of the resulting sequence, one constructs a space $V(X) = \varprojlim B_n$. The following was proved in [DDK, 1979].

(5.4) ***Proposition.*** *There is a one-one correspondence between the components of $V(X)$ and the set $SNT(X)$. If Y is in $SNT(X)$ and $V(X)_Y$ is the component containing the point corresponding to Y, then $V(X)_Y \equiv B_{E(Y)}$.*

We sketch the connection between (5.3) and (5.4). The homotopy groups of the limit space V(X) fit into short exact sequences involving $\varprojlim{}^1$ and $\varprojlim$,

$$* \to \varprojlim{}^1 \pi_{i+1}(B_n) \to \pi_i(V(X)) \to \varprojlim \pi_i(B_n) \to *,$$

(see [DDK, 1979] and [BK, 1972, Chapter XI] for more details). Setting $i = 0$, we obtain a one-one correspondence $\pi_0(V(X)) \approx \varprojlim{}^1 \pi_1(B_n) = \varprojlim{}^1 \mathcal{E}(X^n)$. But $\pi_0(V(X)) \approx SNT(X)$ by (5.4), and so Wilkerson's result follows.

(e) Homotopy Actions of a Group on a Space. Another application of the group of self-homotopy equivalences deals with homotopy actions. A *homotopy action* of a group G on a space X is a homomorphism $\alpha : G \to \mathcal{E}(X)_f$. The homotopy action is called *topological* if α factors through Homeo(X). We call X a *homotopy G-space* in the former case and a *G-space* in the latter case. Note that for a homotopy action α, $\alpha(g)$ is a free homotopy class of a self-homotopy equivalence of X such that $\alpha(g) \circ \alpha(g')$

$= \alpha(gg')$ for all g, $g' \in$ G. Let (X, α) and (Y, β) be homotopy G-spaces and denote by α_g (resp., β_g) any map in the homotopy class $\alpha(g)$ (resp., $\beta(g)$). Then a map $\phi : X \rightarrow Y$ is called a *homotopy-(G)-equivariant* map if $\phi \circ \alpha_g \simeq \beta_g \circ \phi$ for every g $\in$ G. The homotopy G-actions α and β are said to be *equivalent* if there exists a homotopy-equivariant, homotopy equivalence $\phi : X \rightarrow Y$. Consider the following question: When is a homotopy G-action on a space X equivalent to a topological G-action on some space Y? Put another way: When is a homotopy G-action on X *realizable* by a topological G-action? To answer this we first introduce some notation. Let $\pi : E(X)_f \rightarrow \mathcal{E}(X)_f$ denote the mapping of the space $E(X)_f$ onto its path-components, let $E_1(X)_f \subsetneq E(X)_f$ be the path-component of the identity map, and let B denote the classifying space functor. In 1978 Cooke proved the following result.

(5.5) ***Proposition.*** *A homotopy G-action α on X is realizable by a topological G-action if and only if there exists a map $\theta : K(G, 1) \rightarrow B_{E(X)_f}$ such that the following diagram homotopy-commutes*

$$\begin{array}{ccc} & & B_{E(X)_f} \\ & \overset{\theta}{\nearrow} & \downarrow B_\pi \\ K(G, 1) & \xrightarrow{B_\alpha} & K(\mathcal{E}(X)_f, 1). \end{array}$$

Thus the question of when α is equivalent to a topological G-action is just the question of when the map B_α can be lifted to $B_{E(X)_f}$. Since $B_{E_1(X)_f}$ is the fibre of B_π, classical obstruction theory shows that the obstructions to lifting B_α lie in the groups $H^{n+2}(G; \pi_n(E_1(X)_f)) = H^{n+2}(G; \pi_n(X^X, 1))$, $n \geq 1$. Using this fact, Cooke [1978] gave conditions for a homotopy G-action to be realizable by a topological G-action. Two such cases are (1) X is any space and K(G, 1) has dimension less than three (2) X is *l*-local, where *l* is a set of primes not dividing the order of the finite group G. Cooke also gave an example of a homotopy $\mathbb{Z}_2$-action that cannot be realized by a topological action.

We briefly digress to present an unexpected consequence of Cooke's work that was observed by Zabrodsky [1982].

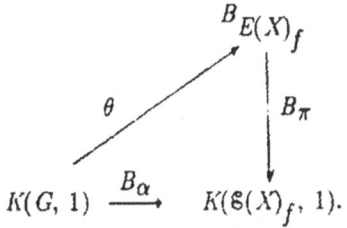

(5.6) *If X is a CW-complex, then there is a space Y of the same homotopy type as X such that every homotopy class in $\mathcal{E}(Y)_f$ contains a homeomorphism.*

Zabrodsky argued as follows. For a given $\mathcal{E}(X)_f$, there is a free group G and an epimorphism $\alpha : G \to \mathcal{E}(X)_f$. By Cooke's result ((1) above), α can be realized by a topological G-action β on a space Y. Thus there is a map $\phi : X \to Y$ that is a homotopy-equivariant, homotopy equivalence. Furthermore, β can be factored through Homeo(Y) as $G \xrightarrow{\beta'} \text{Homeo}(Y) \xrightarrow{\nu} \mathcal{E}(Y)_f$, for some homomorphism β'. Therefore the diagram

$$\begin{array}{ccc} G & \xrightarrow{\beta'} & \text{Homeo}(Y) \\ \alpha \downarrow & & \downarrow \nu \\ \mathcal{E}(X)_f & \xrightarrow{\mathcal{E}(\phi)_f} & \mathcal{E}(Y)_f \end{array}$$

is commutative, where $\mathcal{E}(\phi)_f$ is conjugation by the homotopy class of ϕ. But α is an epimorphism and $\mathcal{E}(\phi)_f$ is an isomorphism, and so ν is an epimorphism. This proves (5.6).

Oprea [1984] investigated homotopy actions within the setting of rational homotopy theory. He defined a homotopy G-action on X to be *rationally elementary equivalent* to a homotopy G-action on Y if there exists a homotopy-equivariant, rational homotopy equivalence $X \to Y$. This generates an equivalence relation called *rational equivalence* (of homotopy G-actions). Let G be a finite group and X a space with finite-dimensional rational homotopy, $\dim(\pi_*(X) \otimes \mathbb{Q}) < \infty$.

(5.7) ***Proposition.*** [*Op*, 1984] *Any homotopy G-action on X is rationally elementary equivalent to a topological G-action.*

Oprea also showed that a homotopy-equivariant, rational self-homotopy equivalence of a G-space is rationally equivalent to an equivariant, rational self-homotopy equivalence.

In 1986 Schwänzl and Vogt defined, for $1 \leq n \leq \infty$, the *n-coherence* of a homotopy action α of G on X. These are increasingly stronger conditions placed on the homotopy action α based on the fact that, for representative maps $\alpha_g : X \to X$ of $\alpha(g)$, $\alpha_{gg'} \simeq \alpha_g \circ \alpha_{g'}$. They are analogous to the coherence conditions for homotopy-associativity of an H-space. It was proved [SV, 1986] that n-coherence is equivalent to the existence of a partial lifting in the diagram of (5.5) defined on the (n+1)-skeleton of K(G, 1). It was also shown that if α is ∞-coherent, then α can be realized by a free, topological G-action. These two results imply (5.5). Furthermore, the authors gave a one-one correspondence between equivalence classes of free realizations of α (as G-actions) and homotopy classes of lifts $K(G, 1) \rightarrow B_{E(X)_f}$ of B_α. Other topics considered in [SV, 1986] are homotopy actions of a topological group G, relative results regarding a group G and a subgroup H, and the realization of a homotopy equivariant G-map by an equivariant G-map.

There is also a connection between homotopy actions and a problem of Steenrod. This problem deals with a group G acting on an abelian group A so that A is a $\mathbb{Z}[G]$-module. Steenrod asked if there is an action of G on a Moore space $X = M(A,n)$ that realizes the $\mathbb{Z}[G]$-module A, that is, such that the reduced homology $\tilde{H}_*(X)$ is isomorphic to A as $\mathbb{Z}[G]$-modules. A negative answer to Steenrod's question was given by Carlsson [1981] for all non-cyclic abelian groups G, by P. Kahn [1982] for $G = GL(r, \mathbb{Z})$ with standard action on $\mathbb{Z}\oplus\cdots\oplus\mathbb{Z}$ (s copies, $4 \leq r \leq s$), by J. Smith [1985] for $G = \mathbb{Z}_2 \oplus \mathbb{Z}_2$ and $A = \mathbb{Z} \oplus \mathbb{Z} \oplus \mathbb{Z}$, and by Assadi [1986] for finite groups G which contain $\mathbb{Z}_p \times \mathbb{Z}_p$ or the generalized quaternion group Q_{2^n} of order 2^n. On the other hand, an obstruction theory was developed by Vogel (see [As, 1986]) for the realization of a $\mathbb{Z}[G]$-module A by a *homotopy* G-action on a Moore space M(A, n). This implied that if Tor $(A, \mathbb{Z}_2) = 0$ or if G has odd order, then any $\mathbb{Z}[G]$-module A can be realized by a homotopy G-action on M(A, n). In addition, Assadi showed that the $\mathbb{Z}[G]$-module he constructed with $G \supseteq \mathbb{Z}_p \times \mathbb{Z}_p$ or $G \supseteq Q_8$ can be realized by homotopy G-actions on a Moore space. From all of these results one obtains many examples of homotopy G-actions on Moore spaces which are not topological.

Finally, Peschke [1987] considered the homotopy action of a group G on a path-connected, group-like space X' defined by a homomorphism $\alpha : G \rightarrow \mathcal{E}_H(X')_f$. In analogy with group theory, the H-semidirect product $X' \rtimes G$ was constructed and proved to be a group-like space. Peschke expressed any group-like space X (not necessarily path-connected) as $X' \rtimes G$, where X' is the path-component of X containing the unit and G is the group of path-components of X, and showed $[A, X]_f \approx [A, X']\rtimes G$, for any A. This enabled him to study the nilpotency of the group $[K, X]_f$ for any finite CW-complex K.

References

Adams, J. F. : Four applications of self-obstruction invariants, Jour. Lond. Math. Soc. 31 (1956), 148-159.

Allaud, G. : On the classification of fiber spaces, Math. Zeit. 92 (1966), 110-125.

Arkowitz, M., Murley, C. P. and Shar, A. O. : The number of multiplications on H-spaces of type (3, 7), Proc. Am. Math. Soc. 50 (1975), 394-398.

Becker, J. C. and Gottlieb, D.H. : Coverings of fibrations, Comp. Math. 26 (1973), 119-128.

Borel, A. : Arithmetic properties of linear algebraic groups, Proc. Int. Cong. of Math. 1962, 10-22.

Bousfield, A. K. and Kan, D. M. : Homotopy Limits, Completions and Localizations, Lecture Notes in Math. 304. Springer-Verlag, 1972.

Brown, E. H. and Copeland, A. C. : An homology analogue of Postnikov systems, Mich. Math. Jour. 6 (1959), 313-330.

Carlsson, G. : A counterexample to a conjecture of Steenrod, Inv. Math. 64 (1981), 171-174.

Curjel, C. R. : On the H-space structure of finite complexes, Comm. Math. Helv. 43 (1968), 1-17.

Dwyer, W. G., Kan, D. M. and Smith, J. H. : Towers of fibrations and homotopical wreath products, Jour. Pure and Applied Algebra 56 (1989), 9-28.

Glover, H. H. and Homer, W. D. : Self maps of flag manifolds, Trans. Amer. Math. Soc. 267 (1981), 423-434.

Gottlieb, D. H. : A certain subgroup of the fundamental group, Amer. Jour. Math. 87 (1965), 840-856.

Gray, B. I. : Spaces of the same n-type, for all n, Top. 5 (1966), 241-243.

Hilton, P. J. : Homotopy Theory and Duality, Gordon and Breach, 1965.

Hoffman, M. : Endomorphisms of the cohomology of complex Grassmannians, Trans. Amer. Math. Soc. 281 (1984), 745-760.

Kahn, P. : Steenrod's problem and k-invariants of certain classifying spaces, Algebraic K-Theory, Lecture Notes in Math. 967 Springer-Verlag (1982), 195-214.

Lewin, J. : A finitely presented group whose group of automorphisms is infinitely-generated, Jour. Lond. Math. Soc. 42 (1967), 610-613.

May, J. P. : Classifying Spaces and Fibrations, Mem. Amer. Math. Soc. 155, 1975.

Mimura, M. : On the number of multiplications on SU(3) and Sp(2), Trans. Amer. Math. Soc. 146 (1969), 473-492.

Mimura, M., Nishida, G. and Toda, H. : On the classification of H-spaces of rank 2, Jour. Math. Kyoto Univ. 13 (1973), 611-627.

Mislin, G. : The homotopy classification of self-maps of infinite quaternionic projective space, Quart. Jour. Math. (2) 38 (1987), 245-257.

Smith, J. R. : Equivariant Moore spaces. II — The low dimensional case, Jour. Pure and Appl. Alg. 36 (1985), 187-204.

Stasheff, J. : A classification theorem for fibre spaces, Top. 2 (1963), 239-246.

Some Research Problems on Homotopy-Self-Equivalences

by

Donald W. Kahn
School of Mathematics
University of Minnesota
Minneapolis, MN 55455

The study of the group of based homotopy classes of homotopy-self- equivalences $\mathcal{E}(X)$ has now been pursued for at least 30 years, and as has been shown by these proceedings, various trends have developed. It is our intention to present some basic research problems in all of these areas, working from the problems proposed by the participants in this conference and others.

As with all branches of mathematics, it is easier to cook-up some slick- sounding problem than it is to pose an intelligent question whose answer would advance the subject in a significant way. For example, since the group of homotopy- self-equivalences of a sphere is clear, one could naively ask for a determination of the groups $\mathcal{E}(X)$ when X is a cell complex with 2 cells in positive dimensions (plus basepoint). But after a little reflection, one sees that this would involve the determination of the self-equivalences of spaces such as

$$X = S^p \vee S^q, \quad 1 < p < q < 2p - 1$$

But a solution of this problem would be tantamount to the determination of the stable homotopy groups of spheres; in fact, the subgroup of $\mathcal{E}(X)$ which fixes homology groups is clearly the homotopy classes of maps $S^q \longrightarrow S^p$.

Before we get into specific categories of problems, let me say that there are two very broad problems - somewhat vague and general - that most workers agree are very important:

A. Calculate the groups $\mathcal{E}(X)$ explicitly in as many cases as possible, and express the known calculations in the most simple and concrete terms.

B. Develop applications of the group $\mathcal{E}(X)$ to other parts of topology (and mathematics in general).

In the following list of problems, I have tried to credit a specific problem to a given mathematician whenever possible. But the lack of a specific name on a problem does not necessarily mean that I am the source of the problem. Finally, I would like to thank Martin Arkowitz for his help and contributions to this chapter.

I. General Problems

The group of homotopy classes of self-equivalences $\mathcal{E}(X)$ is known to be finitely-presented in many cases (see D. Sullivan (1977) and C. Wilkerson (1976)). But in general, $\mathcal{E}(X)$ is infinitely generated, even for finite complexes (see D. Frank - D. Kahn (1977) and D. McCullough (1980)).

Problem 1: Characterize those finite, connected complexes X for which $\mathcal{E}(X)$ is finitely-presented.

There are many more finitely-generated than finitely-presented groups (there are uncountably many groups with 2-generators). The following problem is sometimes called one of "coherence".

Problem 2: Is there a finite, connected complex X , with $\mathcal{E}(X)$ finitely-generated but not finitely- presented?

Spaces X , with $\mathcal{E}(X)$ trivial, are quite rare. An obvious example is $K(Z/2;n)$. The group of based self-equivalences of $K(\pi,n)$ is $Aut(\pi)$, so that one easily sees that - except for $K(Z/2,n)$ - $\mathcal{E}(X)$ cannot be trivial when X is a $K(\pi,n)$. It is known (D. Kahn (1976)) that there is a space X , with non-trivial reduced rational homology, for which $\mathcal{E}(X)$ is trivial. Such spaces - which could be called "homotopically rigid" - might possibly play a role in some (as yet undeveloped) way of decomposing a space.

Problem 3. Determine the spaces (for example connected complexes) X for which $\mathcal{E}(X)$ is trivial.

II. Specific Computations.

There has been some basic work on the determination of the group $\mathcal{E}(X)$ when X is a rank 2 H-space (which is a 1-connected finite complex). See for example Oka (1981 II) and Mimura-Sawashita (1981).

Problem 4 (M. Arkowitz and J.W. Rutter). Complete the calculation of $\mathcal{E}(X)$ for rank 2 H-spaces, specifically determine $\mathcal{E}(G_{2,-2})$ precisely.

Problem 5. (M Arkowitz) Determine the (based) homotopy- self-equivalences for non-simply-connected rank 2 H-spaces. More specifically, calculate the groups

$$\mathcal{E}(RP^i \times S^j), \quad for \quad i = 3,7 \quad and \quad j = 1,3,7$$

and

$$\mathcal{E}(RP^i \times RP^k), \quad for \quad i = 3,7 \quad and \quad k = 3,7.$$

Problem 6 (M. Arkowitz) The determination of $\mathcal{E}(X)$, when X is a real or complex projective space, is classical, and it is known that $\mathcal{E}(HP^2) = Z/2$ and that any self-equivalence of HP^n induces the identity on integral cohomology. C. Curjel and M. Arkowitz (unpublished) have estimated the order of $\mathcal{E}(HP^n)$.

Determine $\mathcal{E}(X)$ when $X = HP^n$, $n > 2$, as well as when X is the Cayley projective plane.

In general, the presence of some well-known structure on a topological space X has not often led to specific calculations of $\mathcal{E}(X)$.

Problem 7 Calculate - as much as is possible - the groups $\mathcal{E}(X)$, when X is a compact Lie group, a Kahler manifold, etc.

There are many questions involving extensions in the determination of, or in calculations involving, $\mathcal{E}(X)$. Two fine examples of this sort are the following problems.

Problem 8 (J.W.Rutter). If $M(\pi, n)$ is a Moore space ($n \geq 2$) with the single homology group π in dimension n , there is an epimorphism

$$\mathcal{E}(M(\pi,n)) \longrightarrow Aut(\pi)$$

whose kernel is known to be $Ext(\pi, \pi_{n+1}(M(\pi,n)))$. (See the papers of M. Barratt on Track Groups (1955)). Calculate the extension for $\mathcal{E}(M(\pi,n))$ precisely.

Problem 9 (J.W. Rutter) There is considerable work on $\mathcal{E}(X)$ for sphere bundles over spheres (see for example, Rutter, (1978), Mimura-Sawashita (1984) & (1986), Sasao (1984) and Yamaguchi (1986) and (1987)). Complete this work by determining the extensions.

III. Special Groups of Homotopy Equivalences

From the very beginning, there has been interest in special kinds of self- equivalences, such as those which preserve some additional structure. For example, if X is a 1-connected finite complex, let $\mathcal{E}_{\#}(X)$ denote the subgroup of classes of self-equivalences which fix homotopy groups, that is $\mathcal{E}_{\#}(X)$ is the kernel of the map

$$\mathcal{E}(X) \longrightarrow Aut(\pi_*(X)).$$

Then $\mathcal{E}_{\#}(X)$ is known to be nilpotent. See Dror-Zabrodsky (1979).

Problem 10 (M. Arkowitz): Relate the nilpotency of the groups $\mathcal{E}_{\#}(X)$ to known numerical invariants of the space X.

In the stable case, the self-equivalences which fix homotopy groups, modulo torsion, and those which fix homology groups, modulo torsion, are the same. In general, we write $\mathcal{E}_*(X)$ for those classes of self-equivalences which fix homology.

Problem 11 (A. Legrand): For reasonable spaces X , what is the relation of the subgroups $\mathcal{E}_{\#}(X)$ and $\mathcal{E}_*(X)$ of the groups $\mathcal{E}(X)$?

There has been some work on the groups of classes of self-equivalences of H- space which are represented by H-maps (see M. Arkowitz and C. Curjel (1967)). Dually, one has the following problem:

Problem 12 (M. Arkowitz): Study the subgroup of self- equivalences of a co-H space, which are represented by co-H maps. Study the "suspension homomorphism" (non-abelian)

$$\mathcal{E}(X) \longrightarrow \mathcal{E}_{co-H}(\Sigma X)$$

Another area with various problems would involve the fibre homotopy equivalences of a fibration $p : E \to B$, as related to the ordinary homotopy equivalences. For example, we have

Problem 13 (K. Tsukiyama): Let $p : E \to B$ be a principal G-bundle, and let $\mathcal{E}_G(E)$ be the group of G-equivariant homotopy-self-equivalences of the fibration. Let $\mathcal{E}(E)_f$ be the group of free (un-based) homotopy self-equivalences of the space E . When is the natural map

$$\mathcal{E}_G(E) \longrightarrow \mathcal{E}(E)_f,$$

which forgets the G-action, a monomorphism?

In this same general direction, there is recent interest in ex-fibrations ($p : X \to B, s : B \to X$), which consist of fibrations with a given section (which might also be required to be a co-fibration). Given another ex- fibration ($q : Y \to B, t : B \to Y$), over the same base B, we call a map $f : X \to Y$ an ex-map if $q \cdot f = p$ and $f \cdot s = t$. The set of ex-homotopy classes of ex-fibre homotopy equivalences of ($p : X \to B, s : B \to X$) to itself forms a group under composition, and we shall write this group $\mathcal{E}(p, s)$. Clearly, we have a natural extension of $\mathcal{E}(X)$ (Take $B = *$).

Problem 14 (P. Booth): Study the group $\mathcal{E}(p, s)$ in general, and connect it to the related problems of the G-equivariant self- equivalences of a free G-space or the self-fibre-homotopy equivalences of a fibration (both of which have been examined in the literature).

As before, let $\mathcal{E}(X)_f$ denotes the group of free homotopy classes of free self-equivalences. If X is 1-connected, or an H-space, then there is no difference between $\mathcal{E}(X)_f$ and $\mathcal{E}(X)$. It is known Becker-Gottlieb (1985) that $\mathcal{E}(RP^{2k})_f$ is trivial, while $\mathcal{E}(RP^{2k}) = Z/2$ (pointed equivalences).

Problem 15 (M. Arkowitz): What is the full relation between the groups $\mathcal{E}(X)_f$ and $\mathcal{E}(X)$? Analyze the obvious map

$$\mathcal{E}(X) \longrightarrow \mathcal{E}(X)_f$$

IV. The Space of Self-Maps

The space of self-maps X^X contains, of course, representatives of all the classes of $\mathcal{E}(X)$. In a few cases, one knows something of the homotopy structure of X^X. For example if X is an abelian topological group, so is X^X, and hence, a connected component is a product of Eilenberg-MacLane spaces. When $X = K(\pi, n)$, see [R. Thom; Colloque de Topologie, Louvain, 1956]. More general results have been obtained in Yamanoshita (1986).

Problem 16 (J.W. Rutter): Determine the homotopy-type of the identity component in X^X, especially in cases where X has more than two non-vanishing homotopy groups.

The late George Cooke (see Cooke (1978)) has studied the question of realizing self-equivalences as homeomorphisms.

Problem 17: In case $\mathcal{E}(X)$ can be realized as homeomorphisms, what additional information can be obtained, for example, about the sapce X^X ?

In closing, let me suggest that groups of classes of homotopy-self- equivalences have enjoyed a rich and diverse history over the last 30 years. Few of the original workers would have predicted - circa 1960 - the many developments which have taken place since then. May the next 30 years be so rich. We can only hope that many of these 17 problems will have been clarified by end of the next three decades.

LIST OF PAPERS ON OR RELEVANT TO GROUPS OF SELF-HOMOTOPY EQUIVALENCES [1]

1958

Barcus, W.D. and Barratt, M.G., *On the homotopy classification of the extensions of a fixed map*, Trans. Amer. Math. Soc. **88** (1958), 57–74.

1963

James, I.M., *The space of bundle maps*, Topology **2** (1963), 45–59.

1964

Arkowitz, M. and Curjel, C.R., *The group of homotopy equivalences of a space*, Bull. Amer. Math. Soc. **70** (1964), 293–296.

Arkowitz, M. and Curjel, C.R., *Groups of homotopy classes*, Springer LNM **4** (1964).

Kahn, D.W., *The group of homotopy equivalences*, Math. Z. **84** (1964), 1–8.

Shih, W., *On the group $\epsilon[X]$ of homotopy equivalence maps*, Bull. Amer. Math. Soc. **70** (1964), 361–365.

1965

Nomura, Y., *A note on fibre homotopy equivalences*, Bull. Nagoya Inst. Tech. **17** (1965), 66–71.

Olum, P., *Self-equivalences of pseudo-projective planes*, Topology **4** (1965), 109–127.

1966

Kahn, P.J., *Self-equivalences of $(n-1)$-connected $2n$-manifolds*, Bull. Amer. Math. Soc. **72** (1966), 562–566.

Nomura, Y., *Homotopy equivalences in a principal fibre space*, Math. Z. **92** (1966), 380–388.

1967

Arkowitz, M. and Curjel, C.R., *On maps of H-spaces*, Topology **6** (1967), 137–148.

Rutter, J.W., *A homotopy classification of maps into an induced fibre space*, Topology **6** (1967), 379–403.

Kudo, Y. and Tsuchida, K., *On the generalized Barcus-Barratt sequence*, Sci. Rep. Hirosaki Univ. **13** (1967), 1–9.

1968

Gottlieb, D., *On fibre spaces and the evaluation map*, Ann. of Math. **87** (1968), 42–55.

1969

Kahn, P.J., *Self-equivalences of $(n-1)$-connected $2n$-manifolds*, Math. Ann. **180** (1969), 26–47.

Rutter, J.W., *Self-equivalences and principal morphisms*, Proc. London Math. Soc. **20** (1970), 644–658.

1970

Rutter, J.W., *Groups of self-homotopy equivalences of induced spaces*, Comm. Math. Helv. **45** (1970), 236–255.

Sieradski, A.J., *Twisted self-homotopy equivalences*, Pacific J. of Math. **3** (1970), 789–802.

---

[1] Compiled from lists prepared by Martin A. Arkowitz, Peter I. Booth and Kouzou Tsukiyama

1971

Metzler, W. and Zimmermann, A., *Selbstäquivalenzen von $S^3 \times S^3$ in quaternionisher Behandlung*, Arch. der Math. **XVII** (1971), 209–213.

Olum, P., *Self-equivalences of pseudo-projective planes II*, Topology **10** (1971), 257–260.

1972

Gottlieb, D.H., *Applications of bundle map theory*, Trans. Amer. Math. Soc. **171** (1972), 23–50.

Kahn, D.W., *A note on H-equivalences*, Pacific J. of Math. **42** (1972), 77–80.

Kahn, D.W., *The group of stable self-equivalences*, Topology **11** (1972), 133–140.

Johnston, P.T., *The stable group of homotopy equivalences*, Quart. J. Math. **23** (1972), 213–219.

Oka, S., *Groups of self-equivalences of certain compexes*, Hiroshima Math. J. **2** (1972), 285–298.

Sieradski, A.J., *Stabilization of self-equivalences of the pseudo-projective spaces*, Michigan Math. J. **19** (1972), 109–119.

Zabrodsky, A., *On the homotopy type of principal classical group bundles over spheres*, Israel J. Math. (1972), 315–325.

1973

Becker, J.C. and Schultz, R.E., *Spaces of equivarinat self-equivalences of spheres*, Bull. Amer. Math. Soc. **79** (1973), 158–162.

Schellenberg, B., *On the self-equivalences of a space with non-cyclic fundamental group*, Math. Ann. **205** (1973), 333–344.

Schellenberg, B., *The group of homotopy self-equivalences of some compact CW-complexes*, Math. Ann. **200** (1973), 253–266.

Sunday, D.M., *The self-equivalences of an H-space*, Pacific J. of Math. **49** (1973), 507–517.

1974

Libermann, G. and Smallen, D.L., *Localization and self-homotopy equivalences*, Duke Math. J. **41** (1974), 183–186.

Oka, S., Sawashita, N. and Sugawa, M., *On the group of self-equivalences of a mapping cone*, Hiroshima Math. J. **4** (1974), 9–28.

Smallen, D., *The group of self-equivalences of certain complexes*, Pacific J.Math. **54** (1974), 269–276.

1975

Sawashita, N., *On the group of self-equivalences of the product of spheres*, Hiroshima Math. J. **5** (1975), 69–86.

Tsukiyama, K., *Note on self-maps inducing the identity automorphism of homotopy groups*, Hiroshima Math. J. **5** (1975), 215–222.

1976

Dyer, M.N., *Homotopy classifications of (π, m)-complexes*, J. Pure Appl. Algebra **8** (1976), 249–282.

Dyer, M.N., *Homotopy trees with trivial classifying ring*, Proc. Amer. Math. Soc. **55** (1976), 405–408.

Kahn, D.W., *Realization problems for the group of homotopy classes of self-equivalences*, Math. Ann. **220** (1976), 37–46.

Sawashita, N., *On the self-equivalences of H-space*, J. Math. Tokushima Univ. **10** (1976), 17–33.

Sieradski, A.J., *Combination isomorphisms and combinatorial homotopy equivalences*, J.Pure Appl. Algebra **7** (1976), 59–95.

Wilkerson, C., *Applications of minimal simplicial groups*, Topology **15** (1976), 115–130.

Wilkerson, C., *Classification of spaces of the same n-type for all n*, Proc. Amer. Math. Soc. **60** (1976), 279–285.

1977

Frank, D. and Kahn, D.W., *Finite complexes with infinitely-generated groups of self-equivalences*, Topology **16** (1977), 189–192.

Sawashita, N., *On H-equivalences of $SU(3), U(3)$ and $Sp(2)$*, J. Math. Tokushima Univ. **11** (1977), 33–47.

Sullivan, D., *Infinitesimal computations in topology*, I.H.E.S. Publ. Math. **47** (1977), 269–331.

Tsukiyama, K., *Note on self-homotopy equivalences of the twisted principal fibrations*, Mem. Fac. Educ. Shimane Univ. **11** (1977), 1–8.

1978

Cooke, G., *Replacing homotopy actions by topological actions*, Trans. Amer. Math. Soc. **237** (1978), 391–406.

Matsuda, T., *On the C_n-equivariant self-homotopy equivalences of spheres*, J. Fac. Sci. Shinshu Univ. **13** (1978), 43–78.

Rutter, J.W., *The group of self-homotopy equivalences of principal three sphere bundles over the seven sphere*, Math. Proc. Camb. Phil. Soc. (1978), 303–311.

1979

Dror, E. and Zabrodsky, A., *Unipotency and nilpotency in homotopy equivalences*, Topology **18** (1979), 187–197.

Dror, E., Dwyer, W.G. and Kan, D.M., *Self-homotopy equivalences of Postnikov conjugates*, Proc. Amer. Math. Soc. **74** (1979), 183–186.

Jajodia, S., *On 2-dimensional CW-complexes with a single 2-cell*, Pacific J. Math. **80** (1979), 191–203.

James, I.M., *On fibre spaces and nilpotency II*, Math. Proc. Camb. Phil. Soc. **86** (1979), 215–217.

Kahn, D.W., *The rigidity problem for stable spaces*, Proc. Amer. Math. Soc. **75** (1979), 139–144.

Matsuda, T., *On the equivariant self-homotopy equivalences of spheres*, J. Math. Soc. Japan **31** (1979), 69–83.

Matsumoto, T., *On homotopy equivalences of $S^2 \times RP^2$ to itself*, J. Math. Kyoto Univ. (1979), 1–17.

1980

Dror, E., Dwyer, W. G. and Kan, D.M., *equivariant maps which are self-homotopy equivalences*, Proc. Amer. Math. Soc. **80** (1980), 670–672.

Dror, E., Dwyer, W.G. and Kan, D.M., *Automorphisms of fibrations*, Proc. Amer. Math. Soc. **80** (1980), 491–494.

McCullough, D., *Finite aspherical complexes with infinitely-generated groups of self-homotopy equivalences*, Proc. Amer. Math. Soc. **80** (1980), 337–340.

Oka, S., *Finite complexes whose self-homotopy equivalences form cyclic groups*, Mem. of Fac. of Sci. Kyushu Univ. Series A. Math. **XXXIV** (1980), 171–181.

Scheerer, H., *Arithmeticity of groups of fibre homotopy equivalence classes*, Manuscripta Math. **31** (1980), 413–424.

Tsukiyama, K., *Self-homotopy equivalences of a space with two nonvanishing homotopy groups*, Proc. Amer. Math. Soc. **79** (1980), 134–138.

Tsukiyama, K., *A remark on fibre homotopy equivalences*, Illinois J. of Math. **24** (1980), 554–559.

1981

Booth, P., Heath, P., Morgan, C. and Piccinini, R., *Remarks on the homotopy type of groups of Gauge transformations*, C.R. Math. Rep. Acad. Sci. Canada **111** (1981), 3–6.

Didierjean, G., *Groupes d'homotopie du monoide des equivalences d'homotopie fibrées*, C.R. Acad. Sci. **292** (1981), 555–558.

Dror, E., Dwyer, W.G. and Kan, D.M., *Self homotopy equivalences of virtually nilpotent spaces*, Comment. Math. Helv. **56** (1981), 599–614.

Hurvitz, S., *The automorphism group of spaces and fibrations*, Pacific J. Math. **96** (1981), 371–388.

Maruyama, K. and Oka, S., *Self-H-maps of H-spaces of type (3,7)*, Mem. of Fac. of Sci. Kyushu Univ. Series A Math. **XXXV** (1981), 375–383.

McCullough, D., *The group of homotopy equivalences for a connected sum of closed aspherical manifolds*, Indiana U. Math. J. **30** (1981), 249–260.

McCullough, D., *Homotopy groups of the space of self-homotopy equivalences*, Trans. Amer. Math. Soc. **264** (1981), 151–163.

McCullough, D., *Connected sums of aspherical manifolds*, Indiana U. Math. J. **30** (1981), 17–28.

Mimura, M. and Sawashita, N., *On the group of self-homotopy equivalences of H-spaces of rank 2*, J. of Math. Kyoto Univ. **21** (1981), 331–349.

Oka, S., *On the group of self-homotopy equivalences of H-spaces of low rank, I, II*, Mem. of Fac. of Sci. Kyushu Univ. Series A Math. **XXXV** (1981), 247–282, 307–323.

Parks, James M., *A note on the monoid of self-equivalences*, Houston J. Math. **7** (1981), 403–406.

Sasao, S., *The stable group of self-homotopy equivalences of sphere bundles over the sphere*, Kodai Math. J. **4** (1981), 231–238.

1982

Ando, Y. and Yamaguchi, K., *On the homotopy self-equivalences of the product $A \times B$*, Proc. Japan Acad. **58** (1982), 323–325.

Brunner, A.M. and Ratcliffe, J.G., *Finite 2-complexes with infinitely-generated groups of self-homotopy equivalences*, Proc. Amer. Math. Soc. **86** (1982), 525–530.

Kolosov, J.E., *Homotopic self-equivalences of highly connected manifolds*, Math. USSR-Sb **41** (1982), 481–494.

Lee, K.B., *Geometric realization of $\pi_0\mathcal{E}(M)$*, Proc. Amer. Math. Soc. **86** (1982), 353–357.

Matsuda, T., *On the unit groups of Burnside rings*, Japanese J. Math. **71–93** (1982), 71–93.

McCullough, D., *Homotopy equivalences of punctured manifolds*, Michigan Math. J. **29** (1982), 457–465.

Meiwes, H., *On fibrations and nilpotency—some remarks upon two articles by I.M. James*, Manuscripta Math. **39** (1982), 263–270.

Plotnick, S., *Homotopy equivalences and free modules*, Topology **21** (1982), 91–99.

Sasao, S. and Ando, Y., *On the group $\epsilon(K(\pi,1)\times X)$ for 1-connected CW-complexes*, Kodai Math. **5** (1982), 65–70.

Sasao, S., *Fibre homotopy self-equivalences*, Kodai Math. J. **5** (1982), 446–453.

Tsukiyama, K., *On the group of fibre homotopy equivalences*, Hiroshima Math. J. **12** (1982), 349–376.

Zabrodsky, A., *On George Cooke's theory of homotopy and topological actions*, in "Current Trends in Algebraic Topology, CMS Conference Proceedings," 1982, pp. 313–317.

1983

Lee, K.B., *Geometric realization of a finite group of $\pi_0\epsilon(M)$*, Proc. Amer. Math. Soc. **87** (1983), 175–178.

Nomura, Y., *Self homotopy equivalences of Stiefel manifolds $W_{n,2}$ and $V_{n,2}$*, Osaka J. Math. **20** (1983), 79–93.

Rutter, J.W., *The group of homotopy self equivalence classes of CW-complexes*, Math. Proc. Camb. Phil. Soc. **93** (1983), 275–293.

Salvetti, M., *Automorphisms of fibre bundles on S^n*, Boll. Un. Math. Ital. D(6) **2** (1983), 99–112.

Sasao, S., *$\epsilon(X)$ for non-simply connected H-spaces*, Kodai Math. J. **6** (1983), 167-173.

Yamaguchi, K., *On the self-homotopy equivalence of the wedge of certain complexes*, Kodai Math. J. **6** (1983), 1–30.

Yamaguchi, K., *On the rational homotopy of $Map(HP^m, HP^m)$*, Kodai Math. J. **6** (1983), 279–288.

1984

Bencivenga, R., *Approximating groups of bundle automorphisms by loop spaces*, Trans. Amer. Math Soc. **285** (1984), 703–715.

Booth, P., Heath, P. Morgan, C. and Piccinini, R., *H-spaces of self-equivalences of fibrations and bundles*, Proc. Lond. Math. Soc. **3** (1984), 111–127.

Maryuma, K. and Mimura, M., *On the group of self-homotopy equivalences of $KP^2\vee S^m$*, Mem. Fac. Sci. Kyushu Univ. **38** (1984), 65–74.

Maryuma, K., *Note on self-H-maps of $SU(3)$*, Mem. Sci. Kyushu Univ. Ser A **38** (1984), 5–8.

McCullough, D., *Compact 3-manifolds with infinitely-generated groups of self-homotopy equivalences*, Proc. Amer. Math. Soc. **91** (1984), 625–629.

Mimura, M. and Sawashita, N., *On the group of self-homotopy equivalences of principal S^3-bundles over spheres*, Hiroshima Math. J. **14** (1984), 415–424.

Oprea, J. F., *Lifting homotopy actions in rational homotopy theory*, J. Pure and Applied Algebra **32** (1984), 177–190.

Sasao, S., *Self-homotopy equivalences of the total spaces of a sphere bundle over a sphere*, Kodai J. Math. **7** (1984), 365–381.

Sawashita, N., *self-H-equivalences of H-spaces with applications to H-spaces of rank 2*, Hiroshima Math. J. **14** (1984), 75–113.

Triantafillou, G., *An algebraic model for G-homotopy types*, Astérisque **113–114** (1984), 312–337.

Yamanoshita, T., *On the spaces of self-homotopy equivalences of certain CW-complexes*, Proc. Japan Acad. **60** (1984), 229–231.

1985

Didierjean, G., *Homotopie de l'espace des équivalences d'homotopie fibrées*, Ann. Inst. Fourier **35** (1985), 33–47.

McCullough, D., *Errata: The group of homotopy equivalences for a connected sum of closed aspherical manifolds*, Indiana U. Math. J. **34** (1985), 201–203.

Roitberg, J., *Residually finite, Hopfian and coHopfian spaces*, Contemp. Math. Amer. Math. Soc. **37** (1985), 131–144.

Sasao, S., *On self homotopy equivalences of S^3-principal bundles over S^n*, Kodai Math. J. **8** (1985), 285–295.

Tsukiyama, K., *Equivariant self-equivalences of principal fibre bundles*, Math. Proc. Camb. Phil. Soc. **98** (1985), 87–92.

Yamanoshita, T., *On the spaces of self-homotopy equivalences for fibre spaces*, Proc. Japan Acad. **61 Ser., A** (1985), 15–18.

Yamanoshita, T., *On the spaces of self-homotopy equivalences of certain CW-complexes*, J. Math. Soc. Japan **37** (1985), 455–470.

Zabrodsky, A., *Endomorphisms in the homotopy category*, Cont. Math. Amer. Math. Soc. **44** (1985), 227-277.

1986

Assadi, A., *Homotopy actions and cohomology of finite groups*, Springer LNM **1217** (1986), 26–57.

Mimura, M. and Sawashita, N., *On p-Sylow subgroups of groups of self homotopy equivalences of sphere bundles over spheres*, Adv. Studies in Pure Math. **9** (1986), 259–271.

Oka, S., *A generalization of the Adam's invariant and applications to homotopy of the exceptional Lie group G_2*, Adv. Studies in Pure Math. **9** (1986), 195–230.

Oshima, H. and Tsukiyama, K., *On the group of equivariant self equivalences of free actions*, Pub. Res. Inst. Math. Sci. Kyoto Univ. **22** (1986), 905–923.

Sawashita, N. and Sugawara, ,M., *On self H-equivalences of an H-space with respect to any multiplication*, Hiroshima Math. J. **16** (1986), 1-20.

Schwänzl, R. and Vogt, R.M., *Coherence in homotopy group actions*, Springer LNM **1217** (1986), 364–390.

Yamaguchi, K., *The group of self-homotopy equivalences of S^3-bundles over S^4, I*, Kodai Math. **9** (1986), 308–326.

Yamanoshita, T., *On the spaces of self-homotopy equivalences for fibre spaces II*, Pub. Res. Inst. Math. Sci. Kyoto Univ. **22** (1986), 43–56.

1987

Maruyama, K., *A remark on the group of self-homotopy equivalences*, Memoirs Fac. Science Kyushu Un. **41** (1987), 81–84.

Mimura, M. and Sawashita, N., *On p-Sylov subgroups of groups of self-homotopy equivalences of sphere bundles over spheres*, in "Homotopy Theory and related topics," North-Holland, Amsterdam-New York, 1987, pp. 259–271.

Peschke, G., *H-semidirect products*, Canad. Math. Bull. **30 (4)** (1987), 402–411.

Sawashita, S. and Sugarwara, M., *On self H-equivalences of homotopy associative H-spaces*, Hiroshima Math. J. **17** (1987), 219–224.

Yamaguchi, K., *The group of self-homotopy equivalences of S^2-bundles over S^4 II, Applications*, Kodai Math. J. **10** (1987), 1–8.

Yamanoshita, T., *On the space of self-homotopy equivalences for spaces*, Kodai Math. **10** (1987), 127–142.

1988

Arkowitz, M., *Formal differential graded algebras and homomorphisms*, J. Pure and Appl. Algebra **51** (1988), 35–52.

Møller, J. M., *Self-maps on twisted Eilenberg-MacLane spaces*, Kodai Math. J. **11** (1988), 372–378.

Rutter, J., *Homotopy classification of maps between pseudo-projective spaces*, Questiones Mathematicae **11** (1988), 409–422.

Rutter, J., *The group of homotopy self-equivalence classes using an homotopy decomposition*, Math. Proc. Cambridge Phil. Soc. **103** (1988), 305–315.

1989

Maruyama, K., *Localization of a certain subgroup of self-homotopy equivalences*, Pacific J. Math. **136** (1989), 293–301.

Morgan, C. and Piccinini, R., *Conjugacy classes of groups of bundle automorphisms*, Manuscripta Mathematica **63** (1989), 233–244.

LECTURE NOTES IN MATHEMATICS
Edited by A. Dold, B. Eckmann and F. Takens

Some general remarks on the publication of proceedings of congresses and symposia

Lecture Notes aim to report new developments - quickly, informally and at a high level. The following describes criteria and procedures which apply to proceedings volumes. The editors of a volume are strongly advised to inform contributors about these points at an early stage.

§1. One (or more) expert participant(s) of the meeting should act as the responsible editor(s) of the proceedings. They select the papers which are suitable (cf. §§ 2, 3) for inclusion in the proceedings, and have them individually refereed (as for a journal). It should not be assumed that the published proceedings must reflect conference events faithfully and in their entirety. Contributions to the meeting which are not included in the proceedings can be listed by title. The series editors will normally not interfere with the editing of a particular proceedings volume - except in fairly obvious cases, or on technical matters, such as described in §§ 2, 3. The names of the responsible editors appear on the title page of the volume.

§2. The proceedings should be reasonably homogeneous (concerned with a limited area). For instance, the proceedings of a congress on "Analysis" or "Mathematics in Wonderland" would normally not be sufficiently homogeneous.

One or two longer survey articles on recent developments in the field are often very useful additions to such proceedings - even if they do not correspond to actual lectures at the congress. An extensive introduction on the subject of the congress would be desirable.

§3. The contributions should be of a high mathematical standard and of current interest. Research articles should present new material and not duplicate other papers already published or due to be published. They should contain sufficient information and motivation and they should present proofs, or at least outlines of such, in sufficient detail to enable an expert to complete them. Thus resumes and mere announcements of papers appearing elsewhere cannot be included, although more detailed versions of a contribution may well be published in other places later.

Contributions in numerical mathematics may be acceptable without formal theorems resp. proofs if they present new algorithms solving problems (previously unsolved or less well solved) or develop innovative qualitative methods, not yet amenable to a more formal treatment. .

Surveys, if included, should cover a sufficiently broad topic, and should in general not simply review the author's own recent research. In the case of such surveys, exceptionally, proofs of results may not be necessary.

§4. "Mathematical Reviews" and "Zentralblatt für Mathematik" recommend that papers in proceedings volumes carry an explicit statement that they are in final form and that no similar paper has been or is being submitted elsewhere, if these papers are to be considered for a review. Normally, papers that satisfy the criteria of the Lecture Notes in Mathematics series also satisfy this requirement, but we strongly recommend that the contributing authors be asked to give this guarantee explicitly at the beginning or end of their paper. There will occasionally be cases where this does not apply but where, for special reasons, the paper is still acceptable for LNM.

§5. Proceedings should appear soon after the meeeting. The publisher should, therefore, receive the complete manuscript (preferably in duplicate) within nine months of the date of the meeting at the latest.

§6. Plans or proposals for proceedings volumes should be sent to one of the editors of the series or to Springer-Verlag Heidelberg. They should give sufficient information on the conference or symposium, and on the proposed proceedings. In particular, they should contain a list of the expected contributions with their prospective length. Abstracts or early versions (drafts) of some of the contributions are helpful.

§7. Lecture Notes are printed by photo-offset from camera-ready typed copy provided by the editors. For this purpose Springer-Verlag provides editors with technical instructions for the preparation of manuscripts and these should be distributed to all contributing authors. Springer-Verlag can also, on request, supply stationery on which the prescribed typing area is outlined. Some homogeneity in the presentation of the contributions is desirable.

Careful preparation of manuscripts will help keep production time short and ensure a satisfactory appearance of the finished book. The actual production of a Lecture Notes volume normally takes 6 -8 weeks.

Manuscripts should be at least 100 pages long. The final version should include a table of contents.

§8. Editors receive a total of 50 free copies of their volume for distribution to the contributing authors, but no royalties. (Unfortunately, no reprints of individual contributions can be supplied.) They are entitled to purchase further copies of their book for their personal use at a discount of 33.3 %, other Springer mathematics books at a discount of 20 % directly from Springer-Verlag. Contributing authors may purchase the volume in which their article appears at a discount of 33.3 %.

Commitment to publish is made by letter of intent rather than by signing a formal contract. Springer-Verlag secures the copyright for each volume.

<u>Addresses:</u>

Professor A. Dold, Mathematisches Institut, Universität Heidelberg,
Im Neuenheimer Feld 288, 6900 Heidelberg, Federal Republic of Germany

Professor B. Eckmann, Mathematik, ETH-Zentrum
8092 Zürich, Switzerland

Prof. F. Takens, Mathematisch Instituut, Rijksuniversiteit Groningen,
Postbus 800, 9700 AV Groningen, The Netherlands

Springer-Verlag, Mathematics Editorial, Tiergartenstr. 17,
6900 Heidelberg, Federal Republic of Germany, Tel.: (06221) 487-410

Springer-Verlag, Mathematics Editorial, 175, Fifth Avenue,
New York, New York 10010, USA, Tel.: (212) 460-1596

Vol. 1259: F. Cano Torres, Desingularization Strategies for Three-Dimensional Vector Fields. IX, 189 pages. 1987.

Vol. 1260: N.H. Pavel, Nonlinear Evolution Operators and Semigroups. VI, 285 pages. 1987.

Vol. 1261: H. Abels, Finite Presentability of S-Arithmetic Groups. Compact Presentability of Solvable Groups. VI, 178 pages. 1987.

Vol. 1262: E. Hlawka (Hrsg.), Zahlentheoretische Analysis II. Seminar, 1984–86. V, 158 Seiten. 1987.

Vol. 1263: V.L. Hansen (Ed.), Differential Geometry. Proceedings, 1985. XI, 288 pages. 1987.

Vol. 1264: Wu Wen-tsün, Rational Homotopy Type. VIII, 219 pages. 1987.

Vol. 1265: W. Van Assche, Asymptotics for Orthogonal Polynomials. VI, 201 pages. 1987.

Vol. 1266: F. Ghione, C. Peskine, E. Sernesi (Eds.), Space Curves. Proceedings, 1985. VI, 272 pages. 1987.

Vol. 1267: J. Lindenstrauss, V.D. Milman (Eds.), Geometrical Aspects of Functional Analysis. Seminar. VII, 212 pages. 1987.

Vol. 1268: S.G. Krantz (Ed.), Complex Analysis. Seminar, 1986. VII, 195 pages. 1987.

Vol. 1269: M. Shiota, Nash Manifolds. VI, 223 pages. 1987.

Vol. 1270: C. Carasso, P.-A. Raviart, D. Serre (Eds.), Nonlinear Hyperbolic Problems. Proceedings, 1986. XV, 341 pages. 1987.

Vol. 1271: A.M. Cohen, W.H. Hesselink, W.L.J. van der Kallen, J.R. Strooker (Eds.), Algebraic Groups Utrecht 1986. Proceedings. XII, 284 pages. 1987.

Vol. 1272: M.S. Livšic, L.L. Waksman, Commuting Nonselfadjoint Operators in Hilbert Space. III, 115 pages. 1987.

Vol. 1273: G.-M. Greuel, G. Trautmann (Eds.), Singularities, Representation of Algebras, and Vector Bundles. Proceedings, 1985. XIV, 383 pages. 1987.

Vol. 1274: N. C. Phillips, Equivariant K-Theory and Freeness of Group Actions on C*-Algebras. VIII, 371 pages. 1987.

Vol. 1275: C.A. Berenstein (Ed.), Complex Analysis I. Proceedings, 1985–86. XV, 331 pages. 1987.

Vol. 1276: C.A. Berenstein (Ed.), Complex Analysis II. Proceedings, 1985–86. IX, 320 pages. 1987.

Vol. 1277: C.A. Berenstein (Ed.), Complex Analysis III. Proceedings, 1985–86. X, 350 pages. 1987.

Vol. 1278: S.S. Koh (Ed.), Invariant Theory. Proceedings, 1985. V, 102 pages. 1987.

Vol. 1279: D. Ieşan, Saint-Venant's Problem. VIII, 162 Seiten. 1987.

Vol. 1280: E. Neher, Jordan Triple Systems by the Grid Approach. XII, 193 pages. 1987.

Vol. 1281: O.H. Kegel, F. Menegazzo, G. Zacher (Eds.), Group Theory. Proceedings, 1986. VII, 179 pages. 1987.

Vol. 1282: D.E. Handelman, Positive Polynomials, Convex Integral Polytopes, and a Random Walk Problem. XI, 136 pages. 1987.

Vol. 1283: S. Mardešić, J. Segal (Eds.), Geometric Topology and Shape Theory. Proceedings, 1986. V, 261 pages. 1987.

Vol. 1284: B.H. Matzat, Konstruktive Galoistheorie. X, 286 pages. 1987.

Vol. 1285: I.W. Knowles, Y. Saitō (Eds.), Differential Equations and Mathematical Physics. Proceedings, 1986. XVI, 499 pages. 1987.

Vol. 1286: H.R. Miller, D.C. Ravenel (Eds.), Algebraic Topology. Proceedings, 1986. VII, 341 pages. 1987.

Vol. 1287: E.B. Saff (Ed.), Approximation Theory, Tampa. Proceedings, 1985–1986. V, 228 pages. 1987.

Vol. 1288: Yu. L. Rodin, Generalized Analytic Functions on Riemann Surfaces. V, 128 pages. 1987.

Vol. 1289: Yu. I. Manin (Ed.), K-Theory, Arithmetic and Geometry. Seminar, 1984–1986. V, 399 pages. 1987.

Vol. 1290: G. Wüstholz (Ed.), Diophantine Approximation and Transcendence Theory. Seminar, 1985. V, 243 pages. 1987.

Vol. 1291: C. Mœglin, M.-F. Vignéras, J.-L. Waldspurger, Correspondances de Howe sur un Corps p-adique. VII, 163 pages. 1987.

Vol. 1292: J.T. Baldwin (Ed.), Classification Theory. Proceedings, 1985. VI, 500 pages. 1987.

Vol. 1293: W. Ebeling, The Monodromy Groups of Isolated Singularities of Complete Intersections. XIV, 153 pages. 1987.

Vol. 1294: M. Queffélec, Substitution Dynamical Systems – Spectral Analysis. XIII, 240 pages. 1987.

Vol. 1295: P. Lelong, P. Dolbeault, H. Skoda (Réd.), Séminaire d'Analyse P. Lelong – P. Dolbeault – H. Skoda. Seminar, 1985/1986. VII, 283 pages. 1987.

Vol. 1296: M.-P. Malliavin (Ed.), Séminaire d'Algèbre Paul Dubreil et Marie-Paule Malliavin. Proceedings, 1986. IV, 324 pages. 1987.

Vol. 1297: Zhu Y.-l., Guo B.-y. (Eds.), Numerical Methods for Partial Differential Equations. Proceedings. XI, 244 pages. 1987.

Vol. 1298: J. Aguadé, R. Kane (Eds.), Algebraic Topology, Barcelona 1986. Proceedings. X, 255 pages. 1987.

Vol. 1299: S. Watanabe, Yu.V. Prokhorov (Eds.), Probability Theory and Mathematical Statistics. Proceedings, 1986. VIII, 589 pages. 1988.

Vol. 1300: G.B. Seligman, Constructions of Lie Algebras and their Modules. VI, 190 pages. 1988.

Vol. 1301: N. Schappacher, Periods of Hecke Characters. XV, 160 pages. 1988.

Vol. 1302: M. Cwikel, J. Peetre, Y. Sagher, H. Wallin (Eds.), Function Spaces and Applications. Proceedings, 1986. VI, 445 pages. 1988.

Vol. 1303: L. Accardi, W. von Waldenfels (Eds.), Quantum Probability and Applications III. Proceedings, 1987. VI, 373 pages. 1988.

Vol. 1304: F.Q. Gouvêa, Arithmetic of p-adic Modular Forms. VIII, 121 pages. 1988.

Vol. 1305: D.S. Lubinsky, E.B. Saff, Strong Asymptotics for Extremal Polynomials Associated with Weights on ℝ. VII, 153 pages. 1988.

Vol. 1306: S.S. Chern (Ed.), Partial Differential Equations. Proceedings, 1986. VI, 294 pages. 1988.

Vol. 1307: T. Murai, A Real Variable Method for the Cauchy Transform, and Analytic Capacity. VIII, 133 pages. 1988.

Vol. 1308: P. Imkeller, Two-Parameter Martingales and Their Quadratic Variation. IV, 177 pages. 1988.

Vol. 1309: B. Fiedler, Global Bifurcation of Periodic Solutions with Symmetry. VIII, 144 pages. 1988.

Vol. 1310: O.A. Laudal, G. Pfister, Local Moduli and Singularities. V, 117 pages. 1988.

Vol. 1311: A. Holme, R. Speiser (Eds.), Algebraic Geometry, Sundance 1986. Proceedings. VI, 320 pages. 1988.

Vol. 1312: N.A. Shirokov, Analytic Functions Smooth up to the Boundary. III, 213 pages. 1988.

Vol. 1313: F. Colonius, Optimal Periodic Control. VI, 177 pages. 1988.

Vol. 1314: A. Futaki, Kähler-Einstein Metrics and Integral Invariants. IV, 140 pages. 1988.

Vol. 1315: R.A. McCoy, I. Ntantu, Topological Properties of Spaces of Continuous Functions. IV, 124 pages. 1988.

Vol. 1316: H. Korezlioglu, A.S. Ustunel (Eds.), Stochastic Analysis and Related Topics. Proceedings, 1986. V, 371 pages. 1988.

Vol. 1317: J. Lindenstrauss, V.D. Milman (Eds.), Geometric Aspects of Functional Analysis. Seminar, 1986–87. VII, 289 pages. 1988.

Vol. 1318: Y. Felix (Ed.), Algebraic Topology – Rational Homotopy. Proceedings, 1986. VIII, 245 pages. 1988.

Vol. 1319: M. Vuorinen, Conformal Geometry and Quasiregular Mappings. XIX, 209 pages. 1988.

Vol. 1320: H. Jürgensen, G. Lallement, H.J. Weinert (Eds.), Semigroups, Theory and Applications. Proceedings, 1986. X, 416 pages. 1988.

Vol. 1321: J. Azéma, P.A. Meyer, M. Yor (Eds.), Séminaire de Probabilités XXII. Proceedings. IV, 600 pages. 1988.

Vol. 1322: M. Métivier, S. Watanabe (Eds.), Stochastic Analysis. Proceedings. 1987. VII, 197 pages. 1988.

Vol. 1323: D.R. Anderson, H.J. Munkholm, Boundedly Controlled Topology. XII, 309 pages. 1988.

Vol. 1324: F. Cardoso, D.G. de Figueiredo, R. Iório, O. Lopes (Eds.), Partial Differential Equations. Proceedings, 1986. VIII, 433 pages. 1988.

Vol. 1325: A. Truman, I.M. Davies (Eds.), Stochastic Mechanics and Stochastic Processes. Proceedings, 1986. V, 220 pages. 1988.

Vol. 1326: P.S. Landweber (Ed.), Elliptic Curves and Modular Forms in Algebraic Topology. Proceedings, 1986. V, 224 pages. 1988.

Vol. 1327: W. Bruns, U. Vetter, Determinantal Rings. VII,236 pages. 1988.

Vol. 1328: J.L. Bueso, P. Jara, B. Torrecillas (Eds.), Ring Theory. Proceedings, 1986. IX, 331 pages. 1988.

Vol. 1329: M. Alfaro, J.S. Dehesa, F.J. Marcellan, J.L. Rubio de Francia, J. Vinuesa (Eds.): Orthogonal Polynomials and their Applications. Proceedings, 1986. XV, 334 pages. 1988.

Vol. 1330: A. Ambrosetti, F. Gori, R. Lucchetti (Eds.), Mathematical Economics. Montecatini Terme 1986. Seminar. VII, 137 pages. 1988.

Vol. 1331: R. Bamón, R. Labarca, J. Palis Jr. (Eds.), Dynamical Systems, Valparaiso 1986. Proceedings. VI, 250 pages. 1988.

Vol. 1332: E. Odell, H. Rosenthal (Eds.), Functional Analysis. Proceedings, 1986–87. V, 202 pages. 1988.

Vol. 1333: A.S. Kechris, D.A. Martin, J.R. Steel (Eds.), Cabal Seminar 81–85. Proceedings, 1981–85. V, 224 pages. 1988.

Vol. 1334: Yu.G. Borisovich, Yu. E. Gliklikh (Eds.), Global Analysis – Studies and Applications III. V, 331 pages. 1988.

Vol. 1335: F. Guillén, V. Navarro Aznar, P. Pascual-Gainza, F. Puerta, Hyperrésolutions cubiques et descente cohomologique. XII, 192 pages. 1988.

Vol. 1336: B. Helffer, Semi-Classical Analysis for the Schrödinger Operator and Applications. V, 107 pages. 1988.

Vol. 1337: E. Sernesi (Ed.), Theory of Moduli. Seminar, 1985. VIII, 232 pages. 1988.

Vol. 1338: A.B. Mingarelli, S.G. Halvorsen, Non-Oscillation Domains of Differential Equations with Two Parameters. XI, 109 pages. 1988.

Vol. 1339: T. Sunada (Ed.), Geometry and Analysis of Manifolds. Procedings, 1987. IX, 277 pages. 1988.

Vol. 1340: S. Hildebrandt, D.S. Kinderlehrer, M. Miranda (Eds.), Calculus of Variations and Partial Differential Equations. Proceedings, 1986. IX, 301 pages. 1988.

Vol. 1341: M. Dauge, Elliptic Boundary Value Problems on Corner Domains. VIII, 259 pages. 1988.

Vol. 1342: J.C. Alexander (Ed.), Dynamical Systems. Proceedings, 1986–87. VIII, 726 pages. 1988.

Vol. 1343: H. Ulrich, Fixed Point Theory of Parametrized Equivariant Maps. VII, 147 pages. 1988.

Vol. 1344: J. Král, J. Lukeš, J. Netuka, J. Veselý (Eds.), Potential Theory – Surveys and Problems. Proceedings, 1987. VIII, 271 pages. 1988.

Vol. 1345: X. Gomez-Mont, J. Seade, A. Verjovski (Eds.), Holomorphic Dynamics. Proceedings, 1986. VII, 321 pages. 1988.

Vol. 1346: O. Ya. Viro (Ed.), Topology and Geometry – Rohlin Seminar. XI, 581 pages. 1988.

Vol. 1347: C. Preston, Iterates of Piecewise Monotone Mappings on an Interval. V, 166 pages. 1988.

Vol. 1348: F. Borceux (Ed.), Categorical Algebra and its Applications. Proceedings, 1987. VIII, 375 pages. 1988.

Vol. 1349: E. Novak, Deterministic and Stochastic Error Bounds in Numerical Analysis. V, 113 pages. 1988.

Vol. 1350: U. Koschorke (Ed.), Differential Topology. Proceedings, 1987. VI, 269 pages. 1988.

Vol. 1351: I. Laine, S. Rickman, T. Sorvali, (Eds.), Complex Analysis, Joensuu 1987. Proceedings. XV, 378 pages. 1988.

Vol. 1352: L.L. Avramov, K.B. Tchakerian (Eds.), Algebra – Some Current Trends. Proceedings, 1986. IX, 240 Seiten. 1988.

Vol. 1353: R.S. Palais, Ch.-l. Terng, Critical Point Theory and Submanifold Geometry. X, 272 pages. 1988.

Vol. 1354: A. Gómez, F. Guerra, M.A. Jiménez, G. López (Eds.), Approximation and Optimization. Proceedings, 1987. VI, 280 pages. 1988.

Vol. 1355: J. Bokowski, B. Sturmfels, Computational Synthetic Geometry. V, 168 pages. 1989.

Vol. 1356: H. Volkmer, Multiparameter Eigenvalue Problems and Expansion Theorems. VI, 157 pages. 1988.

Vol. 1357: S. Hildebrandt, R. Leis (Eds.), Partial Differential Equations and Calculus of Variations. VI, 423 pages. 1988.

Vol. 1358: D. Mumford, The Red Book of Varieties and Schemes. V, 309 pages. 1988.

Vol. 1359: P. Eymard, J.-P. Pier (Eds.), Harmonic Analysis. Proceedings, 1987. VIII, 287 pages. 1988.

Vol. 1360: G. Anderson, C. Greengard (Eds.), Vortex Methods. Proceedings, 1987. V, 141 pages. 1988.

Vol. 1361: T. tom Dieck (Ed.), Algebraic Topology and Transformation Groups. Proceedings, 1987. VI, 298 pages. 1988.

Vol. 1362: P. Diaconis, D. Elworthy, H. Föllmer, E. Nelson, G.C. Papanicolaou, S.R.S. Varadhan. École d'Été de Probabilités de Saint-Flour XV–XVII, 1985–87. Editor: P.L. Hennequin. V, 459 pages. 1988.

Vol. 1363: P.G. Casazza, T.J. Shura, Tsirelson's Space. VIII, 204 pages. 1988.

Vol. 1364: R.R. Phelps, Convex Functions, Monotone Operators and Differentiability. IX, 115 pages. 1989.

Vol. 1365: M. Giaquinta (Ed.), Topics in Calculus of Variations. Seminar, 1987. X, 196 pages. 1989.

Vol. 1366: N. Levitt, Grassmannians and Gauss Maps in PL-Topology. V, 203 pages. 1989.

Vol. 1367: M. Knebusch, Weakly Semialgebraic Spaces. XX, 376 pages. 1989.

Vol. 1368: R. Hübl, Traces of Differential Forms and Hochschild Homology. III, 111 pages. 1989.

Vol. 1369: B. Jiang, Ch.-K. Peng, Z. Hou (Eds.), Differential Geometry and Topology. Proceedings, 1986–87. VI, 366 pages. 1989.

Vol. 1370: G. Carlsson, R.L. Cohen, H.R. Miller, D.C. Ravenel (Eds.), Algebraic Topology. Proceedings, 1986. IX, 456 pages. 1989.

Vol. 1371: S. Glaz, Commutative Coherent Rings. XI, 347 pages. 1989.

Vol. 1372: J. Azéma, P.A. Meyer, M. Yor (Eds.), Séminaire de Probabilités XXIII. Proceedings. IV, 583 pages. 1989.

Vol. 1373: G. Benkart, J.M. Osborn (Eds.), Lie Algebras, Madison 1987. Proceedings. V, 145 pages. 1989.

Vol. 1374: R.C. Kirby, The Topology of 4-Manifolds. VI, 108 pages. 1989.

Vol. 1375: K. Kawakubo (Ed.), Transformation Groups. Proceedings, 1987. VIII, 394 pages, 1989.

Vol. 1376: J. Lindenstrauss, V.D. Milman (Eds.), Geometric Aspects of Functional Analysis. Seminar (GAFA) 1987–88. VII, 288 pages. 1989.

Vol. 1377: J.F. Pierce, Singularity Theory, Rod Theory, and Symmetry-Breaking Loads. IV, 177 pages. 1989.

Vol. 1378: R.S. Rumely, Capacity Theory on Algebraic Curves. III, 437 pages. 1989.

Vol. 1379: H. Heyer (Ed.), Probability Measures on Groups IX. Proceedings, 1988. VIII, 437 pages. 1989